BIOENGINEERING APPROACHES TO PULMONARY PHYSIOLOGY AND MEDICINE

BIOENGINEERING APPROACHES TO PULMONARY PHYSIOLOGY AND MEDICINE

Edited by

Michael C. K. Khoo

University of Southern California
Los Angeles, California

SPRINGER SCIENCE+BUSINESS MEDIA, LLC

Library of Congress Cataloging-in-Publication Data

On file

Proceedings of the 12th Biomedical Simulations Resource Short Course on Bioengineering Approaches to Pulmonary Physiology and Medicine, held May 20, 1995, in Seattle, Washington; and of the Annual Biomedical Engineering Society Conference Symposium on Nonlinear Dynamics and Respiratory Control, held October 6 – 8, 1995, in Boston, Massachusetts

ISBN 978-1-4419-3260-0 ISBN 978-0-585-34964-0 (eBook)
DOI 10.1007/978-0-585-34964-0

Originally published by Plenum Press, New York in 1996
Softcover reprint of the hardcover 1st edition 1996

10 9 8 7 6 5 4 3 2 1

PREFACE

As the current millennium steams towards a close, one cannot help but look with amazement at the incredible amount of progress that has been achieved in medicine in just the last few decades. A key contributing factor to this success has been the importation and blending of ideas and techniques from disciplines outside the traditional borders of medical science. In recent years, the most well-known example is the cross-pollination between molecular biology and medicine. Advances driven by this potent combination have spawned the vision of a future where cures based on gene therapy become commonplace. Yet, as we continue our search for "magic bullets" in the quest to eradicate disease, it is important to recognize the value of other less-heralded interdisciplinary efforts that have laid a large part of the foundation of present-day medicine. In pulmonary medicine, the contribution from the bioengineers (a diverse collection of individuals cross-bred to various degrees in mathematical modeling and experimental physiology) has been larger and more sustained than in many other medical specialties. It is easy to point to the vast array of ventilators, blood-gas analyzers, oximeters, pulmonary function devices, and respiration monitors that are present in any modern clinical setting as solid evidence of the successful synergy between engineering science and pulmonary medicine. However, one must not forget the less tangible, but perhaps more important, contributions that have been derived from mathematical modeling and computer simulation, without which many of these modern instruments would not have come into existence.

From our perspective here and now, it appears that there remains considerable uncharted territory for the pulmonary bioengineer to explore in the era that lies ahead. The quantitative characterization of the respiratory system must move beyond the study of the system in isolation and incorporate *interactions* between respiratory processes and other associated systems, notably the central nervous system. Secondly, the development and application of new quantitative tools will provide us with greater knowledge about the nonlinear properties, time-varying nature, and spatial inhomogeneities of the respiratory system. Thirdly, there is a need to develop models that integrate processes at the organ level with cellular, and even subcellular, processes. This book represents the collective effort of a group of dedicated scientists who have begun to address the first two challenges. The chapters in this book are based primarily on presentations delivered at two meetings held last year. The first was the 12th Short Course sponsored by the Biomedical Simulations Resource at the University of Southern California, a research center funded by the Biomedical Research Program of the Division of Research Resources, National Institutes of Health. The Short Course, which bore the same title as this book, was held on May 20, 1995 in Seattle. The second meeting took the form of a symposium held during the Annual Fall Conference of the Biomedical Engineering Society on October 8, 1995. The symposium, entitled "Nonlinear Dynamics and Respiratory Control," was co-chaired by Dr. Bernard

Hoop and myself. The contributors to this book have taken great pains to give a tutorial flavor to their individual chapters, and I am extremely grateful to them for their efforts. I believe this book should be of greater utility than the usual proceedings volume to researchers and students who wish to learn more about state-of-the-art developments in pulmonary bioengineering.

I wish to thank the Biomedical Research Technology Program of NCRR/NIH for supporting our research activities for the last 11 years as well as for sponsoring this book. I must give due credit to Dr. Bernard Hoop, who did most of the work in organizing the November 1995 symposium. I am grateful to Ms. Stephanie Braun, the BMSR Administrative Assistant, for putting together and running the Short Course flawlessly. My heartfelt thanks also go to Drs. Vasilis Marmarelis and David D'Argenio, co-directors of the Biomedical Simulation Resource, for their encouragement, support, and friendship all these years that I have been at USC. Finally, I would be remiss if I did not acknowledge a contributory role from my family - my wife, Pam, and sons, Bryant and Mason - who patiently endured the many occasions on which I "mysteriously disappeared" to work on this volume.

Michael C.K. Khoo
Los Angeles, California

CONTENTS

Part I: Systems Modeling of Ventilatory Control

Part II: Modeling of the Neural Control of Breathing

Part III: Nonlinear Dynamical Analysis of Respiration

Part IV: Modeling of Pulmonary Mechanics and Gas Exchange

1

UNDERSTANDING THE DYNAMICS OF STATE-RESPIRATORY INTERACTION DURING SLEEP

A Model-Based Approach

Michael C. K. Khoo

Biomedical Engineering Department
University of Southern California
University Park, Los Angeles, California 90089

1. INTRODUCTION

Over past decade, there has been a tremendous increase in the number of pulmonary clinics dedicated to diagnosing and treating primary sleep disorders. An important contributing factor to this phenomenal expansion of the field of sleep medicine is the recognition of the need to better understand the effects of sleep on respiratory control, since it is during sleep that potentially harmful apneas or hypopneas occur in apparently normal individuals. Previous studies have shown that sleep onset leads to a reduction in metabolic rate and the baseline level of ventilation, and an elevation of circulatory delay and average P_{aCO2}[1,2]. These changes work together with the accompanying depression of upper airway muscle tone and chemoresponsiveness to alter the operating set-point of the respiratory control system. In the light stages of sleep, the decrease in damping may be large enough to offset the reduction in controller gain leading to a system that is more susceptible to oscillation[3,4,5]. Under conditions of high loop gain, spontaneous oscillations in ventilation can develop as a result of feedback instability[6]. Alternatively, because of decreased damping, continual random excitation from a variety of sources, including influences from other physiological control systems, can also produce significant periodic and nonperiodic variability in breathing[7]. A number of studies have reported a close association between ventilatory oscillations and fluctuations in sleep state[8,9,10]. These observations suggest a third possibility: that respiratory variability during sleep may be due to primary fluctuations in state[11]. It is likely, however, that more than one of the above three mechanisms are at play in any given instance of sleep-disordered breathing (SDB). Moreover, it is probably reasonable to expect that there will be considerable interplay among the multiple factors and that these relationships will likely be complex and dynamic. Physical intuition and conventional physiological methodology are not the best tools for unraveling the mechanisms underlying such complex interactions. Under these circumstances, the rigorous framework provided by a quantitative *dynamic* model becomes not only useful but, as one could argue, necessary. The theoretical study of Khoo et al.[12] represented a starting point for this kind of model-based approach. However, a number of assumptions were made that did not have firm empirical bases. In this chapter, we consider potential extensions of the model, incorporating data that have been obtained in recent experimental studies. While the specific goal of this study is to enhance our current understanding of the control of breathing during sleep, an equally important objective is to highlight the utility of modeling as a means for hypothesis testing as well as a tool for extracting relevant information from highly variable and dynamic data.

2. THE EXISTING MODEL: LIMITATIONS

We begin by first reviewing how the existing model characterizes the dynamic interactions between state and breathing. A complete description of the model is given in Ref.12. Sleep onset is assumed to be accompanied by the withdrawal or suppression of the "wakefulness drive" that contributes directly to the waking level of ventilation. This represents a large disturbance to the respiratory control system, which is forced to find a new steady state equilibrium level. The reduction in ventilation level leads to an increase in P_{aCO2} and an accompanying decrease in P_{aO2}. Subsequently, these blood gas changes evoke compensatory ventilatory changes through the chemoreflexes. Due to the presence of delays and lags in the feedback control system, rapid changes in wakefulness drive are much more difficult to compensate for than slow changes. Thus, as shown in Fig.1 (*left panel*), the system can attain a stable sleep state with its new ventilatory setpoint if sleep onset takes place gradually. However, with a more abrupt wake-to-sleep transition (but no changes in other parameters), there may be an initial period of apnea (Fig.1, *middle panel*). Subsequently, recovery from the apnea is followed by oscillations in ventilation and blood gases that can persist over several minutes before the system attains its new ventilatory setpoint in sleep. Now, consider the case where the wake-to-sleep transition follows the same rapid time-course; however, the waking drive here is assumed to be 40% larger than in the two previous cases. In this case, sleep onset leads immediately to an apnea that is approximately 30 s long; this is followed by fluctuations in blood gases and ventilation that are large enough to provoke arousal from sleep at ~t=90 s (Fig.1, *right panel*). This produces rapid restoration of the waking drive which, in turn, rapidly reduces P_{aCO2}. At this time, a new sleep onset commences; the concommitant withdrawal of waking drive, coupled with the fall in chemical drive, leads to another period of apnea. These events set the stage for cyclic alternation of apnea and hyperpnea as well as alternations between arousal and sleep. The periodicity of these oscillations in state and ventilation takes on values of approximately 60 s.

In Fig.1 (*right panel*), each hyperpneic phase of the cycle is accompanied by arousal. We have shown that this is not necessarily the case, and depending on the relative combinations of chemoreflex gain and waking drive magnitude, it is also possible to have arousals that do not occur on every hyperpneic phase - instead, these may occur on every *other* crest of the periodic ventilation cycle[12]. However, one important feature of the existing model is that it predicts sustained episodes of apnea or arousals that are highly periodic. Observations of periodic breathing during sleep reveal patterns that can be much less regular. Figure 2 illustrates one such instance. Here, respiratory airflow, arterial O_2 saturation (S_{aO2}) and the electroencephalogram (EEG, C3-A2) were measured in a sleeping healthy subject acclimatized to the simulated equivalent of 6,100 meters above sea level. The segment of data displayed shows sustained periodic breathing with mean inspiratory flow and S_{aO2} oscillating out of phase with each other (Fig.2, *Upper Panel*). The centroid or mean frequency of the EEG corresponding to this segment of data shows significant increases that are closely associated with the hyperpneas of each of the periodic breathing cycles, except for one case at ~34 s. Here, no arousal was detected

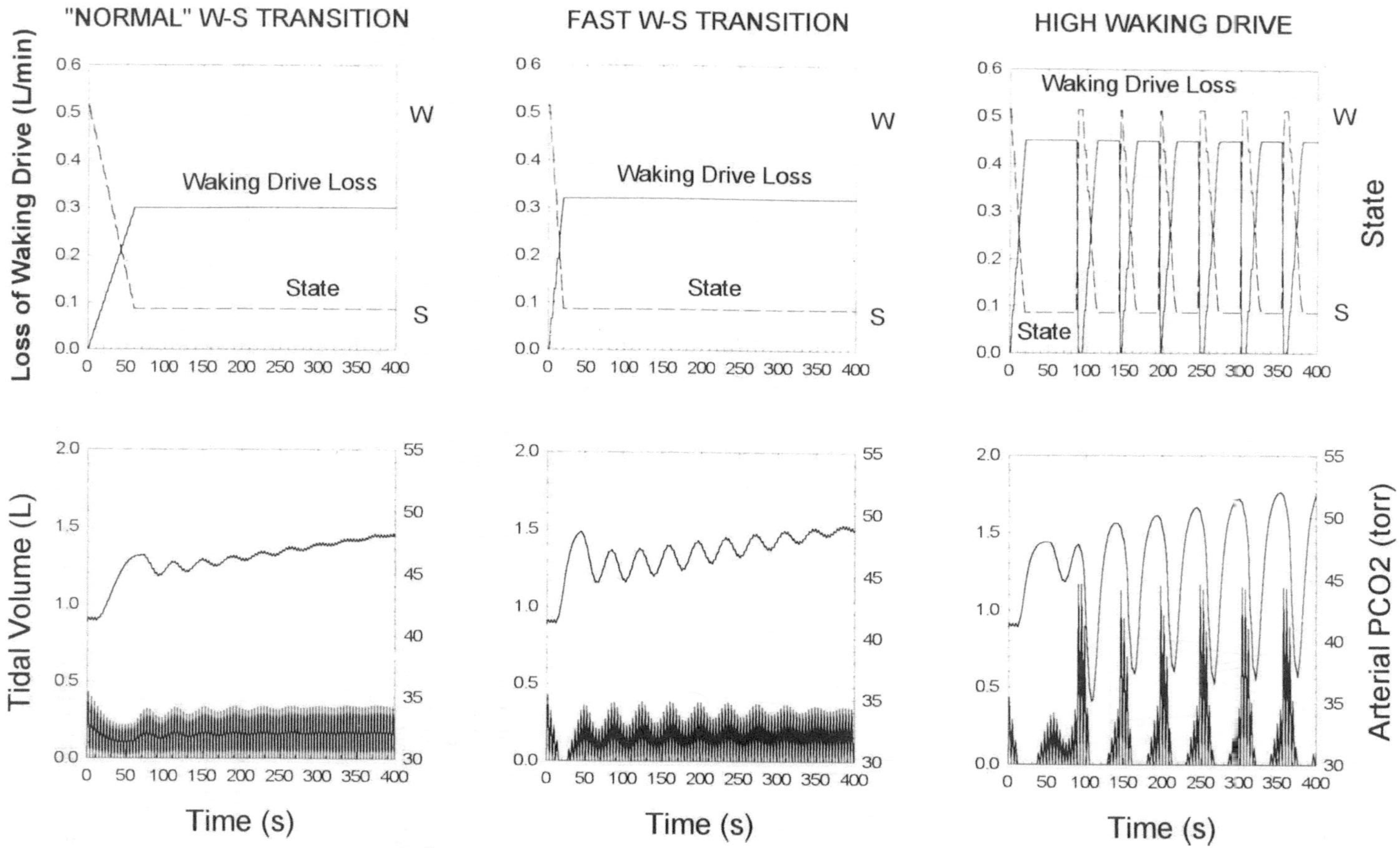

Figure 1. Simulations with the model of Khoo et al.[12], showing how the speed of sleep onset and magnitude of the waking drive affect ventilatory stability. *Left Panel*: "Control" case with "normal" wake-to-sleep transition time and waking drive. Sleep onset produces some minor ventilatory fluctuations which eventually damp out when the next steady state is attained. *Middle Panel*: More abrupt sleep onset leads to initial apnea, followed by ventilatory oscillations lasting several minutes. *Right Panel*: With increased waking drive, abrupt sleep onset can lead to sustained episodes of periodic apnea and arousal.

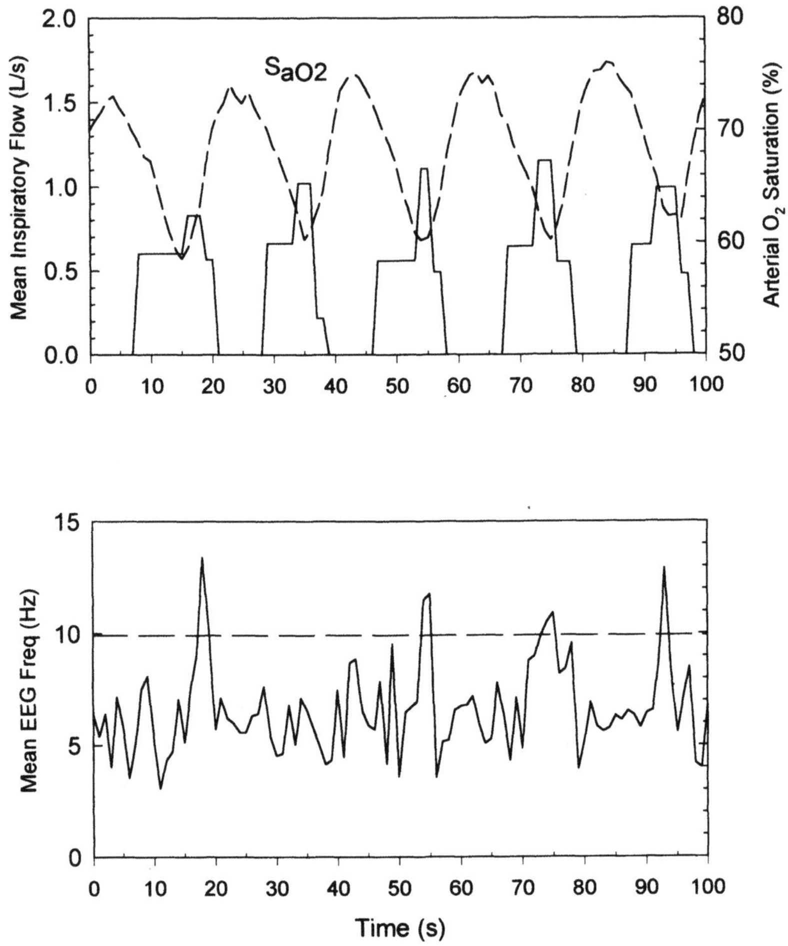

Figure 2. Periodic breathing in a healthy subject during sleep at a simulated altitude of 6,100 m. *Upper Panel*: Time courses of S_{aO2} (broken tracing) and mean inspiratory flow, where the latter is assumed constant for each breath, so that the expiratory portions of the breath cycle are not displayed. Lower Panel: Corresponding time course of the mean EEG frequency, computed from running power spectra (evaluated at 1-second intervals) of the C3-A2 EEG channel. The dashed horizontal line marks the level above which increases in mean EEG frequency are likely to reflect true arousals and unlikely to be due to random fluctuations (5% significance level). See Ref. 13 for more details.

either by visual scoring or by computer analysis. This phenomenon was not specific to this particular segment of data or to this particular individual[13]. Analysis of the data from other subjects also showed a relationship between periodic breathing and arousal that was more irregular than had been predicted by our existing model.

3. EFFECT OF AROUSAL ON STATE AND VENTILATORY OSCILLATIONS

Could the irregularity in the relationship between periodic breathing and arousal be due to inherent stochasticity in the real system? This stochasticity could be due to a variety of sources: fluctuations in blood flow between the lungs and chemoreceptors, fluctuations in cerebral blood flow, time-varying properties of the chemoreceptor cells and the associated neuronal bodies, or time-varying properties of the respiratory centers. Furthermore, the model assumed that arousal would occur once the sum of all chemical drive inputs exceeded a given drive threshold. Could there be direct afferent pathways from the medullary chemoreceptors or carotid bodies to the reticular activating system, which is believed to be responsible for generating arousal, or is the stimulus to arousal derived from the respiratory centers after the chemical inputs have been integrated?

3.1. The Modified Model

In order to address these questions, we made the following modifications to the model: (a) stochastic variability was introduced into the breath pattern by adding white noise (with coefficient of variation of 15%) to the inspiratory duration (T_I) and the expiratory duration (T_E); (b) white noise was also added to the arousal threshold level. Modification "b" may be interpreted as the "lumping together" of all other possible sources of stochasticity in the system that could affect the magnitude and timing of inputs to the reticular activating system. These could include fluctuations in brain blood flow, fluctuations in cardiac output and noise in the associated receptors and neural circuitry. We also explored how different mechanisms for arousal would affect the relative timing of arousals and periodic breathing. As shown in Figure 3, the three possibilities investigated were:

1. arousal due to direct afferents from the medullary chemoreceptors - here, arousal would occur when brain compartment P_{CO2} (P_{BCO2}) exceeds the (stochastic) arousal threshold.
2. arousal due to direct afferents from the carotid-body chemoreceptors - here, arousal would occur when S_{aO2} falls below the (stochastic) arousal threshold.

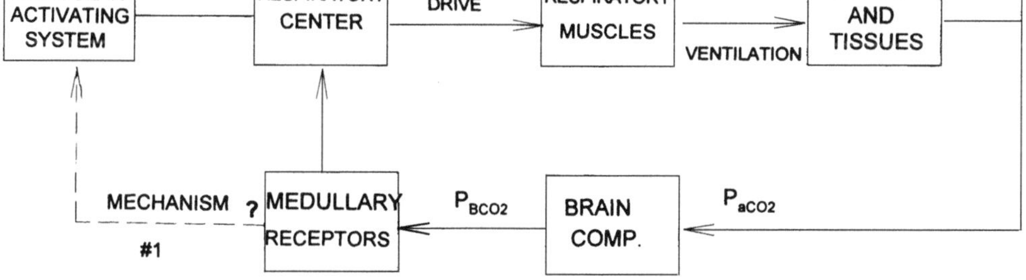

Figure 3. Schematic diagram of three possible mechanisms for respiratory-related arousal. Mechanism #1 represents postulated afferents to the reticular activating system from the medullary chemoreceptors. Mechanism #2 assumes direct arousal inputs from the carotid-body chemoreceptors. Mechanism #3 represents afferents to the reticular activating system from the respiratory centers after chemical drive integration or indirectly from mechanoreceptors in the lungs and chest wall.

3. arousal due to ventilatory drive, regardless of the primary source of stimulation. This could reflect afferents from the respiratory centers or indirectly from mechanoreceptor feedback of ventilatory effort.

3.2. Simulation Results vs Data

As expected, simulations with the modified model showed a rich variety of ventilatory and arousal patterns. Arousal was not always associated with each periodic breathing cycle. The built-in stochasticity led to the appearance of spontaneous arousals, which were unrelated to the periodic breathing cycle. In order to assess how the different mechanisms of arousal might affect the relative timing of arousal and to evaluate model performance vis-a-vis observed data, the following quantities were computed from the model simulations. Following the method employed by Anholm et al.[14], we computed the time (in seconds) between apnea termination and the start of the closest arousal (TAAR) for consecutive cycles of periodic breathing. Arousals that occurred prior to the end of apnea were assigned negative TAAR values, while arousals that happened to coincide with apnea termination would have a TAAR value of zero. Subsequently, we computed from each simulation sequence of approximately 3000 s: (a) the relative incidence of arousals with positive, negative and zero TAAR values; (b) the mean and standard deviation of the positive TAAR values; and (c) the mean and standard deviation of the negative TAAR values.

For each of Mechanisms #1, #2 and #3, the mean arousal threshold and coefficient of variation of the arousal threshold were adjusted until the above derived quantities (a, b and c) were matched as closely as possible with the corresponding values derived from empirical data. Figure 4 shows the result of this exercise. At 6,100 m, Anholm et al.[14] found

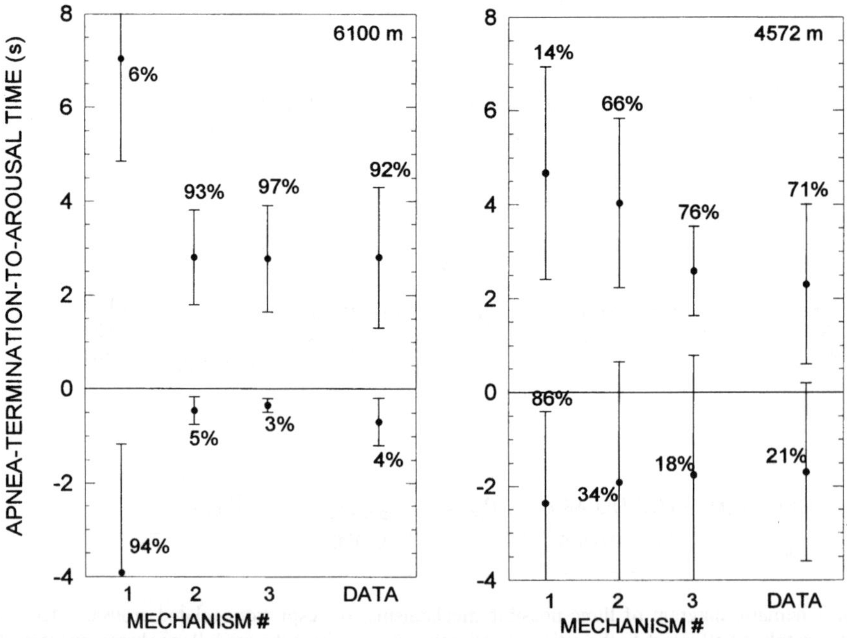

Figure 4. Model predictions of the distribution of TAAR times based on the three postulated mechanisms are compared against corresponding values derived from empirical data for altitudes of: (*Left Panel*) 6,100 m and (*Right Panel*) 4,572 m.

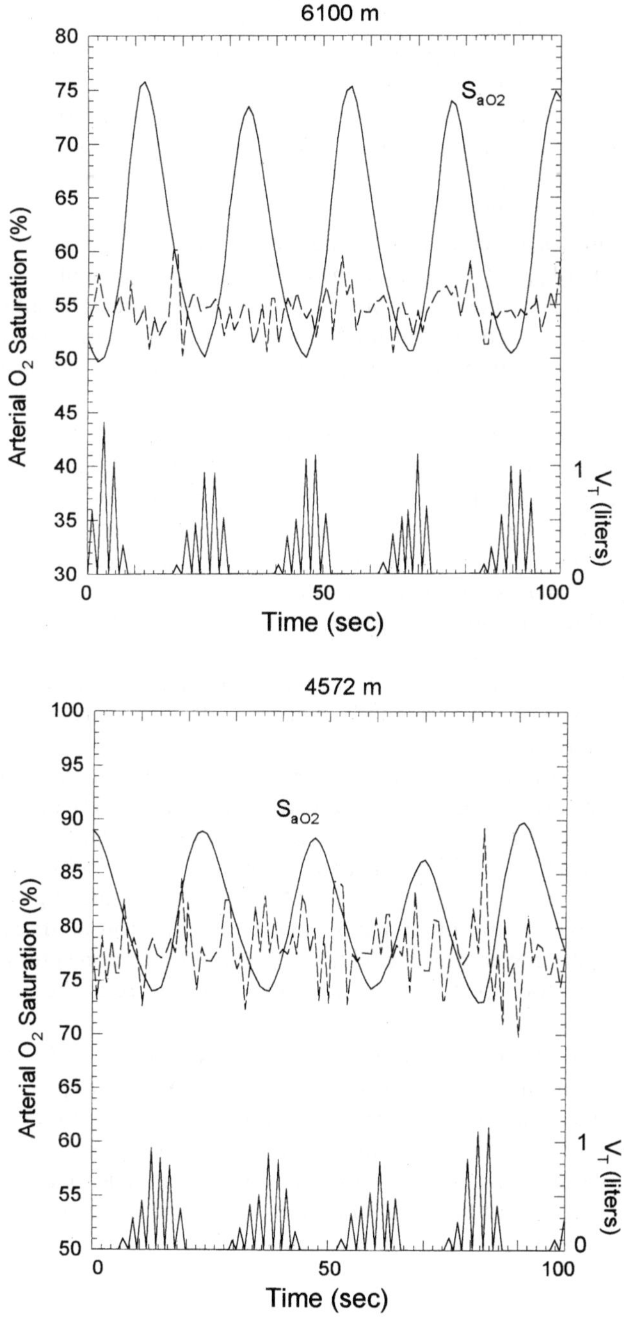

Figure 5. Model simulations of periodic breathing and arousal at 6,100 m (*Left Panel*) and 4,572 m (*Right Panel*) assuming that arousal would be triggered when hypoxia at the level of the carotid-bodies exceeds (or equivalently, when S_{aO2} falls below) a threshold level (Mechanism #2). Note that the mean and coefficient of variation of the arousal threshold (shown as the broken tracing) were selected to produce TAAR distributions that best-matched empirical data.

92% of all associated arousals to occur, on average, 2.8 s after apnea termination (TAAR=2.8 ± 1.5) and only 4% to occur on average 0.7 s before the end of apnea (TAAR=-0.7 ± 0.5). At 4,572 m, the corresponding distributions were 71% for arousals occurring after apnea termination (TAAR=2.3 ± 1.7) and 21% for arousals occurring before the end of apnea (TAAP=−1.7 ± 1.9). At 6,100 m, model predictions assuming Mechanisms #2 and #3 matched the empirical values closely (Fig.4, *Left Panel*); however, at 4,572 m, only the prediction assuming Mechanism #3 fell close to the observed data (Fig.4, *Right Panel*). Thus, the ability of the model to replicate the observed findings was most enhanced if Mechanism #3 was assumed.

At both altitudes, Mechanism #1 generated predictions that were substantially off the mark, suggesting that P_{CO2} at the level of the medullary chemoreceptors is unlikely to be the primary factor in the generation of arousals . We therefore focussed attention on the other two mechanisms. Examples of model simulations based on Mechanism #2 at the two altitudes are displayed in Figure 5. The corresponding simulations assuming Mechanism #3 are shown in Figure 6.

In order to better differentiate the effects of Mechanisms #2 and #3 on TAAR distribution, a comparative sensitivity analysis was conducted. Here, the mean levels and coefficients of variation of the arousal thresholds were independently changed around the "optimal values", and the consequences of these changes on TAAR distribution were recorded. Figure 7 displays the results of the sensitivity analysis assuming Mechanism #2. Here, the sensitivity analysis for 6,100 m was based on nominal values of 54.5% S_{aO2} for the mean arousal threshold and a coefficient of variation of 4.5%, while the corresponding nominal values for 4,572 m werre 77.5% S_{aO2} and 4.5%. In the 6,100 m case, simulations were performed with the mean arousal threshold lowered to 52%, and subsequently, with the threshold increased to 57%. In both instances, coefficient of variation was kept fixed at 4.5%. Then, the threshold level was kept constant at 54.5%, while simulations were conducted with the coefficient of variation decreased to 2% and increased to 7%. The changes in these parameters always produced a greater discrepancy between the model and the observed data. For example, lowering the arousal threshold to 52% S_{aO2} led to the appearance of only arousals that appeared following apnea termination (100% for arousals with positive TAAR); furthermore, the predicted positive TAAR values became significantly larger than the observed values. On the other hand, increasing the arousal threshold to 57% S_{aO2} produced significantly fewer arousals that occurred following apnea termination (46%) and more that occurred before apnea termination (39%), compared to the data, which showed a 92%:4% distribution.

Figure 8 shows the corresponding results for Mechanism #3. Comparing the two sets of results suggests that Mechanism #3 is the more robust of the two possibilities with respect to changes in arousal threshold, ie. deviations away from the "optimal combinations" of mean arousal threshold and coefficient of variation are less likely to substantially affect TAAR distributions at both altitudes. This has led us to conclude that, taking into consideration the limitations imposed by the assumptions embodied in our modified model, arousal is most likely to be generated through Mechanism #3.

4. IMPROVED MODEL OF STATE-CHEMOREFLEX INTERACTION

Our existing model[12] of SDB reflects the widely held belief that sleep produces a modest reduction in chemoresponsiveness. A closer look at the empirical evidence, however, will reveal considerable variability in the magnitude of the reduction across, and even within,

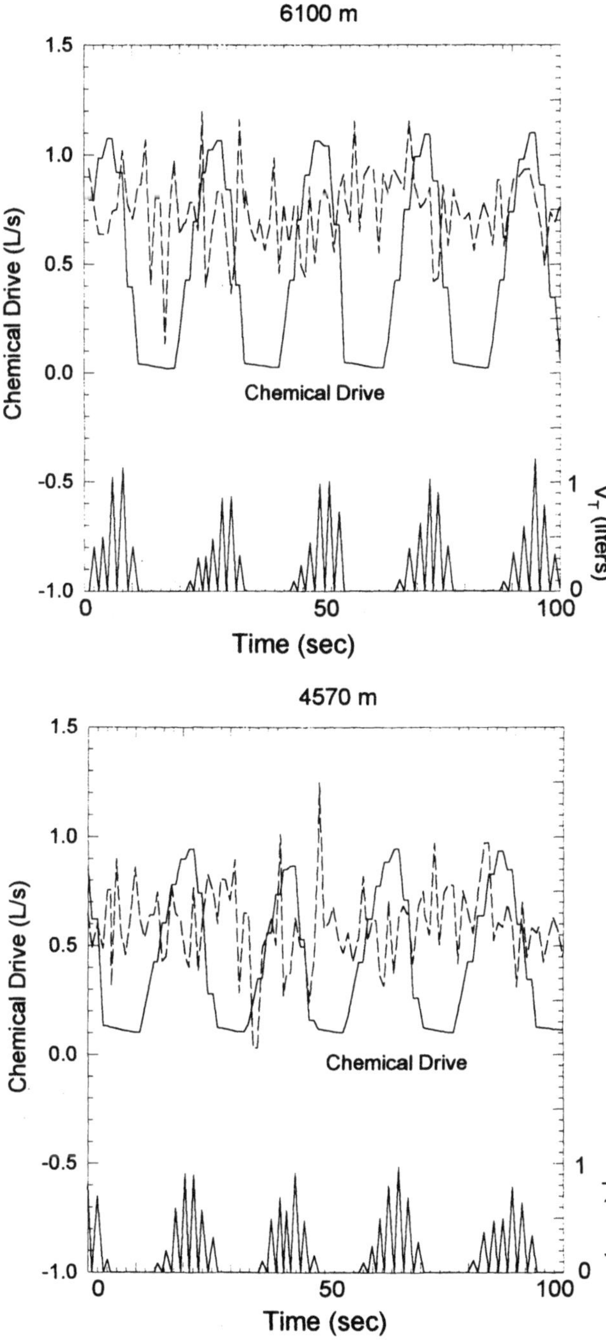

Figure 6. Model simulations of periodic breathing and arousal at 6,100 m (*Left Panel*) and 4,572 m (*Right Panel*) assuming that arousal would be triggered when the integrated neural drive to breathe or feedback from lung and chest-wall mechanoreceptors exceeds a threshold level (Mechanism #3). Note that the mean and coefficient of variation of the arousal threshold (shown as the broken tracing) were selected to produce TAAR distributions that best-matched empirical data.

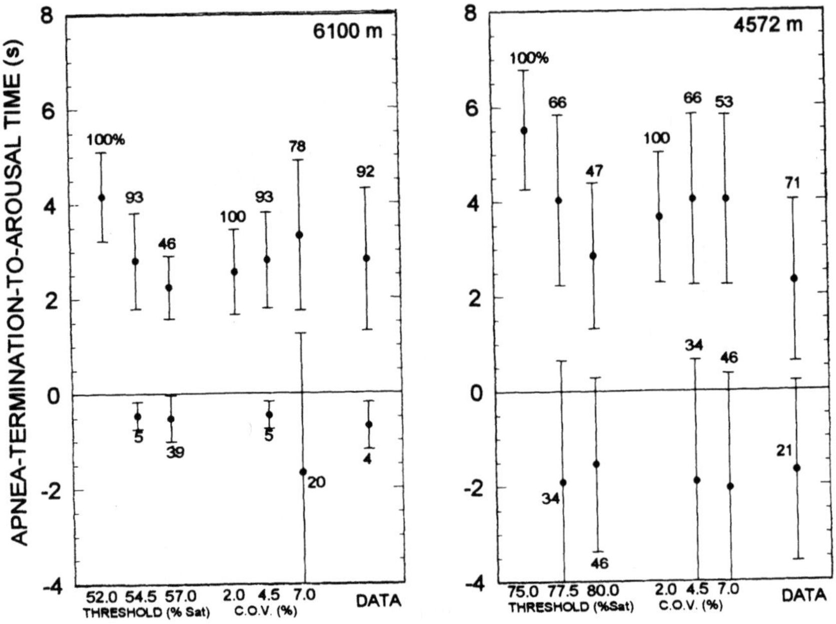

Figure 7. Effects of changes in values of mean arousal threshold level and coefficient of variation on predicted TAAR distributions, assuming Mechanism #2. The "optimal" mean arousal threshold was 54.5% (units of O_2 saturation), and "optimal" coefficient of variation was 4.5% (percentage of the mean level).

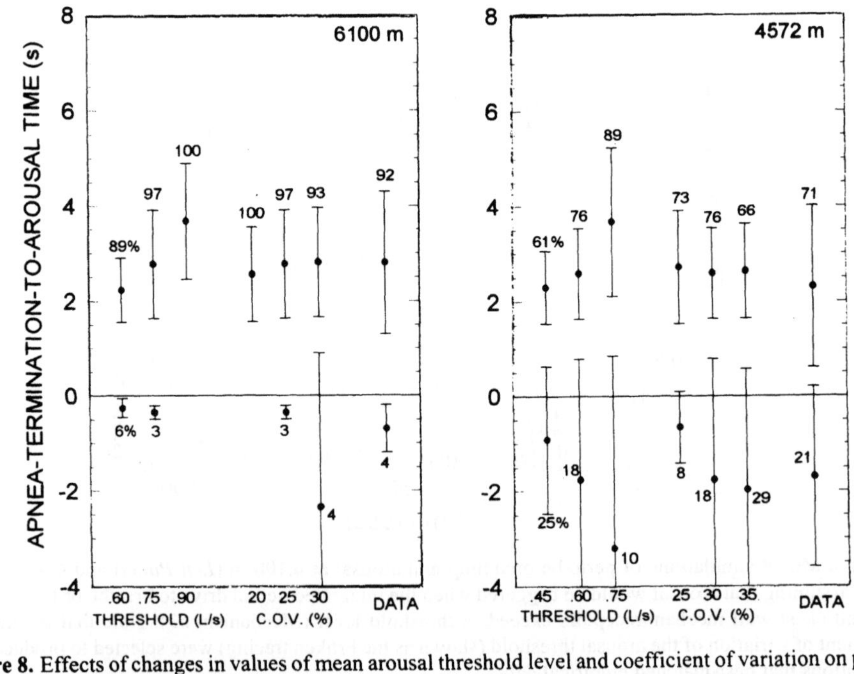

Figure 8. Effects of changes in values of mean arousal threshold level and coefficient of variation on predicted TAAR distributions, assuming Mechanism #3. The "optimal" mean arousal threshold was 0.75 L/s, and the "optimal" coefficient of variation was 25%.

individuals[15,16]. In some instances, increases in chemoresponsiveness have been reported. Furthermore, our current notion that chemoreflex gain is depressed in sleep is derived primarily from inhalation tests that were conducted under steady state or quasi-steady conditions. In order to determine how sleep may affect the *dynamic* characteristics of the respiratory controller, we performed an experimental study in which the ventilatory responses of normal humans to rapidly changing levels of inhaled CO_2 were measured awake and asleep (NREM Stage 2) under normoxic condiitions[17]. We found no significant decrease in hypercapnic gain evaluated in the band of frequencies consistent with periodic breathing, ie. between 0.01 Hz (or equivalently, a cycle time of 100 s) and 0.03 Hz (cycle time of 33 s). A similar study by Modarreszadeh et al.[5], conducted under hyperoxic conditions, revealed a modest reduction in CO_2 gain. Taken together, these studies suggest that in some subjects peripheral chemoresponsiveness may be elevated in the light stages of sleep while central chemoresponsiveness is depressed. The information that can be derived from both studies is limited, however, since they focus on how dynamic chemoresponsiveness changes between two steady (sleep-wake) states. During SDB, there are no real "steady" states: instead, there is the continual punctuation of sleep with transient arousals and the constant alternation between hyperpnea and apnea. How is chemoreflex control altered under such dynamic conditions? In the existing model, we assume that arousal would lead to an immediate switch from the reduced chemoreflex gain consistent with sleep to the higher chemoreflex gain associated with wakefulness. At the same time, there would also be complete restoration of the wakefulness drive. How realistic are these assumptions? We undertook empirical studies to address this question and to obtain a more quantitative characterization of ventilatory control during arousal.

4.1. Experimental Data Selection and Preprocessing

To investigate how arousal affects the chemoreflex controller, we analyzed selected segments of data collected in our previous study of dynamic chemoresponsiveness[17]. In that study, overnight polysomnographic recordings were made of healthy adult volunteers who breathed through a face mask. The inspiratory port of the mask was connected to a three-way inflatable balloon valve that enabled the inhaled gas composition to contain either air or a mixture of 5%CO_2, 21% O_2 and 74% N_2. Throughout most of the night, the subjects breathed air. However, at various times during the study, the inhalate was switched on a breath-by-breath basis between air and the CO_2 mixture. The sequence of switches was determined by a controlling computer that employed a five-digit shift register algorithm for generating pseudorandom numbers[18]. Each test sequence consisted of a total of 62 breaths. In addition to the standard polysomnographic variables, we measured respiratory airflow and end-tidal P_{CO2} (P_{ETCO2}).

From the airflow and exhaled CO_2 signals, tidal volume (V_T), inspiratory (T_I) and expiratory durations, and P_{ETCO2} were determined on a breath-by-breath basis using a computerized algorithm. We used the mean inspiratory flow (V_T/T_I) to represent ventilatory drive. Since the breath duration was variable from breath to breath, the V_T/T_I and P_{ETCO2} time-series were converted into equivalent uniformly-spaced sequences using the resampling method of Waggener et al.[19]. Thus, assuming that each breath parameter remains constant over a given breath, the average value of the parameter was calculated over successive intervals of fixed duration equal to the mean breath period (T_T) of the data segment. The power spectra from successive segments of the EEG data (central lead) were computed using an autoregressive spectral estimation approach[20]. From each power spectrum, we calculated the "EEG high-frequency power", ie. the combined power of the alpha (8–12 Hz) and beta (14–25 Hz) bands expressed as a percentage of the total EEG power between 1 and 25 Hz. These computations were performed over EEG data segments of duration T_T so that the state

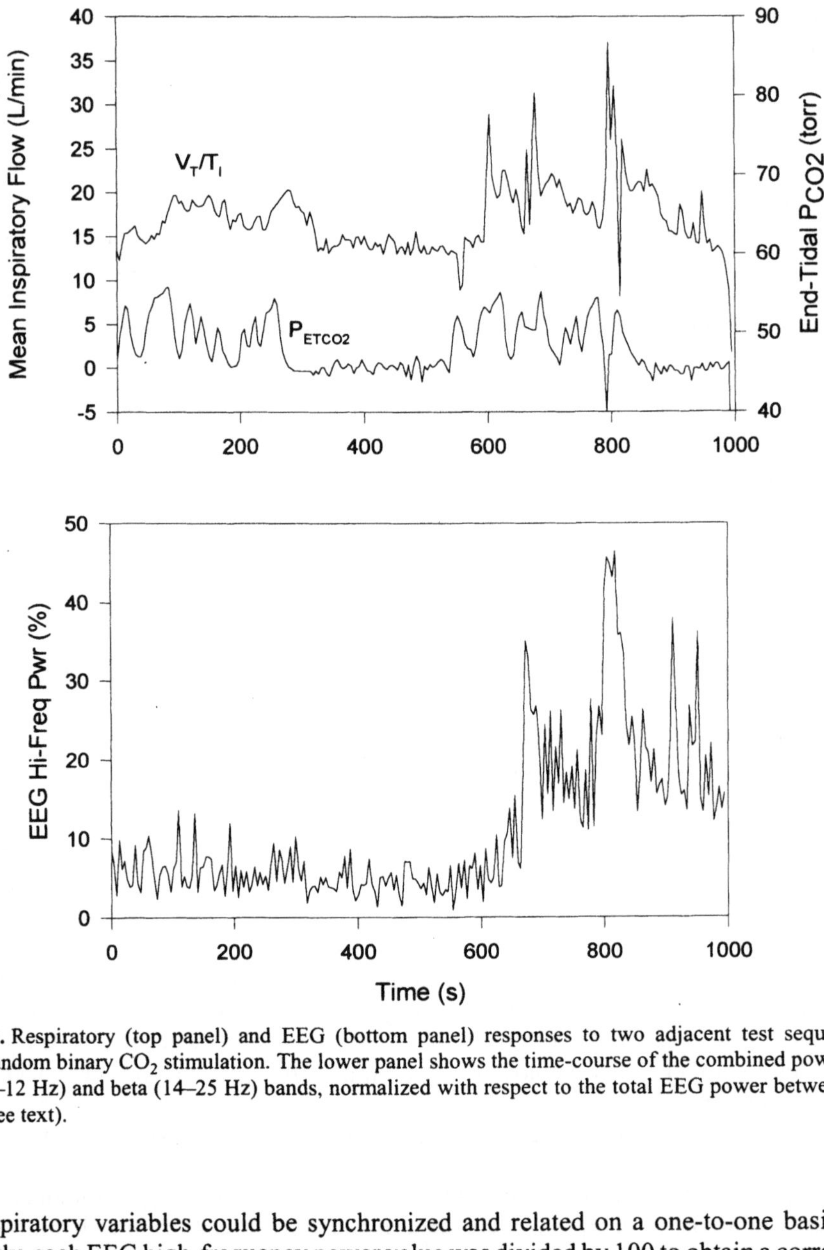

Figure 9. Respiratory (top panel) and EEG (bottom panel) responses to two adjacent test sequences of pseudorandom binary CO_2 stimulation. The lower panel shows the time-course of the combined power of the alpha (8–12 Hz) and beta (14–25 Hz) bands, normalized with respect to the total EEG power between 1 and 25 Hz (see text).

and respiratory variables could be synchronized and related on a one-to-one basis. Subsequently, each EEG high-frequency power value was divided by 100 to obtain a corresponding "state index", S(n).

During administration of the test sequence, the appearance of transient arousals was not infrequent, particularly towards the end of the test. Since the concentration of CO_2 employed was significantly lower than the levels previously reported to induce arousal, we speculate that the abrupt changes in state may have been due to stimulation of airway CO_2 or irritant receptors by the rapid changes in CO_2 content, or stimulation of the olfactory receptors. The data segments chosen for analysis generally contained the responses to two tests conducted a few minutes apart. The first response would contain no arousal while, during the second test sequence, arousal would occur approximately near the middle of the test. An example of one the segments analyzed is shown in Figure 9. In this experiment, the subject was in Stage 2 sleep. The top panel of Figure 9 shows P_{ETCO2} and V_T/T_I during the administration of two test sequences, while the lower panel displays the time-course of the EEG high-frequency power. During the first test sequence (0 to 280 s), there was a clear modulation of V_T/T_I by the pseudorandom changes in inhaled CO_2, but the EEG showed no discernible changes in sleep stage. However, soon after the start of the second test sequence (530 to 820 s), large and abrupt increases in ventilation appeared. The accompanying large increases in EEG high-frequency power confirm that these ventilatory changes reflect arousal from sleep. However, in this and most other cases, the change in state was generally not sustained and the subject reverted back to a stable sleep stage soon after the test sequence was terminated.

4.2. Model Development

In order to determine the relative contributions of chemical drive and state-related drive in the arousal responses that we measured, we first considered the mathematical characterization of the steady-state effects of sleep on chemoreflex control. Sleep is known to affect the slope and position of the steady state ventilatory response to hypercapnia. This observation can be expressed mathematically as:

$$\Delta D = (G_{ch} + G_i[S]) \cdot \Delta P_{ACO2} + G_S \cdot \Delta S \tag{1}$$

where ΔD, ΔP_{ACO2} and ΔS represent changes in ventilatory drive, alveolar (or equivalently, arterial) P_{CO2} and state, reespectively, from some nominal control level. The term $G_S \cdot \Delta S$ represents an additive state-related component to total ventilatory drive, or the steady state "wakefulness drive". The chemoreflex gain is composed of state-related (G_i) and non-state-related (G_{ch}) components. G_i determines the magnitude of the effect of state on chemoresponsiveness.

A straightforward extension of the steady state model given by Equation (1) to incorporate linear dynamics is:

$$\Delta D(n) = \sum_{k=0}^{M} [h_{ch}(k) + h_i(k) \cdot \Delta S(n)] \cdot \Delta P_{ACO2}(n - k - N_D) + G_S \cdot \Delta S(n) + e(n) \tag{2}$$

where $h_{ch}(k)$ $\{0 \le k \le M\}$ and $h_i(k)$ $\{0 \le k \le M\}$ represent the non-state-dependent and the state-dependent impulse responses, respectively, of the respiratory controller. Thus, there are 3 postulated components of the total change in drive. First, there is a non-state-dependent, purely chemical component $\Delta D_{ch}(n)$, which in Equation (2) is given by the convolution of $h_{ch}(k)$ with delayed values of ΔP_{ACO2}. Secondly, there is a chemoreflex component which depends on and is modulated by changes in state, which we will refer to as the "interaction component", $\Delta D_i(n)$. In Equation (2), this component is produced by the convolution of $h_i(k)$ with delayed values of ΔP_{ACO2}, but the amplitude of $h_i(k)$ is modulated by the current value of $\Delta S(n)$. Thus, here, the assumption is that any change in state at "breath" n will affect ventilatory drive within the same breath but will have no effect on drive in future breaths. This is simply a direct extension of the steady state notion that changes in hypercapnic ventilatory response slope accompany changes in sleep-wake state. Finally, there is the direct additive contribution to drive that is proportional to the measured change in state, $\Delta D_s(n) = G_S \Delta S(n)$.

Equation (2) uses a breath-number time-base (n), which implicitly assumes that each "breath" is of uniform duration T_T. Another assumption built into Equation (2) is the lung-to-chemoreceptor delay of N_D breaths. The last term in Equation (2), e(n), represents the discrepancy (or residual) between the observed drive, which we assume is measured in the form of V_T/T_I, and the predicted value at "breath" n.

4.3. Parameter Estimation

In order to determine the dynamics of state-respiratory interaction using the controller model described by Equation (2), we need to estimate all the unknown parameters (or coefficients) from the data. The unknown parameters to be identified total 2M+3, where M represents the duration (in"breaths") of the controller impulse response. If the longest time constant of the respiratory chemoreflexes is 100 s, then assuming an average breath duration of 4 s, M should be on the order of 75. Since our selected data segments were no longer than 250 breaths, the problem is clearly overparametrized. Solving Equation (2) by least-squares would yield meaningless parameter estimates. Thus, in order to circumvent this difficulty, it is necessary to constrain the "forms" of the impulse responses. One way of doing this is to express the impulse responses as the sum of a finite number of weighted basis functions. In our case, we chose the Laguerre set, since they have a "built-in" exponential that enables the constructed impulse response to mimick the long-tailed distributions which have been deduced in previous studies of breath-to-breath CO_2 responses[5,18]. The Laguerre expansion technique has also proven useful in characterizing the dynamics of other physiological systems[21]. Thus, $h_{ch}(k)$ and $h_i(k)$ are now expressed as follows:

$$h_{ch}(k) = \sum_{j=0}^{q} a_j L_j(k) \tag{3}$$

$$h_i(k) = \sum_{j=0}^{q} b_j L_j(k) \tag{4}$$

where $L_j(k)$ $\{0 \le k \le M\}$ represents the j-th order discrete-time orthonormal Laguerre function, and a_j and b_j $(0 \le j \le q)$ are unknown coefficients to be estimated. For each order, j, the values of the function at all times k are defined, and given by[22]:

$$L_o(k) = \sqrt{\alpha^k(1-\alpha)}\,, \qquad 0 \le k \le M \tag{5}$$

$$L_j(k) = \sqrt{\alpha}\; L_j(k-1) + \sqrt{\alpha}\;\; L_{j-1}(k) - L_{j-1}(k-1) \qquad 1 \le j \le q, \;\; 1 \le k \le M \tag{6}$$

The parameter α $(0<\alpha<1)$ determines the rate of exponential decline of the Laguerre function, and is selected such that, for given M and q, the values of the constructed impulse response become insignificant as k approaches M. Laguerre functions of orders 0 to 4 are displayed in Figure 10.

Substituting Equations (3) through (6) into Equation (2) enables us to obtain an expression relating change in drive to new derived variables $\{u_j(n), v_j(n)\}$ with far fewer unknown parameters $(2q+3 \ll 2M+3)$:

$$\Delta D(n) = \sum_{j=0}^{q} a_j \cdot u_j(n) + \sum_{j=0}^{q} b_j \cdot v_j(n) + G_s \cdot \Delta S(n) + e(n) \tag{7}$$

where

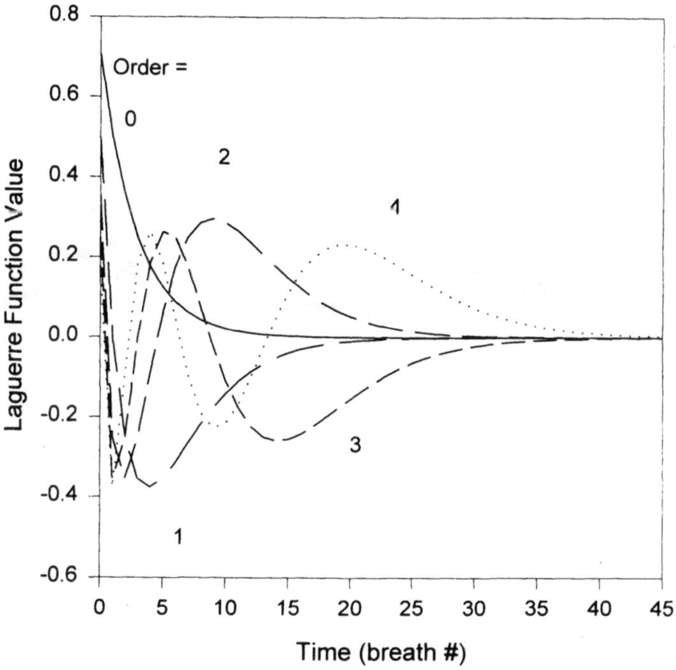

Figure 10. Examples of Laguerre functions of orders 0 (solid tracing), 1 (long dashes), 2 (medium-length dashes), 3 (short dashes) and 4 (dotted tracing), as generated by Equations (5) and (6). The impulse responses to be estimated are constructed from the sum of weighted combinations of *q* such functions.

$$u_j(n) = \sum_{k=0}^{M} L_j(k) \cdot \Delta P_{ACO2}(n - k - N_D) \tag{8}$$

$$v_j(n) = \sum_{k=0}^{M} L_j(k) \cdot \Delta P_{ACO2}(n - k - N_D) \tag{9}$$

Equation (7) thus becomes a linear relation that can be solved by least-squares minimization.

In addition to the parameters a_j $\{0 \leq j \leq q)$, b_j $\{0 \leq j \leq q\}$ and G_S, the number of Laguerre functions to be used (q) and the delay (N_D) also needed to be determined. This was achieved in the following way. Least squares minimization was performed to solve Equation (7) for a range of values of q and N_D. For each combination of q and N_D, the Akaike Information Criterion[20] (AIC) was computed. Finally, a global search was performed to determine the combination of q and N_D that produced the lowest AIC. Values of N_D from 1 to 5, which were considered physiologically feasible, were explored. The value of N_D=0 was omitted since this would yield a "false minimum" - a solution that reflected the effect of ventilation on P_{ACO2} rather than the other way around. We have found from previous studies that this simple constraint circumvents the problem of feedback, since we are really dealing with a closed-loop control system but making measurements of variables within the loop[17,23,24,25]. Finally, since the approximate duration of the ventilatory controller impulse response is known from previous studies[5,17,18,25], the parameter α could be determined for any given combination of q and M using an algorithm developed by Marmarelis[22].

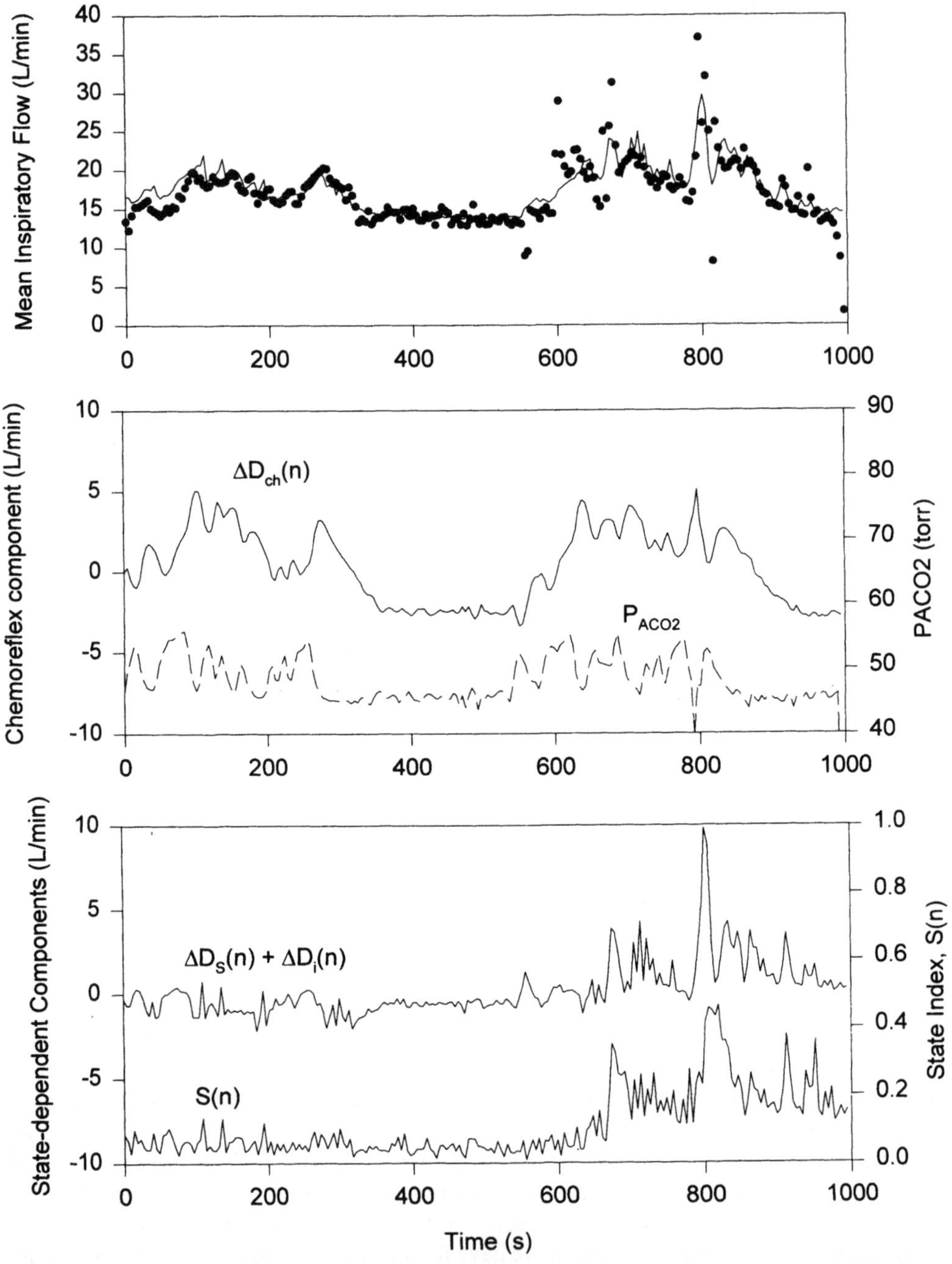

Figure 11. Example of model performance in one dataset (Subject #1). *Top Panel*: Predicted (solid tracing) vs. measured (closed circles) V_T/T_I. *Middle Panel*: Measured P_{ACO2} and corresponding prediction of the non-state-related chemical component, $\Delta D_{ch}(n)$. *Bottom Panel*: Measured EEG high-frequency power, displayed as the state index S(n), and the sum of the state-dependent components of drive, $\Delta D_S(n)$ and $\Delta D_i(n)$.

4.4. Results

An example of how well the model predicts the data is shown in Figure 11 (top panel). Here, the V_T/T_I sequence displayed in Figure 9 is shown as closed circles. The corresponding best-fit model prediction is shown as the solid tracing. With the exception of a few points in the latter half of the sequence, there is excellent agreement between model prediction and data. The middle panel of Figure 11 shows measured P_{ETCO2} ($=P_{ACO2}$) and the predicted non-state-related chemoreflex component of the changes in ventilatory drive, $D_{ch}(n)$. Accordingly, the responses to both test sequences of pseudorandomly modulated P_{CO2} are similar, and unaffected by the changes in state that begin to appear at ~620 s. The bottom panel shows the state index, S(n), and the sum of the state-dependent components, $\Delta D_S(n)$ and $\Delta D_i(n)$. $\{\Delta D_S(n)+\Delta D_i(n)\}$ shows little activity in the first two-thirds of the experimental trial, but displays abrupt increases during the arousal period.

Figure 12 shows the cumulative contributions of the 3 components to the model prediction as one progressively incorporates each component. In the top panel of Figure 12, with both state-dependent components excluded, the model accounts for only 18.3% of the total variance about the mean. Including only the additive state-dependent component leads to a substantial reduction of the residual variance. Now, the model is able to account for 45.0% of the total variance about the mean. Finally, with the incorporation of the multiplicative interaction term, 59.5% of the total variance can be explained by the model. Thus, in this subject, the additive state-dependent component accounted for most (26.7%) of the changes in ventilatory drive, followed by the purely chemical component (18.3%) and the interaction component (14.5%).

In the data selected from 6 subjects, the fraction of total variance about the mean that could be explained by the model was 50.9 ± (SD)12.2 %, with values ranging from 36.0% to 68.9%. These indices can also be expressed in terms of the correlation coefficient, r, between predicted ΔD and measured ΔD. The mean (±SD) value of r was 0.70±0.11, with the corresponding range being 0.51 to 0.83. The best-fit parameters for each dataset are displayed in Table 1. N_D ranged from 1 to 3, while the number of Laguerre functions needed to produce the optimal estimate of the impulse responses $h_{ch}(k)$ and $h_i(k)$ ranged from 7 to 9.

It is difficult to glean any kind of insight from the values of the Laguerre coefficients in Table 1. However, they can be combined to synthesize the corresponding impulse response functions $h_{ch}(k)$ and $h_i(k)$, using Equations (3) and (4). An example of this kind of synthesis is shown in Figure 13. Here, the estimated time-varying kernel $h_{ch}(k)+h_i(k)\cdot\Delta S(n)$ was estimated from the data of Subject #1 over the time interval 450 to 750 s. Thus, at each "breath" n, we computed the corresponding impulse response, but only 11 of these impulse responses, uniformly spaced in time, are shown in Figure 13 (top panel) for the sake of clarity. The bottom panel shows the EEG time course and predicted and measured changes in V_T/T_I corresponding to this segment of data. There is a noticeable increase in amplitude of the impulse response after 660 s, when the arousal begins to occur.

4.5. Limitations

Although we were able to account for much of the state-respiratory interaction in the data with the use of the model presented here, the characterization is by no means complete. The use of the EEG for deriving a template of dynamic changes in state can be criticized on a number of fronts. First, our state index was based on a central lead (C3-A2). This channel may not be the best source for arousal detection. A closer look at Figure 13 (lower panel) certainly suggests this possibility: at ~610 s, there is a transient but large increase in V_T/T_I, but the state index does not show any significant change at all. On the other hand, it may be

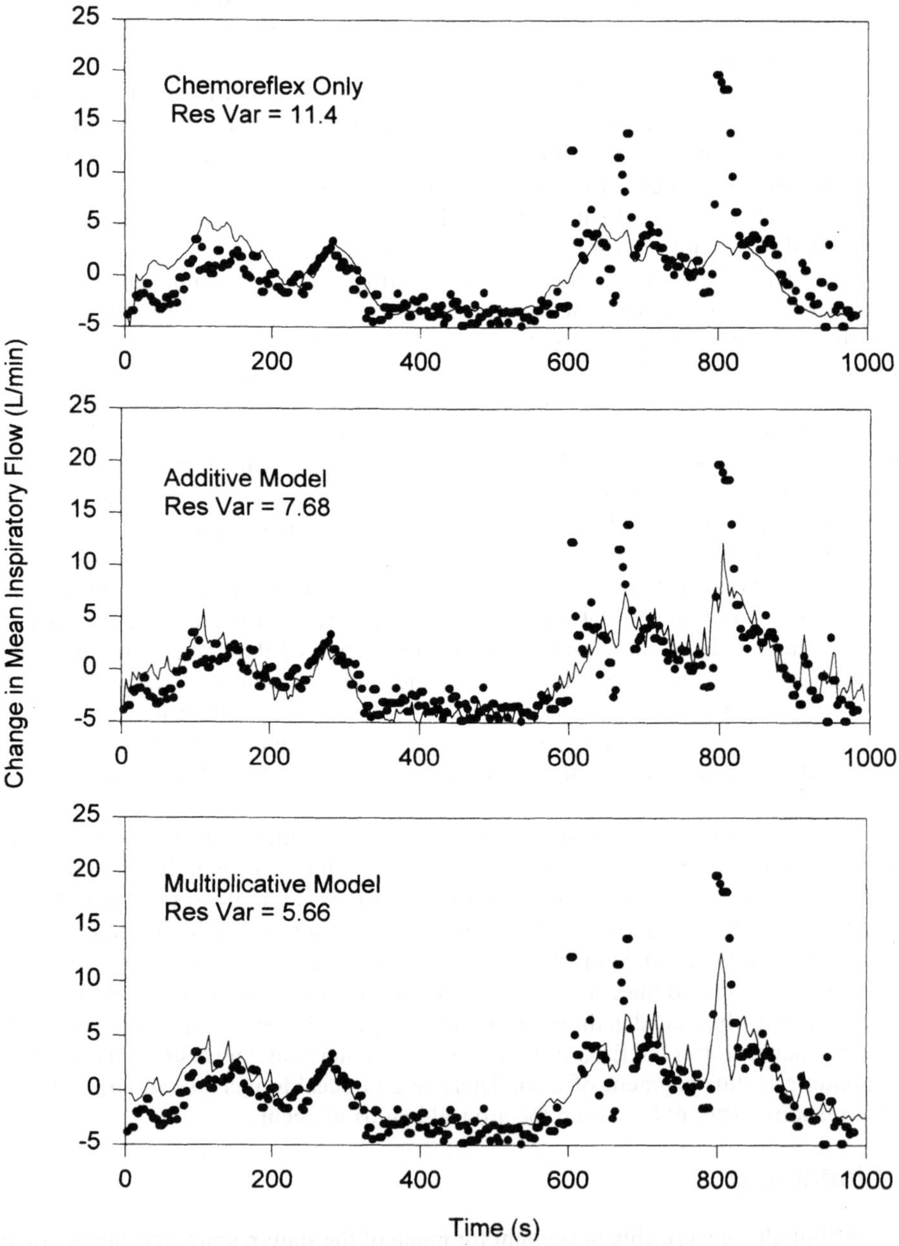

Figure 12. Measured "breath-to-breath" changes in mean inspiratory flow (closed circles) compared against model predictions (solid tracing) for the following cases: (*Top Panel*) inclusion of only the non-state-dependent, purely chemical component; (*Middle Panel*) inclusion of the chemical component and the additive state-dependent component; (*Bottom Panel*) inclusion of all three components. Note how significantly the residual variance decreases at each of the stages.

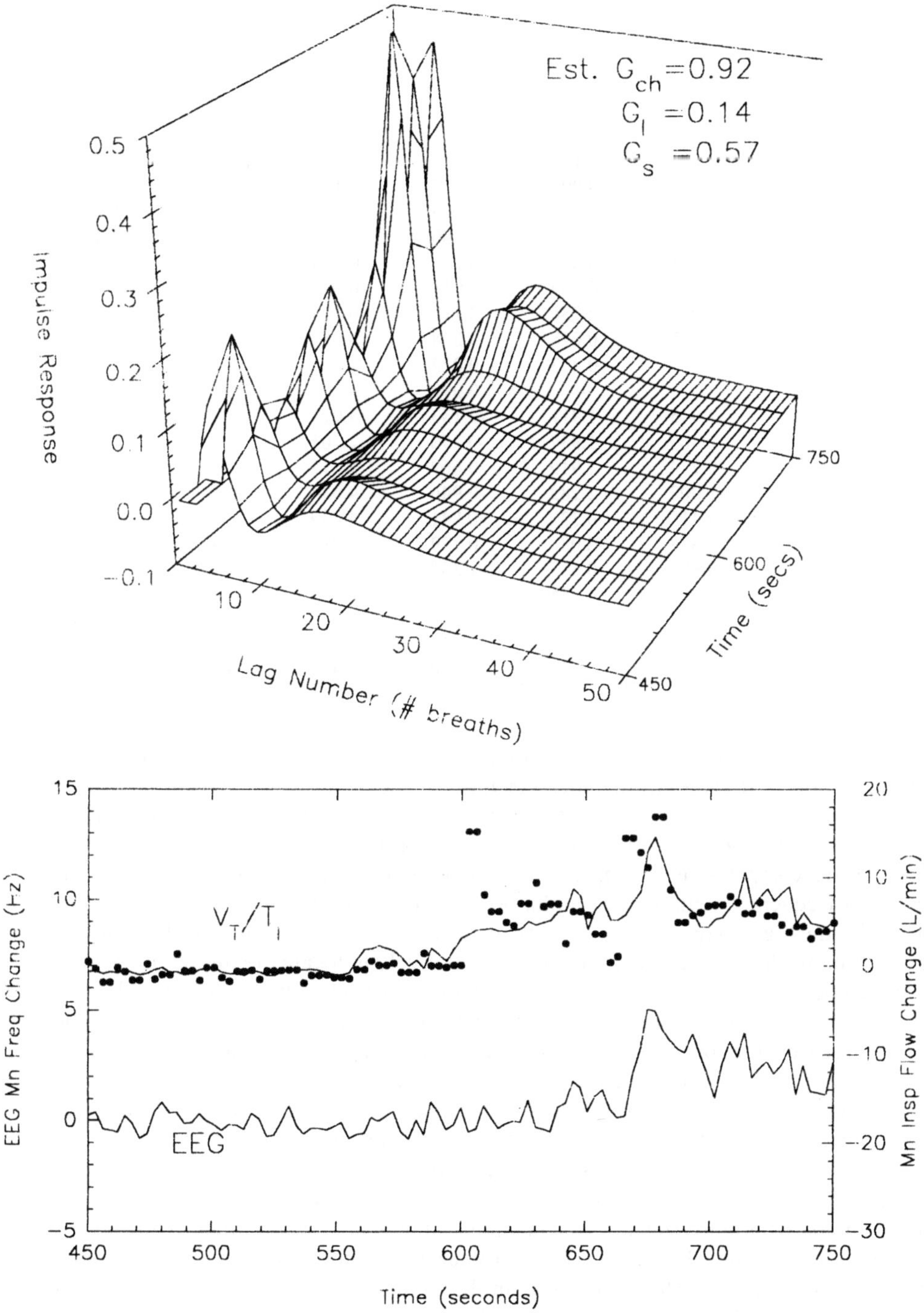

Figure 13. Top Panel: Behavior of the time-varying kernel $h_{ch}(k)+h_i(k)\cdot\Delta S(n)$, over a selected segment of data from Subject #1. Note that, for the sake of clarity, the kernel is shown only at selected intervals. *Bottom Panel*: Time-courses of predicted (solid tracing) and measured (filled circles) V_T/T_I and mean EEG frequency corresponding to the same segment of data.

Table 1. Best-fit values of the parameters deduced from the data. See text for details

	Subject					
Parameter	1	2	3	4	5	6
Avg.TT (s)	4.4	4.2	4.7	4.4	3.7	3.3
N_D	1	3	3	2	1	1
α	0.4	0.4	0.3	0.4	0.4	0.4
a_0	-0.110	0.191	0.233	-0.092	0.186	0.373
a_1	-0.422	-0.093	-0.170	-0.132	-0.157	0.381
a_2	-0.107	0.054	0.005	-0.026	0.134	0.099
a_3	-0.295	0.024	0.280	-0.100	0.106	-0.104
a_4	-0.093	0.031	0.179	-0.034	0.056	0.115
a_5	-0.070	-0.139	0.038	-0.024	-0.001	-0.148
a_6	-0.070	0.236	0.112	0.104	-0.049	0.034
a_7	-0.037	-0.028	0.228	0.054	0.041	-0.008
a_8	—	-0.203	0.022	—	—	0.121
a_9	—	—	-0.090	—	—	—
b_0	-2.645	1.911	-0.133	3.011	0.867	1.799
b_1	-2.469	1.937	-0.250	2.701	-0.907	2.020
b_2	-0.274	1.919	-0.669	8.217	0.622	-0.632
b_3	-1.007	2.282	-0.946	4.634	-0.469	-1.821
b_4	-0.415	-0.332	-0.841	4.039	-0.592	0.339
b_5	-0.838	-2.088	-0.129	1.990	-0.471	-0.911
b_6	-0.859	-0.577	-1.337	3.394	-1.299	0.451
b_7	0.061	-4.285	-0.598	2.375	0.857	1.071
b_8	—	-3.116	1.141	—	—	0.029
b_9	—	—	-0.154	—	—	—
G_s	13.16	45.59	6.390	27.42	14.04	10.42

possible also that this increase in respiration was not state-related. However, a comparison of state indices derived from other electrode placement sites should prove valuable to improving the quality of data selected to be one of our model inputs. The use of the "EEG high-frequency power" to represent the state index may also be a limitation. However, we have made comparisons among several possible frequency-based features (eg. mean or centroid frequency, delta power, alpha power, etc.) that one might extract from the EEG. Thus far, the high-frequency power appears to be the most sensitive indicator. An alternative approach, in which the EEG is used only qualitatively to locate the timing of the arousal but not included explicitly into the model, is described in a companion chapter in this volume[26].

The assumption that changes in state affect ventilatory drive "instantaneously" appears reasonable, but should be explored further. Our model is intrinsically a linear one. Thus, at any given level of chemical drive, the same change in state occurring at a later time should yield the same increase in state-dependent ventilatory drive. However, data from a recently completed study[27] suggest that this assumption may not be valid. In this study, healthy adult volunteers during Stage 2 sleep were exposed to loud 5-second tones in order to provoke arousal. The ventilatory time-coursse before and immediately following the tone were compared. Figure 14 shows an example of this kind of experiment. Time 0 s is defined as the instant at which the tone began (black bar in top panel of Fig.14). Immediately following the start of the tone, there was an abrupt increase in the EEG high-frequency power (top panel), signalling the start of an arousal. The corresponding ventilatory response, which was rapid and quite substantial, is depicted in the second panel in the form of tidal volume (only the inspiratory portions are shown), and in the third panel as P_{100}, the mouth occlusion pressure measured 100 ms after the start of each inspiration[28]. End-tidal P_{CO2} (displayed in the bottom panel) showed only a small variation. However, following the first arousal

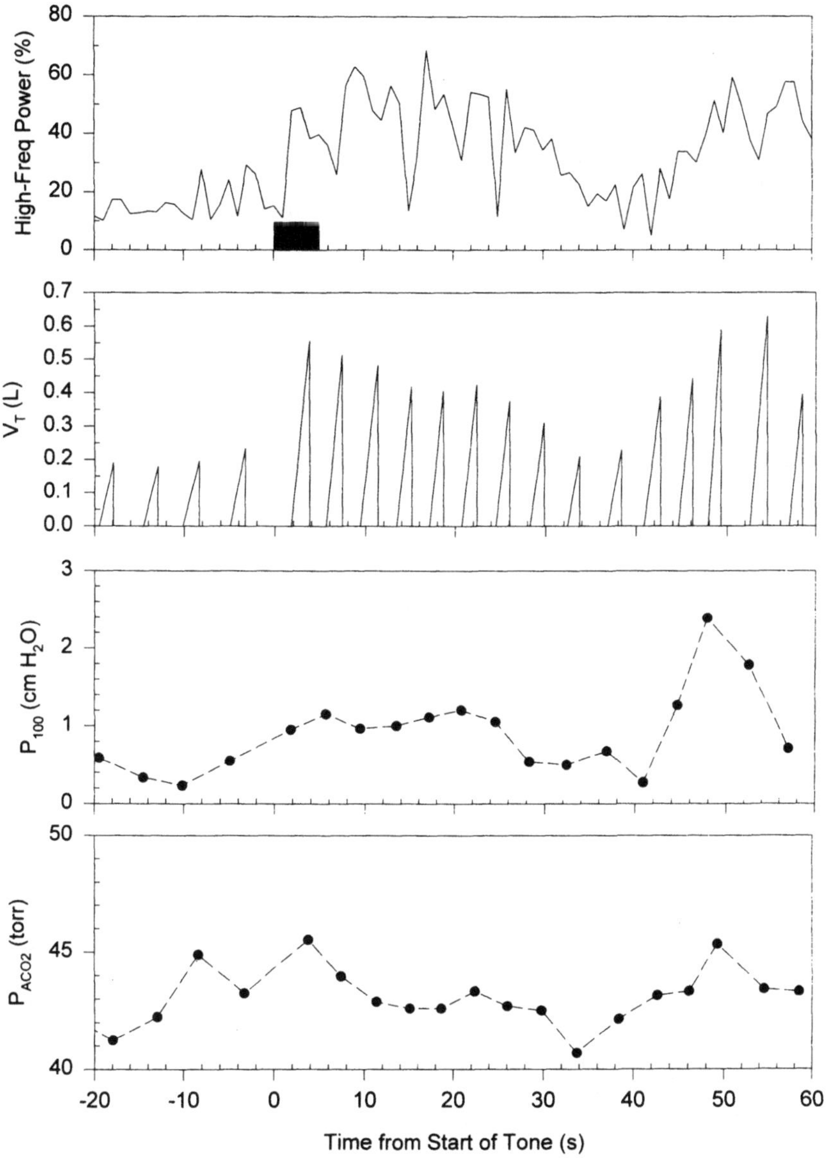

Figure 14. Ventilatory response to transient arousal induced by acoustic stimulation. *Top Panel:* Time course of EEG high-frequency power; the filled bar represents the duration over which the tone is administered. *Second Panel:* Tidal volume (only inspiratory portion) time course. *Third Panel:* Breath-to-breath values of P_{100}. *Bottom Panel:* Time-course of end-tidal P_{CO2}.

"wave", another transient bout sets in at ~40 s. Although the EEG high-frequency changes in the second wave are similar in magnitude to the first, the ventilatory response (particularly in the case of P_{100}) in the second is noticeably larger. This may reflect a change in state that is inadequately represented by the EEG high-frequency changes. Alternatively, there may be a sensitization of the ventilatory response to arousal immediately following an initial episode. Such a feature can only be adequately described with the use of a model with time-varying parameters or the inclusion of an appropriate memory-type nonlinearity.

5. CONCLUDING REMARKS

The primary purpose of this chapter has been to illustrate the approaches we have adopted recently to extend our original computational model of sleep-disordered breathing. It is clear that a major stumbling block for further development is simply the lack of quantitative data about the dynamics of ventilatory and state fluctuations during sleep. This is precisely the reason for our heavy emphasis on empirically-based parameter estimation in the second part of this work. There are many important aspects that we have not included in this work. For instance, there remains much to be done to develop a good predictive model of upper airway dynamics. The effect of arousal on the respiratory pattern generator and the many issues surrounding the relative timing of arousal and breath generation - these have all not been addressed thus far. Nevertheless, from a more utilitarian perspective, it may not be necessary to incorporate all the details or to attain a complete understanding of all the mechanisms involved, if the utlimate goal is to be able to "customize" models of sleep-disordered breathing for individual subjects so that more precise modes of therapy can be prescribed.

ACKNOWLEDGMENTS

This work was supported in part by NHLBI Research Career Development Award HL-02536, an American Lung Association Career Investigator Award, and Division of Research Resources (NIH) Grant RR-01861.

REFERENCES

1. Phillipson, E.A., and G. Bowes. Control of breathing during sleep. In: *Handbook of Physiology, Section 3: The Respiratory System II*, edited by N.S. Cherniack and J.G. Widdicombe. Bethesda, MD: American Physiological Society, 1986, p.649–690.
2. Bulow, K. Respiration and wakefulness in man. *Acta Physiol. Scand.* 59(Suppl. 209):1–110, 1963.
3. Chapman, K.R., E.N. Bruce, B. Gothe and N.S. Cherniack. Possible mechanisms of periodic breathing during sleep. *J. Appl. Physiol.* 64:1000–1008, 1988.
4. Khoo, M.C.K. Periodic breathing. In: *The Lung: Scientific Foundations, 2nd Edition*, edited by R.G. Crystal, J.B. West, E. Weibel and P.J. Barnes. Philadelphia: Lippincott–Raven, 1996, p.137.1–137.13.
5. Modarreszadeh, M., E.N. Bruce, H. Hamiltonn, and D.W. Hudgel. Ventilatory stability to CO_2 disturbances in wakefulness and quiet sleep. *J. Appl. Physiol.* 79:1071–1081, 1995.
6. Khoo, M.C.K., R.E. Kronauer, K.P. Strohl, and A.S. Slutsky. Factors inducing periodic breathing in humans: a general model. *J. Appl. Physiol.* 53:644–659, 1982.
7. Bruce, E.N., and J.A. Daubenspeck. Mechanisms and analysis of ventilatory stability. In: *Regulation of breathing*, edited by J.A. Dempsey. New York: Marcel Dekker, 1994, p.285–313.
8. Anholm, J.D., A.C.P. Powles, R. Downey III, C.S. Houston, J.R. Sutton, M.H. Bonnet, and A. Cymerman. Operation Everest II: Arterial oxygen saturation and sleep at extreme simulated altitude. *Am. Rev. Respir. Dis.* 145:817–826, 1992.
9. Trinder, J., W. Whitworth, A. Kay, and P. Wilken. Respiratory instability during sleep onset. *J. Appl. Physiol.* 73:2462–2469, 1992.
10. Pack, A.I., M.F. Cola, A. Goldszmidt, M.D. Ogilvie, and A. Gottschalk. Correlation between oscillations in ventilation and frequency content of the electroencephalograph. *J. Appl. Physiol.* 72:985–992, 1992.
11. Pack, A.I. Sleep state and periodic ventilation. In: *Modeling and Parameter Estimation in Respiratory Control*, edited by M.C.K. Khoo. New York: Plenum Press, 1989, p.181–192.
12. Khoo, M.C.K., A. Gottschalk, and A.I. Pack. Sleep-induced periodic breathing and apnea: a theoretical study. *J. Appl. Physiol.* 70:2014–2024, 1991.

13. Khoo, M.C.K., J.D. Anholm, S.W. Ko, R. Downey III, A.C.P. Powles, J.R. Sutton, and C.S. Houston. Dynamics of periodic breathing and arousal during sleep at extreme altitude. Respir. Physiol., 1996 103:33-43, 1996.
14. Anholm, J.D., R. Downey III, J. MacQuarrie, A.C.P. Powles, and C.S. Houston. Periodic breathing and sleep disruption at high altitude: Operation Everest II. In: *Hypoxia and Mountain Medicine*, edited by J.R. Sutton, G. Coates and C.S. Houston. Burlington, VT: Queen City Printers, 1991 (abstract).
15. Hedemark, L.L., and R.S. Kronenberg. Ventilatory and heart rate responses to hypoxia and hypercapnia during sleep in adults. *J. Appl. Physiol.* 53:307–312, 1982.
16. Honda, Y., F. Hayashi, A. Yoshida, Y. Ohyabu, Y. Nishibayashi, and H. Kimura. Overall "gain" of the respiratory control ssystem in normoxic humans awake and asleep. *J. Appl. Physiol.* 55:1530–1535, 1983.
17. Khoo, M.C.K., F. Yang, J.W.. Shin, and P.R. Westbrook. Estimation of dynamic chemoreflex gain in wakefulness and NREM sleep. *J. Appl. Physiol.* 78:1052–1064, 1995.
18. Sohrab, S., and S.M. Yamashiro. Pseudorandom testing of ventilatory response to inspired carbon dioxide in man. *J. Appl. Physiol.* 49:1000–1009, 1980.
19. Waggener, T.B., A.R. Stark, B.A. Cohlan, and I.D. Frantz III. Apnea duration is related to ventilatory oscillation characteristics in newborn infants. *J. Appl. Physiol.* 57:536–544, 1984.
20. Akay, M. *Biomedical Signal Processing*. New York: Academic Press, 1994.
21. Chon, K.H., N.H. Holstein-Rathlou, D.J. Marsh, and V.Z. Marmarelis. Parametric and nonparametric nonlinear modeling of renal autoregulation dynamics. In: *Advanced Methods of Physiological System Modeling*, Vol.3, edited by V.Z. Marmarelis. New York: Plenum Press,1994, p.195–210.
22. Marmarelis, V.Z. Identification of nonlinear biological systems using Laguerre expansions of kernels. *Ann. Biomed. Eng.* 21:573–589, 1993.
23. Khoo, M.C.K., and V.Z. Marmarelis. Estimation of peripheral chemoreflex gain from spontaneous sigh responses. *Ann. Biomed. Eng.* 17:557–570, 1989.
24. Khoo, M.C.K. Non-invasive tracking of peripheral ventilatory response to carbon dioxide. *Int. J. Biomed. Comput.* 24:283–295, 1989.
25. Yang, F., and M.C.K. Khoo. Ventilatory response to randomly modulated hypercapnia and hypoxia in humans. *J. Appl. Physiol.* 76:2216–2223, 1994.
26. Koh, S.W., R.B. Berry, J.W. Shin, and M.C.K. Khoo. Estimation of changes in chemoreflex gain and 'wakefulness drive' during abrupt sleep-wake transitions. In: *Bioengineering Approaches to Pulmonary Physiology and Medicine*, edited by M.C.K. Khoo. New York: Plenum Press, 1996 (this volume).
27. Khoo, M.C.K., S.S.W. Koh, J.J.W. Shin, P.R. Westbrook, and R.B. Berry. Ventilatory dynamics during transient arousal from NREM sleep: Implications for respiratory control stability. *J. Appl. Physiol.* 80:1475-1484, 1996.
28. White, D.P. Occlusion pressure and ventilation during sleep in normal humans. *J.Appl. Physiol.* 61:1279–1287, 1986.

2

THE NORTH CAROLINA RESPIRATORY MODEL

A Multipurpose Model for Studying the Control of Breathing

Frederic L. Eldridge

Departments of Medicine and Physiology
University of North Carolina
Chapel Hill, North Carolina 27599

1. INTRODUCTION

Computer-based mathematical models of respiration have been used for some decades to study respiratory control. Such models can mimic both steady state and dynamic behaviors to a wide variety of forcings[1] and have been particularly useful in gaining understanding of some of the chemical and circulatory factors underlying unstable, periodic breathing of the Cheyne–Stokes type[2-4].

Most of the models have dealt primarily with lung-blood-tissue gas transport and exchange, with chemoreceptor-based feedback mechanisms for control. Mechanical factors such as lung and thoracic compliance and airway resistance and body weight (obesity) have usually not been included. Neither have strictly neural mechanisms not involving chemoreceptor feedback, such as input to the controller from suprapontine brain (voluntary, the "central command" input associated with exercise[5]), input from working muscle (peripheral neural) during exercise, the neural input associated with wakefulness, neuronal depression due to hypoxia of the brain, and especially the mechanism of short-term potentiation (STP) of respiratory neurons which has been postulated as a potential stabilizer of respiration. It should be noted that Khoo and his coworkers[6,7] have recently incorporated "a wakefulness" drive in their model for the purpose of studying periodic breathing and apnea and show that arousals may be an important factor in destabilizing breathing during sleep. The question of hypoxic neuronal depression has also been modeled[8] and shown to be potentially destabilizing. I am not aware of any model that contains the mechanism of short-term potentiation.

The intent behind the present model was to incorporate most of these factors in a general model of respiratory gas exchange and control. Because there is currently no usable model of the neuronal oscillator in the medulla (but see J. Smith in this volume), a standard limit-cycle oscillator (the Fitzhugh–Van der Pol–Bonhoeffer[9]) has been used to represent the medullary controller. "Output" generated by the oscillator is used to activate respiratory muscles (with attention to length-tension properties) and upper airway muscles, which in

Bioengineering Approaches to Pulmonary Physiology and Medicine, edited by Khoo
Plenum Press, New York, 1996

turn lead to volume changes and tidal gas exchange. Activities of central (increased by ↑CO_2) and peripheral chemo-receptors (increased by ↑CO_2, by ↓O_2 and by ↑ potassium) act as feedbacks to the controller. Input to the controller from the cortex (voluntary), that associated with the wakefulness effect on the reticular formation, the "central command" signal during exercise, and short-term potentiation of respiratory neurons are incorporated in the model.

2. THE MODEL

2.1. Central Controller

For the central neural oscillator, the model uses the Fitzhugh–Bonhoeffer–Van der Pol (BVP) limit-cycle equations[9]:

$$dx/dt = \varepsilon(y + x - x^3/3 + z) \text{ and} \quad (1)$$

$$dy/dt = -(x - a + by)/\varepsilon \quad (2)$$

where $1 - 2b / 3 < a < 1$; $0 < b < 1$; and $b < \varepsilon^2$

Equation 1 has been modified to include a variable (D):

$$dx/dt = (y + x - (x^3/D)/3 + z \quad (3)$$

where magnitude of D is equivalent of respiratory "drive" and has the effect of changing the size of the limit-cycle[10,11]. An example of the unmodified Fitzhugh BVP limit-cycle (where $\varepsilon = 3$, $a = 0.7$, $b = 0.8$, $z = 0.6$ and $D = 1$) is shown in Fig. 1. Figure 1 also shows a threshold level (y_{th} at $y = 0.3$) which divides portions of the cycle defined for the purposes of the model as inspiratory and expiratory.

Figure 2 shows examples of the effects of changing D, i.e., "respiratory drive." Note that at D values below 0.3 the limit-cycle shows no oscillation and stabilizes at fixed points, falling farther below threshold (y_{th}, unlabelled horizontal lines) as drive becomes smaller.

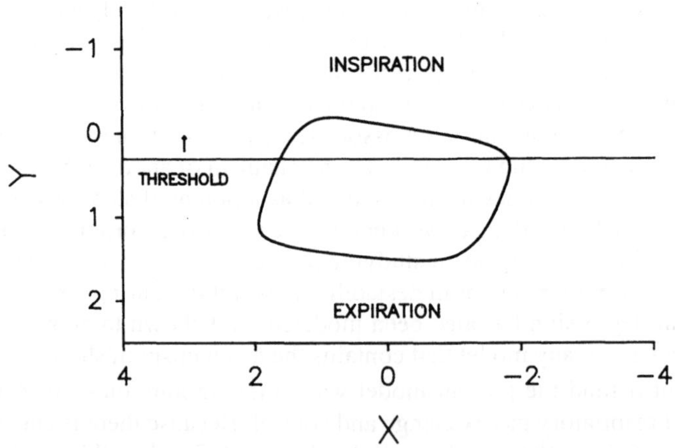

Figure 1. Limit-cycle generated by Fitzhugh- BVP equations. For the purposes of the respiratory model, threshold value (y_{th}) at $y = 0.3$ divides the inspiratory and expiratory portions of the cycle. "Respiratory drive" (D) in this example is 1.0.

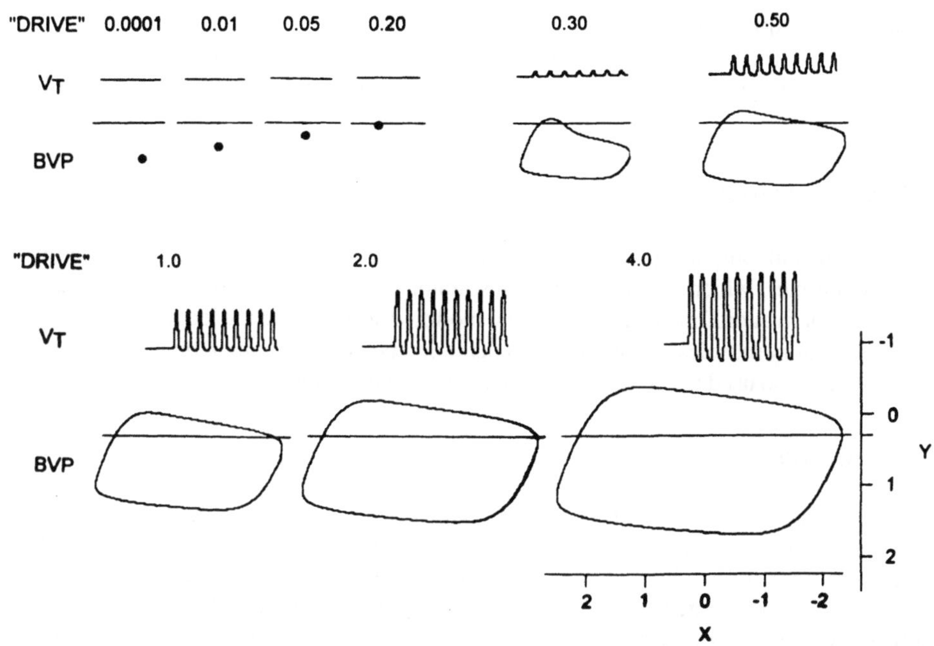

Figure 2. Limit-cycles generated by Fitzhugh BVP oscillator model at different "respiratory drives" (D in equation 3). Horizontal line in each limit-cycle represents threshold for transition from expiration to inspiration. Tidal volumes generated by the model for each level of drive are shown above the limit-cycle. Note that "drives" below 0.3 are below threshold and generate no rhythm.

At a drive level of 0.3, regular oscillations develop and the limit-cycle (BVP) becomes progressively larger as D becomes larger.

Stochastic processes ("noise"), created from the computer's random number generator, can be added during the running of the model to x and y values with each iteration of

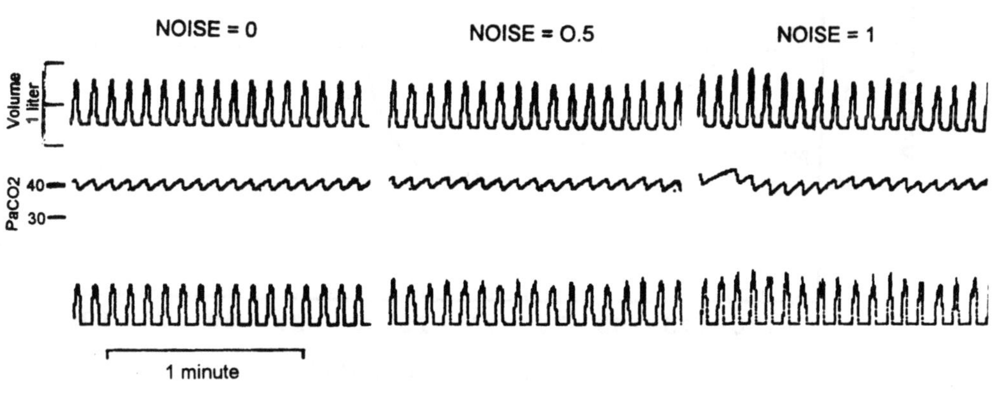

Figure 3. Example of effects of simulated stochastic processes ("noise") in the model. Bottom tracing represents inspiratory neural output of controller. Increasing noise increases irregularity of timing and magnitude of all variables.

the BVP equations, which results in random small changes in magnitude and timing of inspiration and expiration (Fig. 3)

Neural outputs of the controller are [ABS($y - y_{th}$)] for inspiration and [($y - y_{th} \cdot f$)] for expiration, f being an empirical factor to yield an appropriate magnitude of expiratory neural output.

2.2. Muscle Pressures

The neural outputs generated by the controller alternately activate inspiratory and expiratory muscles, which in turn produce inspiratory and expiratory muscle pressures. For a given "neural activation," muscle pressures are modified according to the length-tension relationships of respiratory muscles[12] (Fig. 4), and depending upon the volume of the thorax. Inspiratory neural output is also used to activate upper airway respiratory muscles, e.g., genioglossus m.

2.3. Ventilation

The generated muscle pressures act against elastic recoils of the lungs and thorax (Fig. 5), and overall airway resistance, which includes lower airway resistance and that of pharynx and nose, to produce airflow and tidal gas exchange (see Figs. 2 and 7). Elastic recoil curves are generated according to the exponential relations between pressure and volume shown by Salazar and Knowles[13] for the lungs. Similar equations are used for chest wall recoil in the opposite direction. The stable volume at FRC is 1.5 liters above RV when default curve shapes and elastances are used and body weight is a normal 150 lbs., but there is an additional mechanical factor related to increasing obesity which causes FRC to become closer to RV as weight increases.

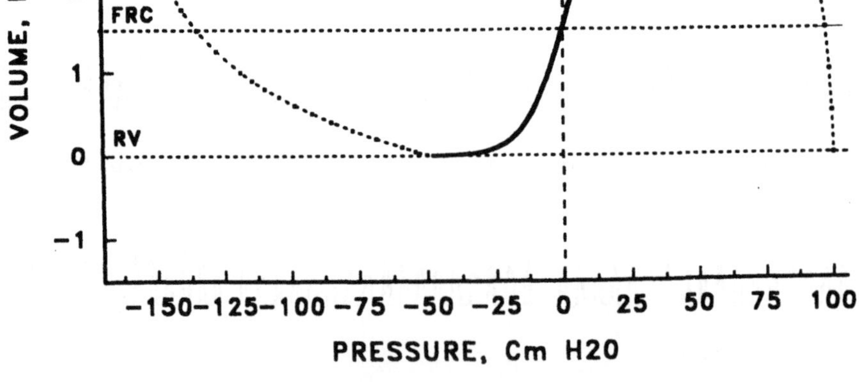

Figure 4. Static pressure-volume diagram of values generated in model for static elastic properties of respiratory system (Prs, solid line) and maximum inspiratory and expiratory muscles pressures (dashed lines). The diagram shows the effect of lung volume on generation of muscle pressure.

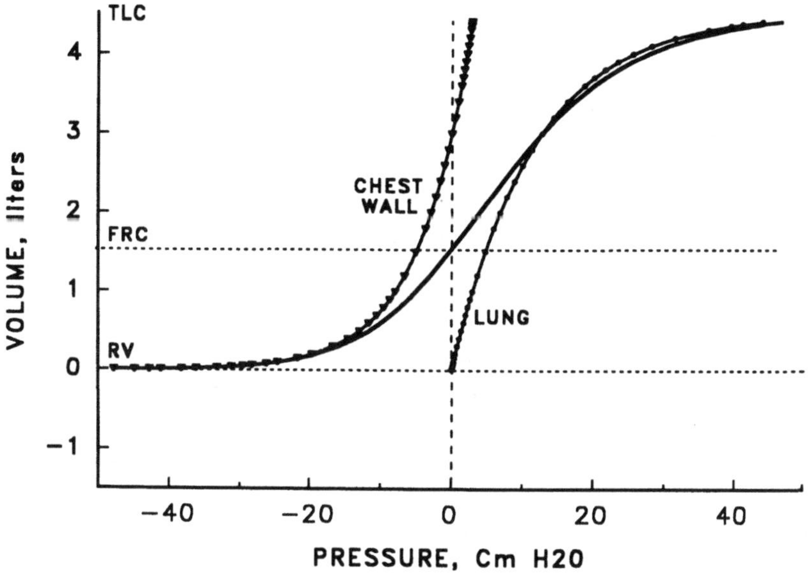

Figure 5. Static pressure-volume plots for lung, chest wall, and respiratory system (P_{rs}, solid line), generated with default parameters in model.

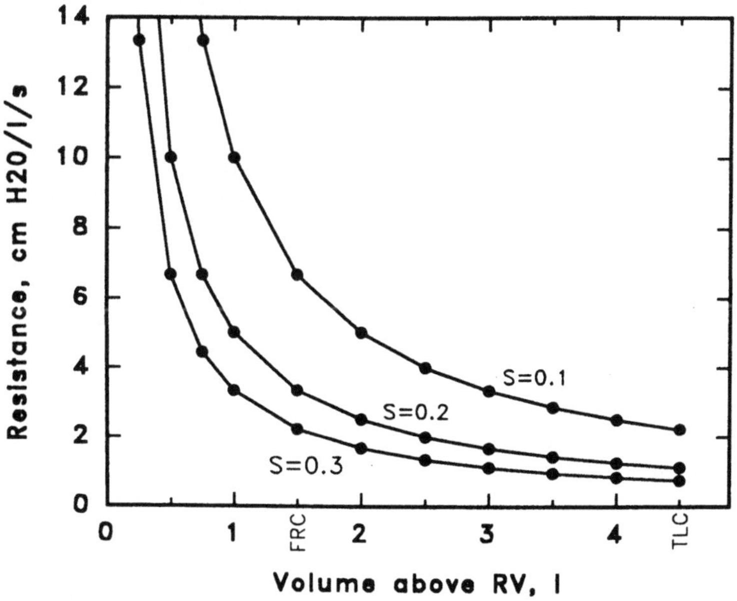

Figure 6. Plot of the effect of lung volume on lower airway resistance. Default slope (S) of relation of conductance vs. lung volume is 0.3 and yields a resistance of 2.0 cm H_2O/l per sec at FRC.

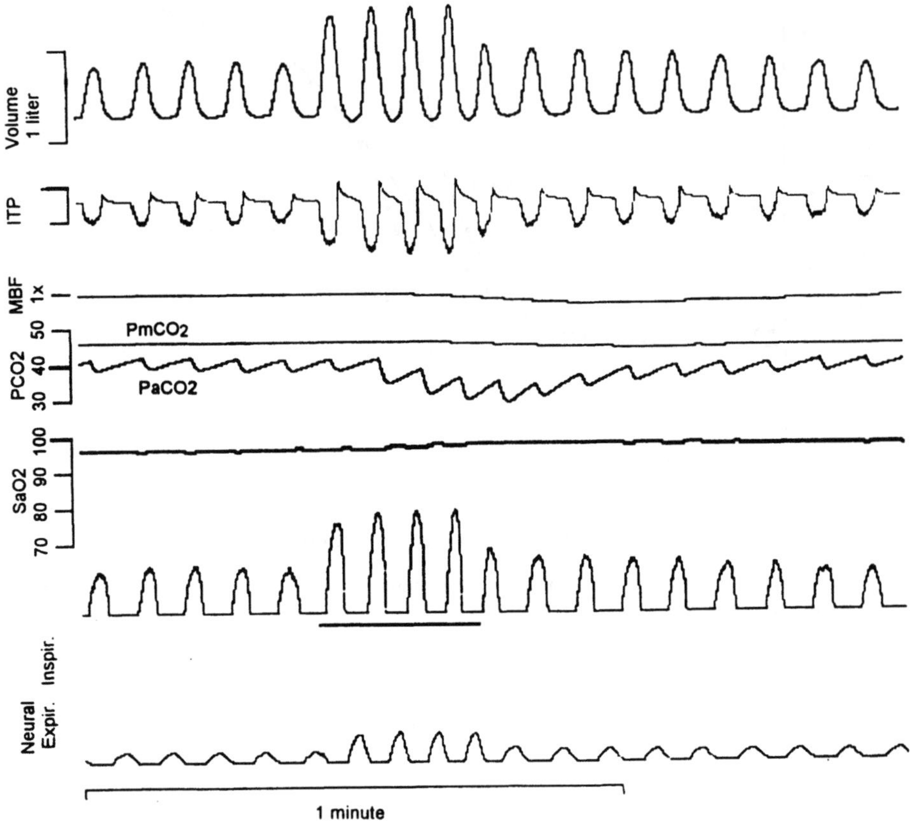

Figure 7. Example of model tracings on monitor screen. Quiet breathing at left, then 20 sec of neurally driven increased breathing, equivalent to voluntary hyperventilation. Tidal volume at top; ITP = intrathoracic pressure (calib. = 5 cm H_2O); MBF = medullary blood flow factor; $PmCO_2$ = medullary PCO_2.

Pharyngeal airway resistance is calculated from pharyngeal area, with assumed mechanical characteristics, including maximum area, static closing pressure[14-15], level of neural activity acting on upper respiratory muscles[16] and a factor converting neural activity to muscle pressure.

Lower airway resistance increases as lung volume falls below FRC and decreases slightly as volume increases. The values at any lung volume can be calculated because the reciprocal of resistance (conductance) is linearly related to lung volume[17]. In the model it is assumed that resistance is infinite at RV. When the slope (S) of that relationship is 0.3, the resistance at FRC is 2.0 cm H_2O/l per sec (default value). Fig. 6 shows this and also shows the effect of changing lung volume and the effect of smaller values of S which can be selected by the model's operator.

2.4. Gas Exchange

Mass balance equations are used for CO_2 and O_2 exchange; they are similar to those reported by Khoo et al.[3,6] and similar parameters are used for distribution volumes. Ventilatory dead space is defaulted at 0.15 l but any value can be selected. At rest $\dot{V}O_2$ is defaulted at 0.25 l/min, RQ at 0.8 and cardiac output at 5 l/min. Pulmonary shunt is defaulted at 2%

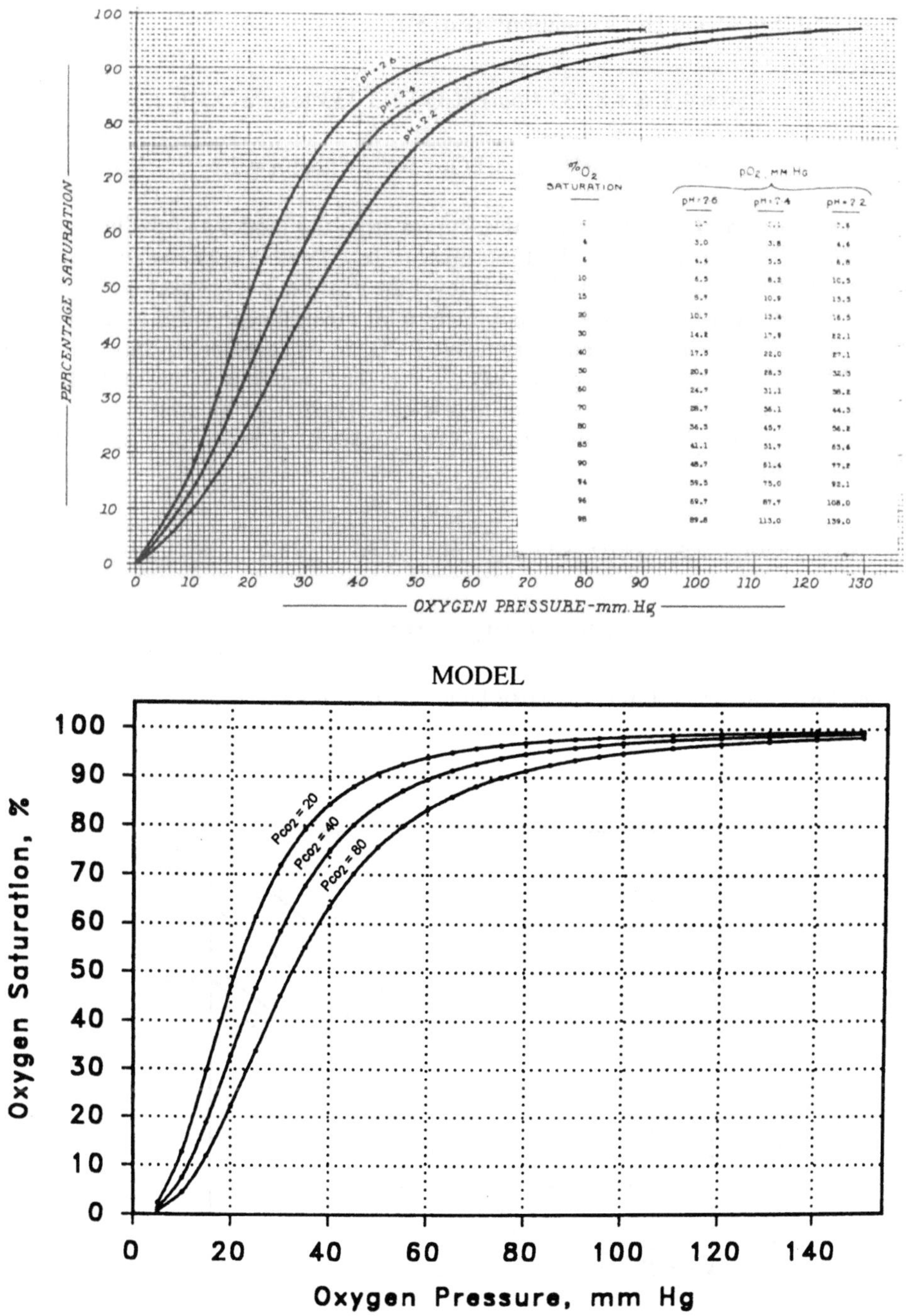

Figure 8. Oxyhemoglobin dissociation curves at different pH/PCO_2 values. *Top:* for human blood from Dill[19]. *Bottom:* from model using calculated P_{50} values and Hill's equation. Only below 40% SaO_2 do the model's curves differ much from the Dill curves.

but any value can be selected. In addition, dissolved O_2 is incorporated at all levels of the circulation.

Conversions to O_2 saturation from PaO_2, and vice versa, are made with use of a calculated P50 (Eq. 4) and Hill's equation[18] (Eq. 5):

$$P50 = 4.333 + 8.333 \cdot \log(PaCO_2) + 3.333 \cdot \log(PaCO_2)^2 \qquad (4)$$

$$SaO_2 = (P)_2/P50)^n/(1 + PO_2/P50)^n \text{, where } n = 2.6 \qquad (5)$$

A comparison of the O_2 dissociation curves from the model and the human blood studies of Dill[19] shows extremely high correlation (r >0.9995) between the two above saturation values of 40% (Fig. 8). Below this level the relation is less exact.

A separate set of equations and parameters for medullary metabolic rate (MR_{med}) and blood flow (MBF) allows medullary tissue PCO_2 to be calculated. Baseline default value of MR_{med} is 0.032 l/min·kg and RQ_{med} = 1. The baseline value of MBF is 0.8 l/min·kg). MBF is modulated by a factor that is calculated from an empirically derived (based on limited data[20]) sigmoidal interactive function of medullary arterial PCO_2 and SO_2 (Fig. 9):

$$\text{MBF factor} = \exp(\text{ATN}((PaCO_{2med} - 40) \cdot .03)) + (96.5 - SaO_{2med}) \cdot .02)) \qquad (6)$$

$$\text{MBF} = \text{MBF}_{base} \cdot \text{MBF factor} \qquad (7)$$

where the lower and upper limits for MBF factor are 0.4 and 3.0 respectively.

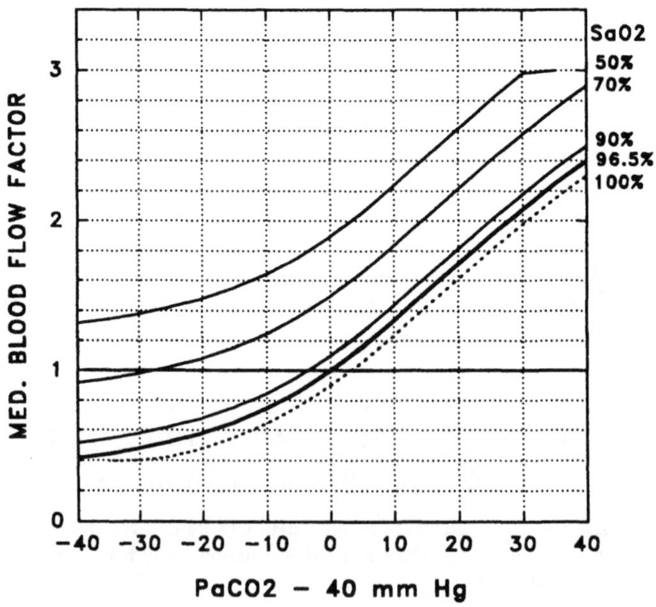

Figure 9. Plots of model's medullary blood flow factor as function of arterial PCO_2 and SaO_2 at the medulla.

Appropriate delays of blood gases from lungs to carotid bodies (normal = 6 sec) and to medulla (normal = 8 sec) are incorporated into the program as default values but can be changed if desired.

2.5. Inputs to the Controller

Inputs, which act as neural inputs from various sources (see Fig. 10), change the value of D in the Fitzhugh-BVP equations, thereby changing the oscillator's magnitude which in turn changes inspiratory and expiratory outputs They include the following:

2.5.1. Input from Carotid Bodies. In the carotid bodies, there are multiplicative interactions between PaO_2 and $PaCO_2$ (Fig. 11), as well as the mutiplicative interaction between PaO_2 and potassium concentration $[K^+a]$ at the chemoreceptor[21] (Fig. 12) when $[K^+a]$ increases above a normal resting level of 4 meq / l during exercise.

2.5.2. Input from Central Chemoreceptors Responding to Medullary PCO_2. There is a threshold for central chemoreceptor responses at 45 mm Hg. The form of the relation is an empirical 2nd order equation that yields appropriate ventilatory responses to CO_2.

2.5.3. Wakefulness Input. There is a "neural" input associated with the awake state[22], which is removed during sleep. This input leads to more ventilation during wakefulness than during sleep, i.e., a shift to the left of about 5 mm Hg $PaCO_2$ during wakefulness, and is also involved in the "dog-leg" phenomenon[23] when $PaCO_2$ is below 40 mm Hg. It leads to a multiplicative interaction with the other drives so that the ventilatory response to CO_2 is steeper in the awake state than in the sleep state (see Fig. 14).

2.5.4. Other Non-Chemical Inputs. Inputs due to other neural but non-chemoreceptor related mechanisms that can be added or subtracted from D as desired by the program's operator. Such positive neural drive inputs would be the equivalent of mechanisms such as the "central command" from suprapontine brain that is associated with exercise[24], other peripheral (neural) mechanisms, and voluntary hyperventilation

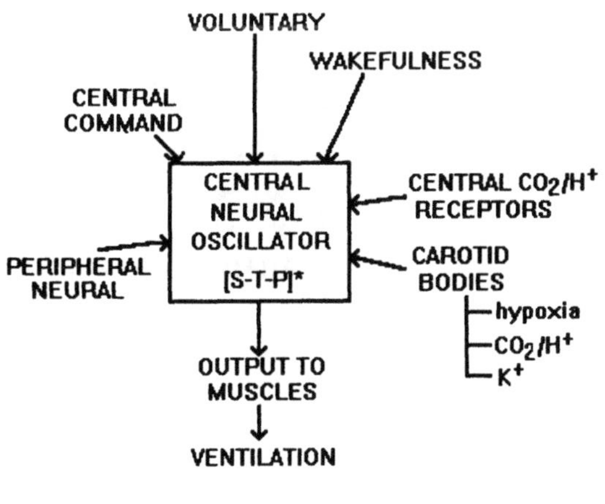

Figure 10. Schematic diagram of central oscillator and the various inputs used in the model. *S-T-P represents short-term potentiation of neural pathways in medulla. See text.

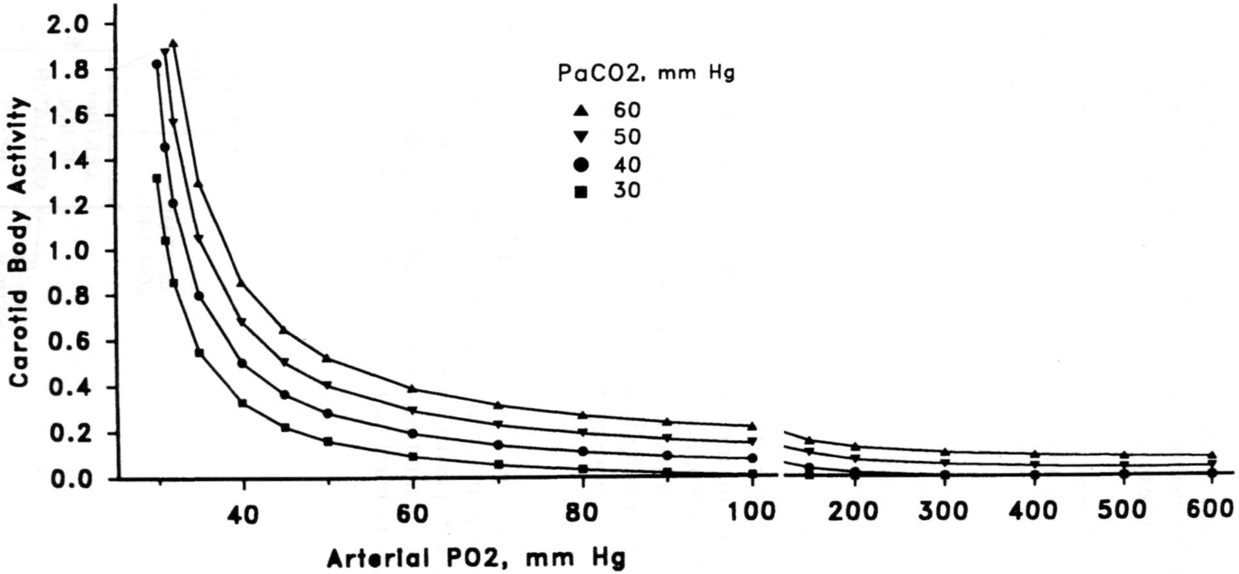

Figure 11. Relations between arterial PO_2 and "neural" activity of the carotid body at different PaO_2 and $PaCO_2$ levels. Carotid body activity is in D units that act in the Fitzhugh-BVP oscillator equations (see Eq. 3).

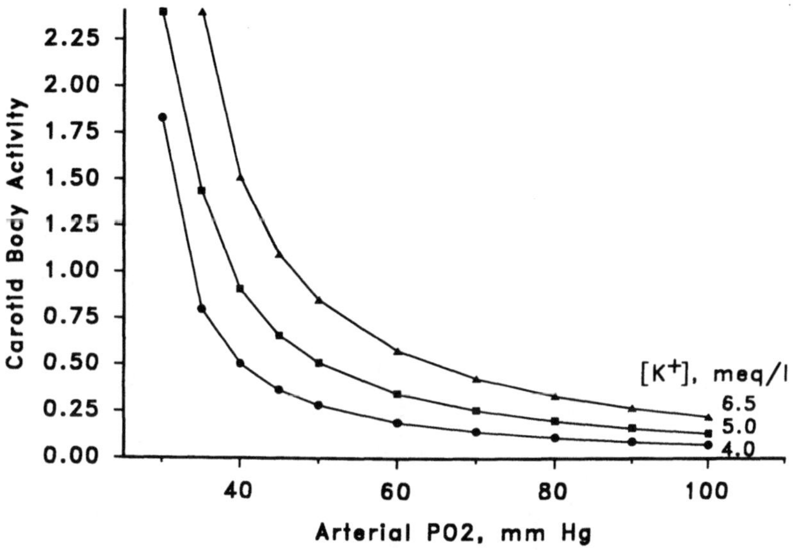

Figure 12. Relations between arterial PO_2 and "neural" activity of the carotid body at different concentrations of arterial potassium [K+]. Activity is in D units that act on the Fitzhugh-BVP (see Eq. 3).

driven from the cortex. Negative values would be the equivalent of an input with an inhibitory effect on breathing.

2.5.5. Short-Term Potentiation. In addition, there are equations that reflect the respiratory oscillator's intrinsic neural dynamics due to short-term potentiation (STP) of neuronal pathways[25]. STP is responsible for a slow component of the changes of respiratory output at the onset of a stimulus and the even slower changes at the offset, i.e., the respiratory afterdischarge[26]. Some amount of STP is probably present at every level of respiratory drive. It is thought to contribute to the ventilatory response to exercise[5] and to be important in preventing hypocapnic post-hyperventilation apnea[27] and periodic breathing[28,29]. The STP time constants are defaulted in the model at t_{on} = 10 sec, t_{off} = 40 sec[25] but they can be assigned any desired values. In the model the level of STP is determined continuously and added as an input to the controller. Figure 13 shows examples of development and decay of STP with several modes of stimulation in the cat and also shows that the model develops the same behavior.

2.6. Operation of the Model

The computer program is written in QBasic but is compiled so it runs fast enough on a PC 486-33Mhz or faster computer. At the default resolution (dt = 0.03), 333 iterations of the equations are performed per model-breath (67 iterations per model-sec), but other resolutions can be used.

Tracings of tidal volume, intrathoracic pressure, medullary blood flow factor, medullary PCO_2, $PaCO_2$ and SaO_2 at the carotid body, neural inspiratory activity and neural expiratory activity, along with time markers for each model-minute and a marker indicating whether there is wakefulness or sleep, are displayed on the computer's monitor screen. Numerical values of $PaCO_2$, medullary $PmCO_2$, PaO_2 and SaO_2, respiratory frequency and tidal volume are also displayed on the screen.

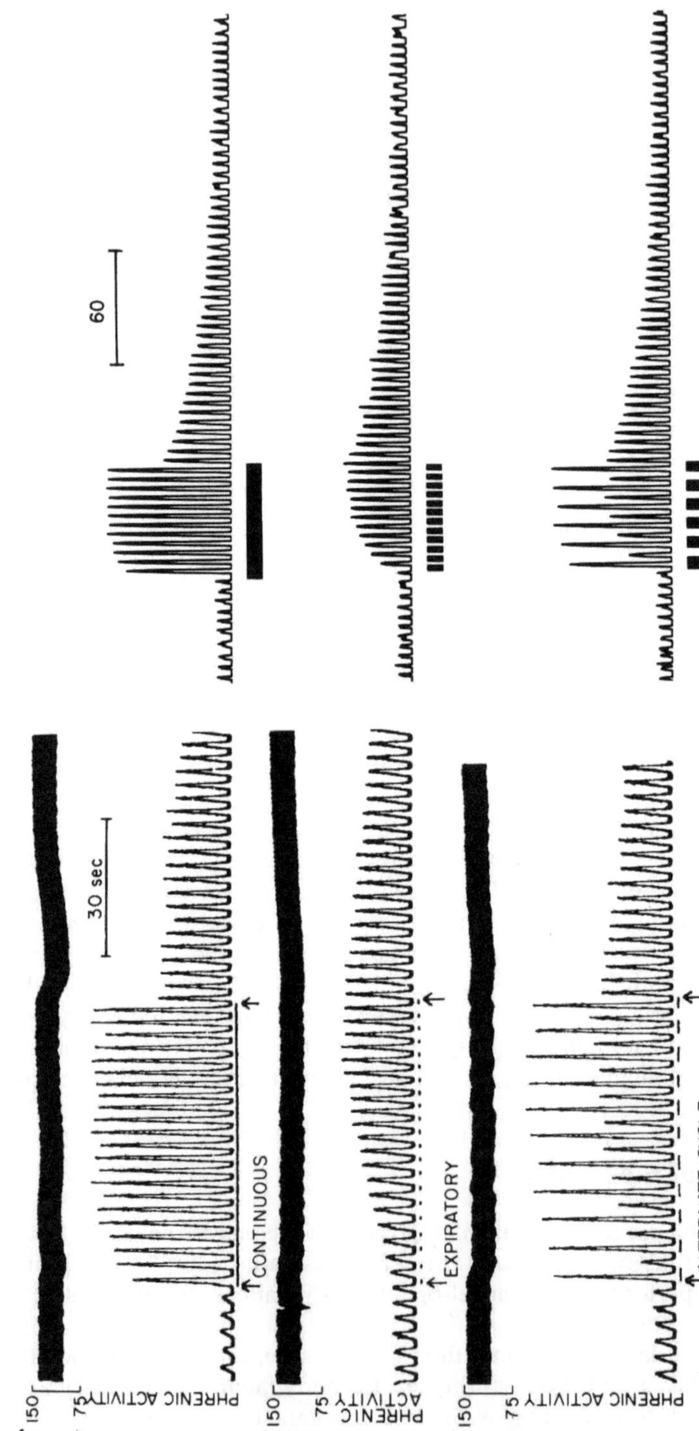

Figure 13. In cat (left, phrenic activity), different patterns of carotid sinus nerve stimulation all lead to development of short-term potentiation (STP), and then to afterdischarges during its decay. Development of STP can be clearly seen in expiratory-only and alternate breath stimulation panels. Fitzhugh-BVP model (right) with the equivalent of STP shows same behavior.

Theseandothernumericalbreath-by-breathvariables,includingminuteventilation,end-tidal, end-inspiratory and venous PCO_2 can be sent to a Laserjet printer when desired, and tracings generated on the monitor can be printed by means of a "screendump" program, such as HPSCREEN (distributed by Hewlett-Packard) activated by <shift-printscreen> keys.

Although the model can use entirely default parameter values, chosen to mimic the normal adult, values and screen tracing speed and appearance can be selected by the operator. In addition, two dozen of the parameters and operating controls can be changed independently during the running of the model by mean of F keys or Control- keys. They include: level of neural drive (see Section 2.5.4 above); level of "noise;" central CO_2 threshold; gain of peripheral and central chemoreceptive mechanisms; lung-to-carotid body and lung-to-medulla delay times; mechanical, muscular and neural parameters that affect upper airway (nose and pharynx) function; percentage pulmonary shunt; dead space; changes of $PICO_2$ and PIO_2, and factors that affect wake breathing. Some of the keys can be used to change screen tracing speed and tracing amplitude, to cause breath-by-breathing printing of results, and to reset screen.

Blood gases and/or MBF can be clamped at fixed levels if desired. The model can be run during wake or sleep and changed from one to the other at any time. It can also be run with or without short-term potentiation to visualize the role STP plays. "Exercise" can be mimicked by changing metabolic rate, cardiac output and A-V gas differences and introducing a factor representing changes of $[K^+a]$ and a factor representing the "central command" neural drive.

3. RESULTS

The model allows study of a wide variety of processes and questions of which the following are examples. Figures are given for some of them.

1. CO_2 and O_2 Ventilatory Responses, Awake and Asleep(Figs. 14 and 15)

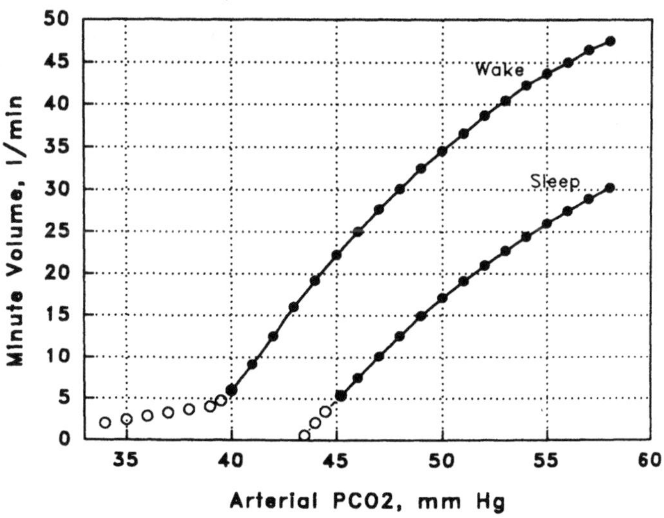

Figure 14. Ventilatory CO_2 response curves generated (with default values) by model during wakefulness and during sleep. Close circles at and above level of quiet breathing; open circles represent responses below level of quiet breathing. "Dogleg" disappears during sleep.

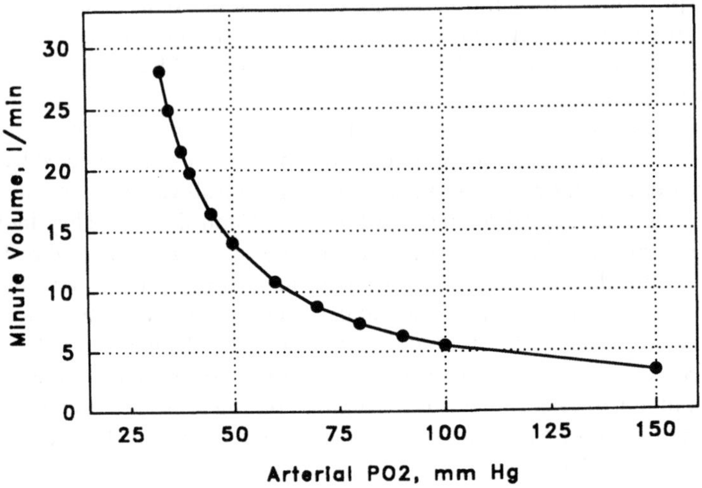

Figure 15. Model's ventilatory response to hypoxia (default parameters).

2. Effects of Changes of Dead Space on Gas Exchange and Ventilation
3. Effects of Varying Pulmonary Shunt on PaO_2 and SaO_2. The effects of different inspired PO_2 levels on these values can also be studied (Fig. 16).
4. Effects of Obesity on Lung Volume, Shunts and Gas Exchange
5. Effects of Carotid Body Ablation, at Rest Awake, during Sleep, and during Exercise
6. Effects of Ondine's Curse Syndrome, Ie. Very Small Central and Peripheral Gains
7. Effects of Changing from Wakefulness to Sleep (Fig. 17) and Vice Versa. See Phillipson and Bowes[22].
8. Effects of Different Circulatory Delays. Changes of timing of oscillating blood gases at the carotid bodies, due to different circulatory delays, can produce significant effects on central neural respiratory responses (Fig. 18). The findings are due to the differences in central effects of carotid body output depending on

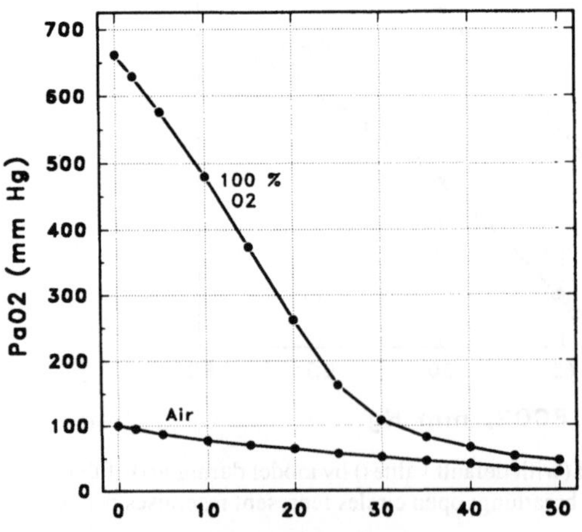

Figure 16. Relation determined in model between magnitude of pulmonary shunt and arterial O_2 pressure (PaO_2) during air and 100% O_2 breathing.

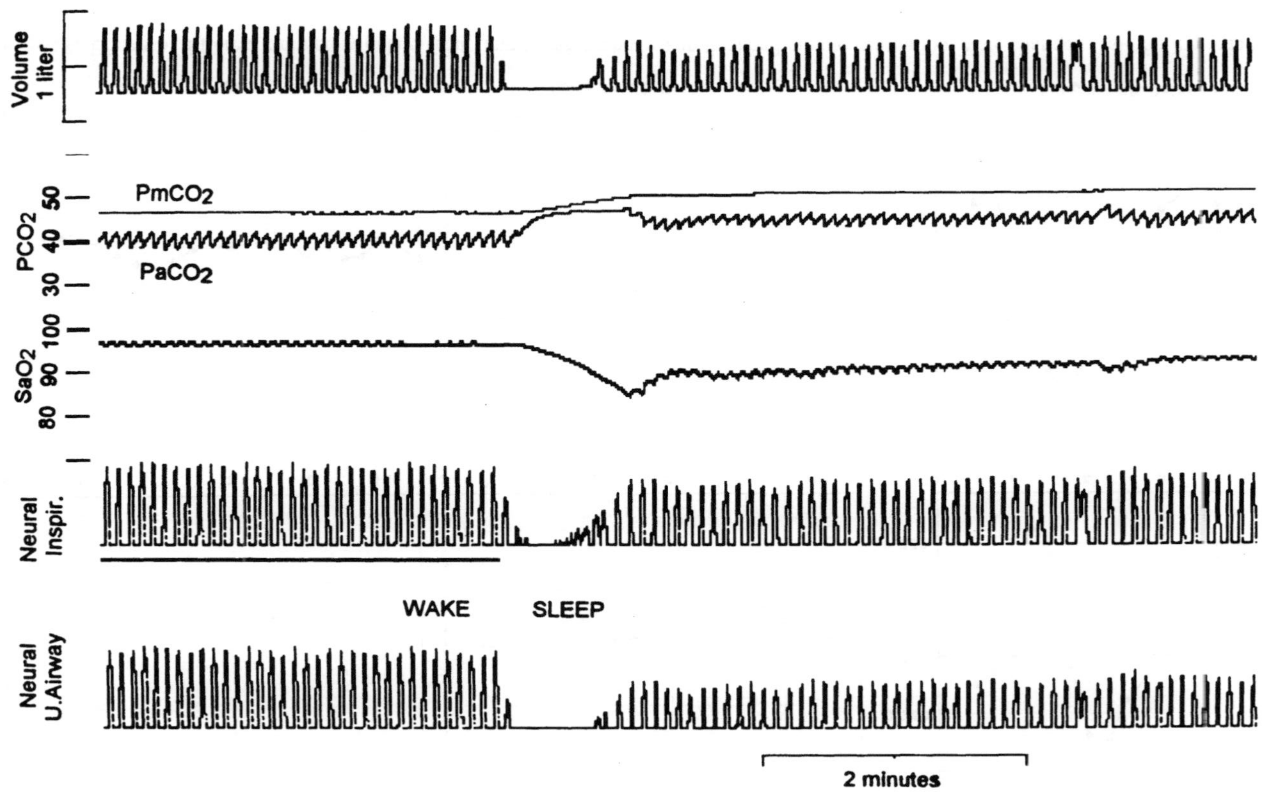

Figure 17. Screen tracings of model during wakefulness (at left) and change to sleep (at right). Loss of wakefulness leads to short neural apnea, a brief period of oscillating respiratory output and then stabilization at a lower level of ventilation and a rise of $PaCO_2$ and $PmCO_2$ of about 5 mm Hg. The slight irregularities are due to stochastic processes included in the model.

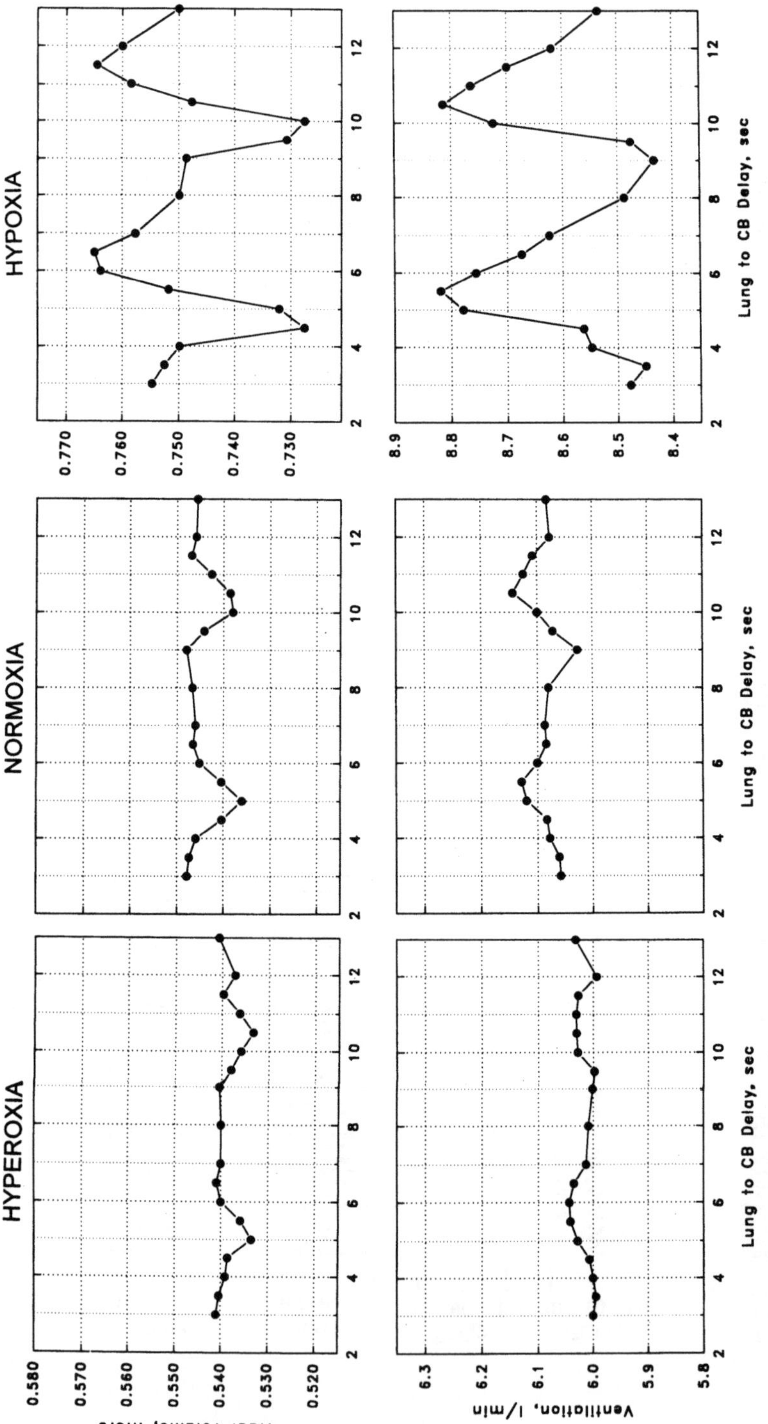

Figure 18. Effects of different circulatory delays from lung to carotid bodies on tidal and minute ventilation. Note changed scale for hypoxic panels.

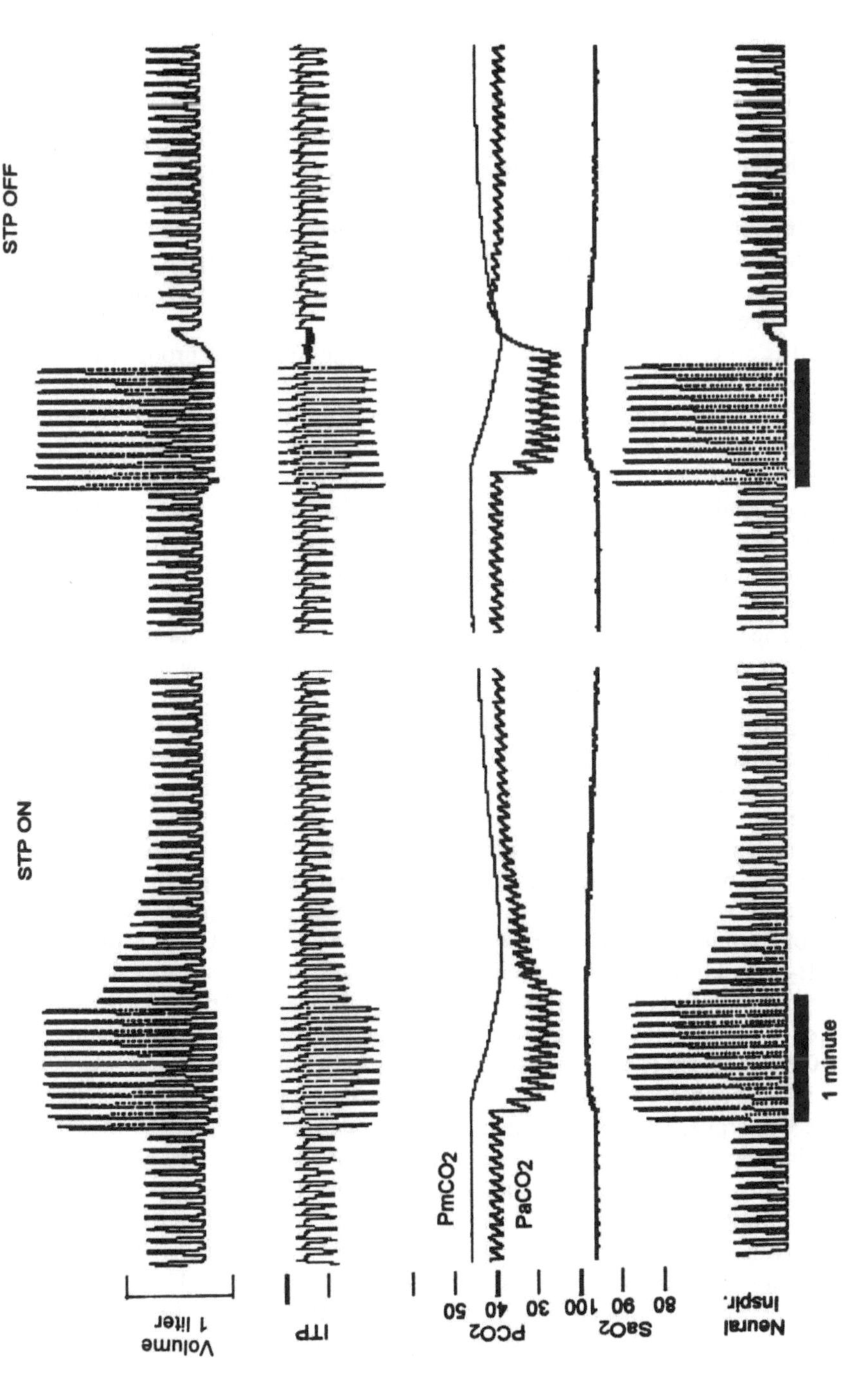

Figure 19. Model (wake) during hyperventilation of 1 min duration with STP (left) and without STP (right). Note that STP prevents posthyperventilation apnea.

the phase of central neural respiratory activity[30]. That it is the timing at carotid bodies that is important is shown by increasing oscillation of tidal volume and ventilation with increasing hypoxia. When carotid bodies are ablated in the model (not shown), there are no oscillations of central output.

9. Short-Term Potentiation. The model also allows a comparison of the effects of presence or absence of short-term potentiation (STP) on stability of breathing, on post-hyperventilation breathing (Fig., 19), on periodic breathing (Fig. 20), and on exercise hyperpnea.
10. Development of Cheyne–Stokes and Obstructive Periodic Breathing and Apnea. The likelihood of development of periodic breathing can be raised by increasing central gain, by prolonging lung-to-chemoreceptor circulation times, and by intermittent upper airway obstruction[31]. It has also been suggested that the loss of short-term potentiation (STP), the mechanism that leads to the respiratory afterdischarge and stabilizes respiration[25,28], should be important in allowing periodic breathing to develop[28,29].

    Figure 20 is an example of Cheyne–Stokes periodic breathing produced in the model during sleep by increasing central neural gain and eliminating the mechanism that leads to STP but retaining normal circulatory delays. Typical Cheyne–Stokes breathing and apnea are present (left) and lead to oscillations of blood gases, medullary $PaCO_2$ and medullary blood flow. The periodic breathing is primarily dependent upon the $PaCO_2$ oscillations since SaO_2 and PaO_2 oscillations are rather small. The addition of STP in the middle of the tracing leads to slow stabilization of breathing, confirming that it is an important factor in the prevention of periodic breathing.

    The top panel of Fig. 21 shows an example of obstructive apnea during sleep caused by upper airway (pharyngeal) narrowing and closure which has been produced by altering mechanical characteristics of the pharynx in the model. In the bottom panel of Fig. 21, the central neural gain has also been increased but no other parameters changed. This leads to so-called "mixed" central and obstructive apneas and supports the idea that pure obstructive and "mixed" apneas are merely part of a continuum and not fundamentally different in their mechanisms.

    The effects of continuous positive airway pressure (CPAP) on obstructive sleep apnea can also be studied. Figure 22 (left panel) shows an example of obstructive apnea caused, as in Fig. 21, by mechanical narrowing of upper airway in the absence of continuous positive airway pressure (CPAP). When CPAP is increased to 3 cm H_20 (middle panel), the upper airway is expanded and resistance decreases but not to normal, and FRC increases; the apneas disappear but periodic breathing continues due to what is the equivalent of the Upper Airway Resistance Syndrome. With a CPAP of 6 cm H_20 (right panel), the airway is fully opened and breathing stabilizes—the slight irregularities are due to stochastic processes in the model.
11. Exercise Hyperpnea. As well as the traditional feedback mechanisms involving carbon dioxide and oxygen, the model includes those processes proposed by Eldridge and Waldrop[5] to be important in the genesis of exercise hyperpnea: they include rapidly acting neural mechanisms (central command and input from peripheral musculature), short-term potentiation of central neurons in the respiratory pathway, and the rise of serum $[K^+a]$ during exercise (t of 45 sec) and fall during recovery (t of 30 sec)[32]. Figure 23 (top panel) shows the findings at rest ($\dot{V}O_2$ of 0.25 l/min), during exercise ($\dot{V}O_2$ of 1.0 l/min) for 6 min, and during recovery in the intact model which has all of the processes noted above. It shows ventilation (l/min), and end-tidal, mean arterial and medullary PCO_2 (mm Hg). It can be seen that the pattern of the ventilatory response to exercise is much like

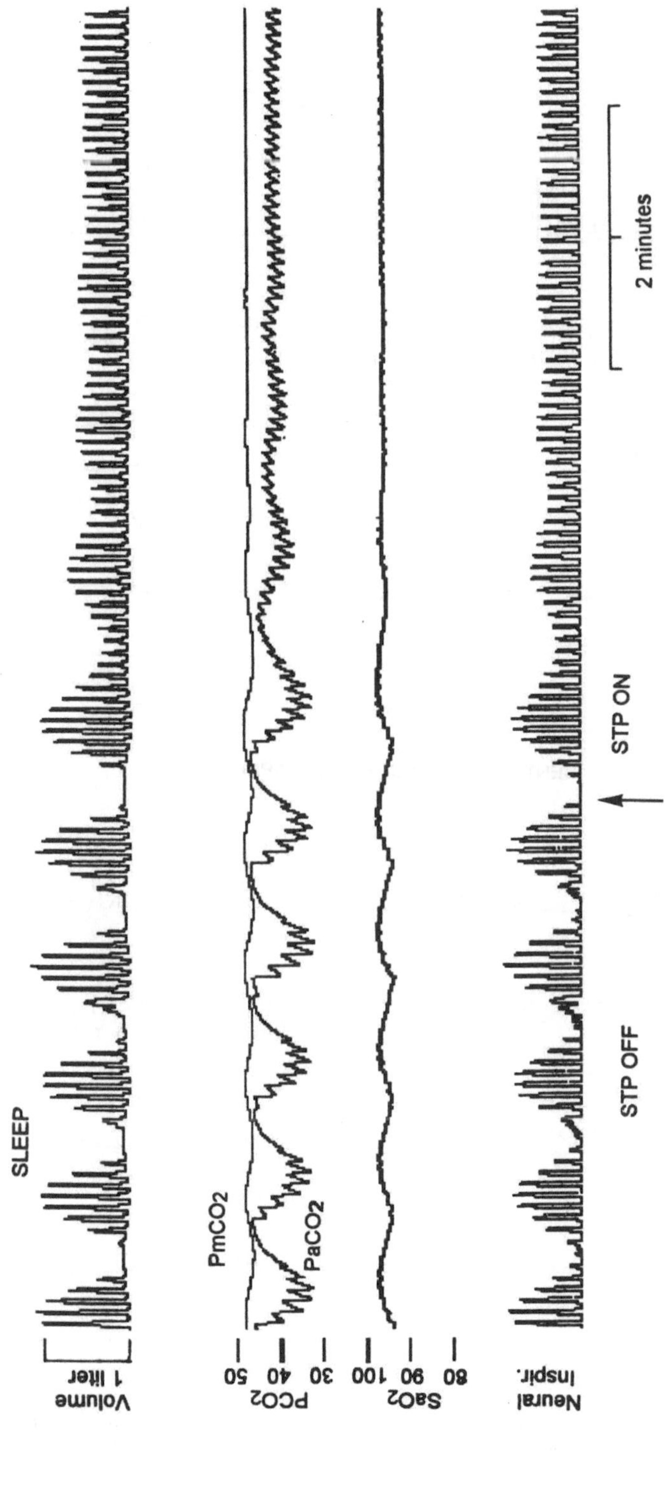

Figure 20. Cheyne-Stokes-like periodic breathing (left) in model (sleep, increased central neural gain) when STP is not active. Breathing is stabilized (right) when STP is added.

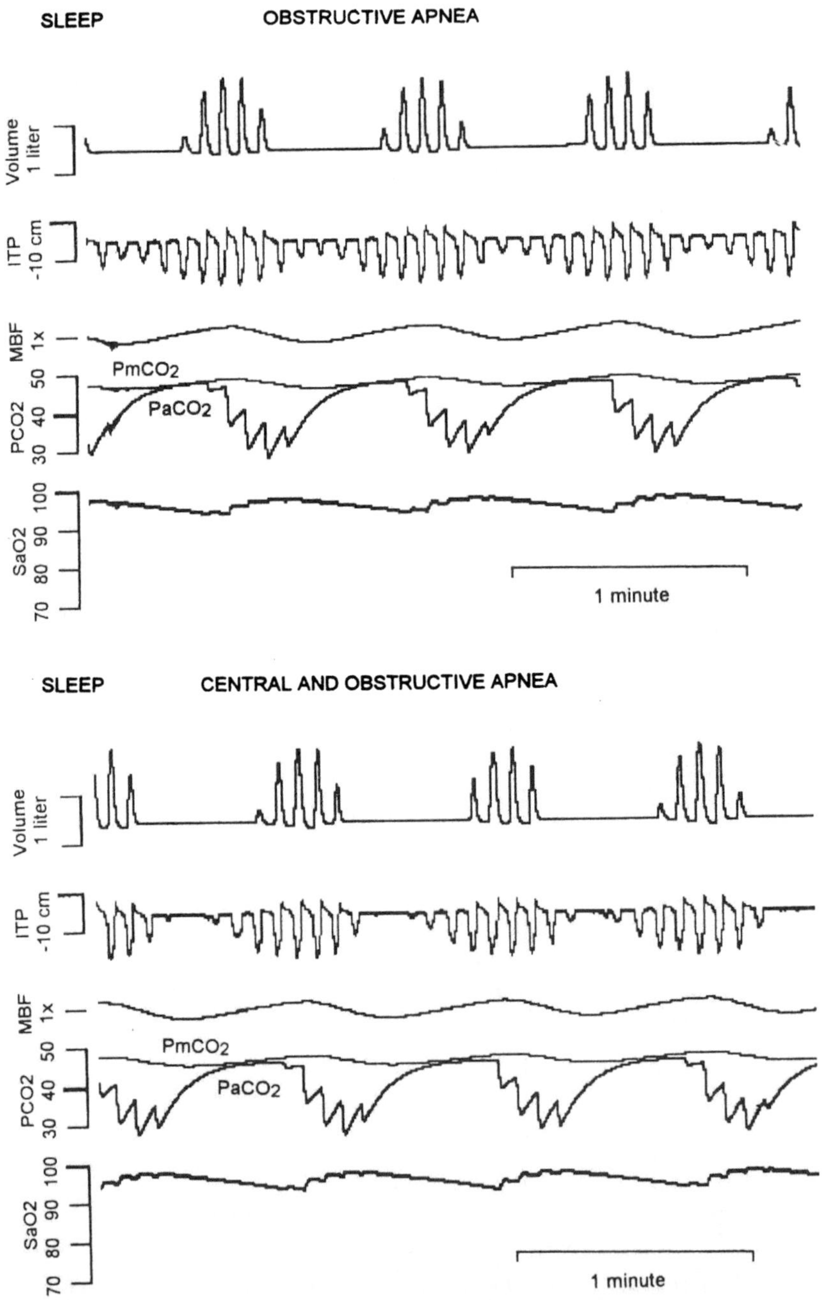

Figure 21. Top panel: Obstructive sleep apnea with no central apnea. *Bottom panel:* Combined central and obstructive apnea. ITP = intrathoracic pressure, cm H_2O; MBF = medullary blood flow factor; $PmCO_2$ = medullary PCO_2.

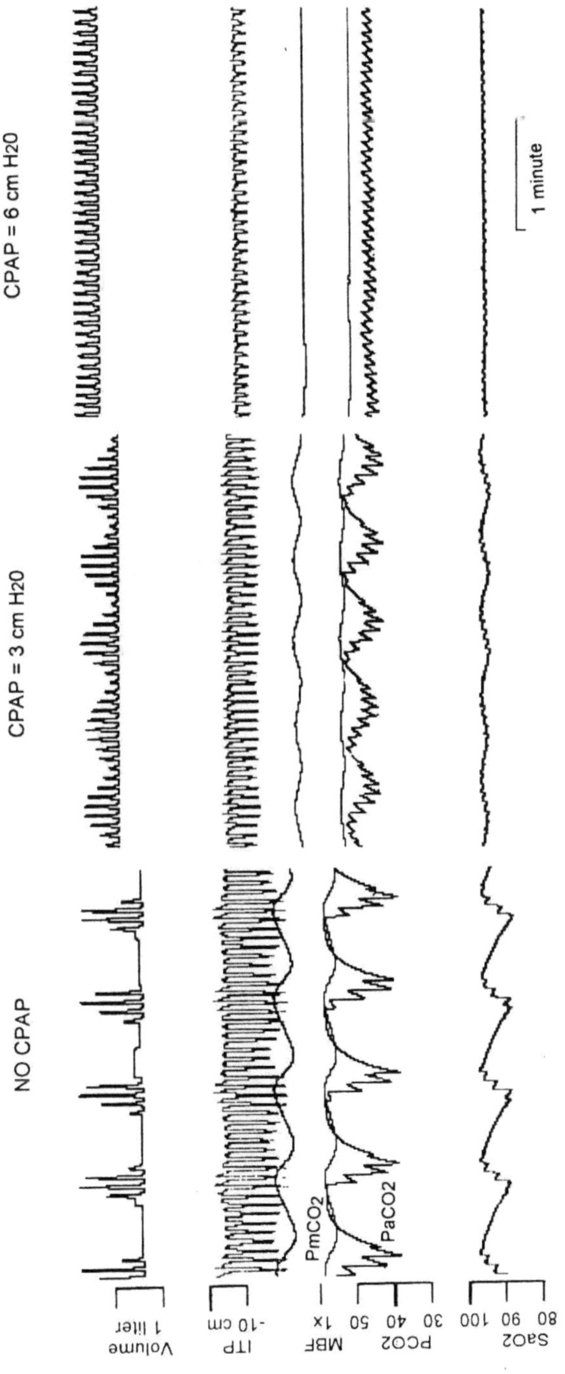

Figure 22. Effect of several levels of continuous positive airway pressure (CPAP) on obstructive sleep apnea.

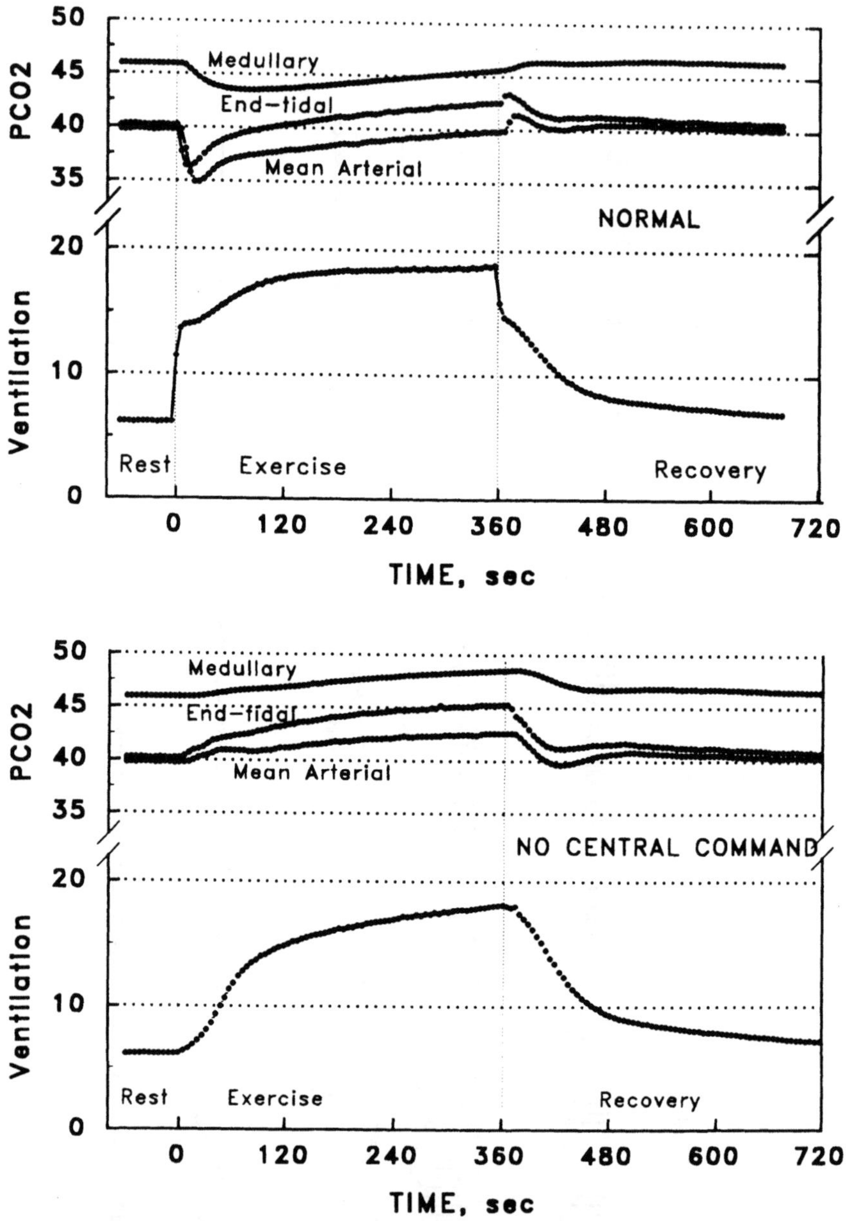

Figure 23. Breathing during rest ($\dot{V}O_2 = 0.25$ l/min) and moderate exercise ($\dot{V}O_2 = 1.0$ l/min). *Top:* Model has normal parameters including central neural command, STP, and rising [K^+a]. *Bottom:* same parameters except for absence of central command.

the well-known studies of Dejours[33] and diagram of Whipp[34]. There are transient changes of arterial and medullary PCO_2 but they have the same value at 6 min of exercise as they did at rest.

The bottom panel of Fig. 23 shows the same rate of exercise in the model which has been set up so that all parameters are the same, except that the central command has been removed. The difference from normal is obvious, a very different time course, a slightly lower ventilation at 6 min, (18.1 vs 18.8 l/min), and definitely higher arterial (42.6 vs 39.8 mm Hg) and medullary (48.5 vs 45.5 mm Hg) PCO_2 levels. The model thus supports the idea that the neural central command is necessary for the rapid ventilatory response to exercise and for the maintainance of isocapnia in the steady-state. In the absence of the central command, increased stimulation of the chemoreceptors, replacing input that normally comes from the central command, is necessary to produce the hyperpnea.

The importance of other potential factors in exercise hyperpnea can easily be tested by omitting them or changing their parameters.

4. COMMENT

An attempt has been made in the model described here to incorporate not only gas exchange and chemical feedback mechanisms, but also respiratory and mechanical processes involved in breathing, as well as the neural mechanisms that the author considers important in the control of respiration.

There are several deficiencies in the present model:

a. It does not have much of a frequency response when respiratory drive and tidal volume increase. In this respect it acts more like a vagotomized preparation. For example, the ventilatory response to CO_2 is mostly due increase of tidal volume. In the latest version of the model (see below in Acknowledgment), however, this deficiency has been corrected and respiratory frequency can increase about 3-fold during the highest levels of respiratory drive, as in exercise.
b. The model does not develop respiratory acidosis and its effect on chemoreceptors during heavy exercise. This means that the excessive breathing, relative to metabolic rate, and resulting hypocapnia do not occur in the current model.
c. The model does not incorporate calculation of tissue oxygenation in the medulla and does not develop the central neuronal depression of hypoxia due to release of inhibitory neurotransmitters, such as adenosine. Such a model has been constructed[8]; from that model it has been suggested hypoxic depression of medullary neurons may increase the likelihood of developing periodic breathing.
d. The Fitzhugh-VDP limit-cycle oscillator almost certainly does not exactly mimic the true limit-cycle of the medulla's oscillator. If an oscillator, based on neuronal function in the medulla, could be developed (see J. Smith in this volume), it might be possible to replace the Fitzhugh oscillator with such a neuronally based oscillator.

Despite the deficiencies, the model has proved to be quite useful in analyzing a number of elements involved in respiratory control, and allows comparison with published experimental results. Because so many of the various factors in the program can be changed while it is running, it is possible to study the importance of these factors. Finally, the model has proved useful in the teaching of the physiology of respiratory control.

ACKNOWLEDGMENTS

Supported in part by NIH Merit Award Grant Hl-17689. The author is prepared to make available a limited number of copies of the model's program in the form of compiled executable code (*.EXE), along with instructions for its use on an IBM-PC type computer, to individuals interested in using the model for purposes of teaching or research. Requests will be accepted only via e-mail at the following address: *eldridge.phys@mhs.unc.edu*

REFERENCES

1. Grodins, F.S., J. Buell and A.J. Bart. Mathematical analysis and digital simulation of the respiratory control system. *J. Appl. Physiol.* 22: 260-276, 1967.
2. Longobardo, G.S., N.S. Cherniack and A.P. Fishman. Cheyne-Stokes breathing produced by a model of the human respiratory system. *J. Appl. Physiol.* 21:1839-1846, 1966.
3. Khoo, M.C.K., R.E. Kronauer, K.P. Strohl and A.S. Slutsky. Factors inducing periodic breathing in humans: a general model. *J. Appl. Physiol.* 53: 644-659, 1982.
4. Rapoport, D.M., R.G. Norman and R.M. Goldring. CO_2 homeostasis during periodic breathing: predictions from a computer model. *J.Appl. Physiol.* k75: 2302-2309, 1993.
5. Eldridge, F.L., and T.G. Waldrop. Neural control of breathing during exercise. In: *Exercise, Pulmonary Physiology and Pathophysiology*. New York, Dekker, 1991, pp. 309-370.
6. Khoo, M.C.K., A. Gottschalk and A.I. Pack. Sleep-induced periodic breathing and apnea: a theoretical study. *J. Appl. Physiol.* 70: 2014-2024, 1991.
7. Khoo, M.C.K., F. Yang, J.J.W. Shin and P.R. Westbrook. Estimation of dynamic chemoresponsiveness in wakefulness and non-rapid-eye-movement sleep. *J. Appl. Physiol.* 78: 1052-1064, 1995.
8. Takahashi, E., and K. Doi. Destabilization of the respiratory control by hypoxic ventilatory depressions: A model analysis. *Jap. J. Physiol.* 43: 599-612, 1993.
9. Fitzhugh, R. Impulses and biological states in theoretical models of nerve membrane. *Biophysical J.* 1:445-446. 1961.
10. Eldridge, F.L. The limit-cycle oscillator: A model for the respiratory oscillator. *Biomed. Simulations Resource Workshop on Models of Complex Respiratory Dynamics of Sleep and Wakefulness.* Anaheim, May 11, 1991.
11. Eldridge, F.L. Phase resetting of respiratory rhythm - experiments in animals and models. In: *Springer Series in Synergetics, Vol. 55. Rhythms in Physiological Systems.* Berlin, Springer-Verlag, 1991, pp. 165-175.
12. Rochester, D.F., and N.M.T. Braun. The respiratory muscles. *Basics of RD.* Vol. 6: No. 4, 1978.
13. Salazar, E., and J. H. Knowles. An analysis of pressure-volume characteristics of the lungs. *J. Applied Physiol.* 19: 97-104, 1964.
14. Isono, S., D.L. Morrison, S.H. Launois, T.R. Feroah. W.A. Whitelaw and J.E. Remmers. Static mechanics of the velopharynx of patients with obstructive sleep apnea. *J. Appl. Physiol.* 75: 148-154.
15. Morrison, D.L., S.H. Launois, S. Isono, T.R. Feroah, W.A. Whitelaw and J.E. Remmers. Pharyngeal narrowing and closing pressures in patients with obstructive sleep apnea. *Am. Rev. Respir. Dis.* 148: 606-611, 1993.
16. Leiter, J.C., S.L. Knuth and D. Bartlett, Jr. Dependence of pharyngeal resistance on genioglossal EMG activity, nasal resistance, and airflow. *J. App. Physiol.* 73: 584-590, 1992.
17. Briscoe, W.A., and A.B. DuBois. The relationship between airway conductance and lung volume in subjects of different age and body size. *J. Clin. Invest.* 37:1279-1285, 1958.
18. Altman, P.L., and D.S. Dittmer (Eds). *Respiration and Circulation.* Fed. Am. Soc. for Experimental Biology. Bethesda, MD, 1971, p. 74.
19. Dill, D.B. In: *Handbook of Respiratory Data in Aviation*, Committee on Medical Research. Washington, 1944.
20. Santiago, T.V., and N.H. Edelman. Brain blood flow and control of breathing. In: *Handbook of Physiology, The Respiratory System, Control of Breathing.* Bethesda, MD: Am. Physiol. Soc., 1986, sect. 3, Vol. II, pt. 1, Chapt. 6, pp. 163-179.
21. Patterson, D.J. Potassium and ventilation in exercise. *J. Appl. Physiol.* 72: 811-820, 1992.
22. Phillipson, E.A., and G. Bowes. Control of breathing during sleep. In: *Handbook of Physiology, The Respiratory System, Control of Breathing.* Bethesda, MD: Am. Physiol. Soc., 1986, sect. 3, Vol. II, pt. 2, Chapt. 19, pp. 649-689.
23. Cunningham, D.J.C., P.A. Robbins and C.B. Wolff. Integration of respiratory responses to changes in alveolar partial pressures of CO_2 and O_2 and in arterial pH. In: *Handbook of Physiology, The Respiratory System, Control of Breathing.* Bethesda, MD: Am. Physiol. Soc., 1986, sect. 3, Vol. II, pt. 2, Chapt. 15, pp. 475-528.

24. Eldridge, F.L., D.E. Millhorn, J.P. Kiley and T.G. Waldrop. Stimulation by central command of locomotion, respiration and circulation during exercise. *Respir. Physiol.* 59: 313-337, 1985.
25. Wagner, P.G., and F.L. Eldridge. Development of short-term potentiation of respiration. *Respir. Physiol.* 83: 129-140, 1991.
26. Eldridge, F.L., and P. Gill-Kumar. Central neural respiratory drive and afterdischarge. *Respir. Physiol.* 40: 49-63, 1980.
27. Tawadrous, F.D, and F.L. Eldridge. Posthyperventilation breathing patterns after active hyperventilation in man. *J. Appl. Physiol.* 37: 353-356, 1974.
28. Eldridge, F.L. Central neural respiratory stimulatory effect of active respiration. *J. Appl. Physiol.* 37: 723-735, 1974.
29. Younes, M. The physiological basis of central apnea and periodic breathing. *Curr. Pulmonol.* 10: 265-326, 1989.
30. Eldridge, F.L, and D.E. Millhorn. Oscillation, gating, and memory in the respiratory control system. In: *Handbook of Physiology, The Respiratory System, Control of Breathing.* Bethesda, MD: Am. Physiol. Soc., 1986, sect. 3, Vol. II, pt. 1, Chapt. 3, pp. 93-114.
31. Khoo, M.C.K. Periodic breathing. In: *The Lung: Scientific Foundations.* New York, Raven Press, Ltd., 1991, pp. 1419-1431.
32. Conway, J., D.J. Patterson, E.S. Peterson and P.A. Robbins. Changes in arterial potassium and ventilation in response to exercise in humans. *J. Physiol. (Lond.)* 399: 36P, 1988.
33. Dejours, P. Control of respiration in muscular exercise. In: *Handbook of Physiology, Sect. 3. Respiration, Vol. I.* Washington, DC: Am. Physiol. Soc., 1964, pp. 631-648.
34. Whipp, B.J. The control of exercise hyperpnea. In: *Regulation of Breathing, Part I.* New York: Marcel Dekker, Inc., 1981, pp. 1069-1139.

3

THE APPLICATION OF ARX MODELLING TO VENTILATORY CONTROL

System Characterization and Prediction

Yoshitaka Oku

Department of Clinical Physiology
Chest Disease Research Institute
Kyoto University
Kyoto 606, Japan

1. INTRODUCTION

Instability of the respiratory control system causes periodic breathing, and life-threatening hypoxia may also occur during this time. Simple analysis of the steady-state ventilatory response to hypercapnia or hypoxia is not sufficient to evaluate the stability of this system, because the respiratory control system has a number of negative feedback loops, as shown in Figure 1.

In determining the stability of the negative feedback system, the *timing* as well as the magnitude of the system's corrective actions are critical parameters, since a corrective action occurring out of phase may further increase the error. There are several factors that promote the instability of the system[1]. To predict the behavior of the system, the characteristics of the ventilatory controller are obviously important. However, the characteristics of the plant are equally important. Therefore, it would be advantageous to be aware of the *dynamic* characteristics of the ventilatory system (both controller and plant) of *each individual.*

The general strategy to identify the characteristics of the system is to impose an exogenous input, and then assess the response of the system. However, in most cases the system has many inputs, and thus the response of the system may be affected by changes in other inputs. To eliminate the influence of these other inputs, one must control all other inputs and keep them constant. Clearly, this is extremely difficult because the respiratory control system is a typical multi-input system, as shown in Figure 2.

Suppose that hypoxic stimuli are imposed on the respiratory system. If the arterial PCO_2 level is uncontrolled, then the increase in ventilation due to hypoxia would result in hypocapnia, which would in turn decrease the ventilation. Therefore, the net ventilatory response depends on the combined effects of hypoxia and hypercapnia. During exercise, many factors are involved in the control of ventilation[2], such as corollary inputs from

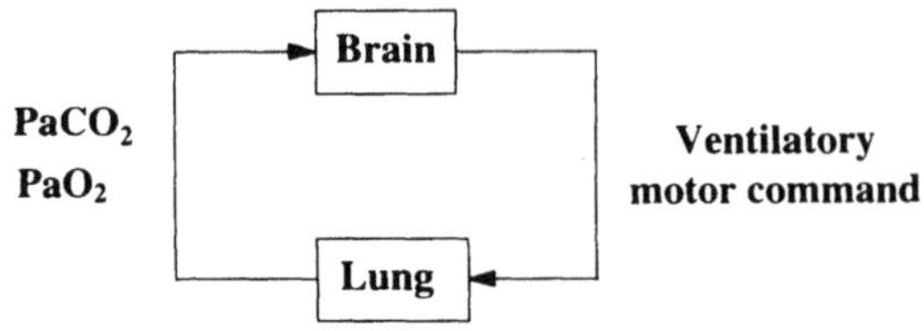

Figure 1. Schematic representation of the chemical feedback control system. Inherent background noise in the ventilatory output is transmitted through the feedback loop and affects fluctuation in the blood gas components. Therefore, fluctuations in the blood gas components are not independent of the inherent noises in the ventilatory output, and the cross correlation method does not yield an accurate frequency response characteristic.

locomotive centers, chemoreceptor afferent inputs, inputs from mechanoreceptors in the exercising limbs, the potassium concentration, cardiodynamic signals, etc. Clearly, it is not technically feasible to analyze the effect of a single input by keeping all other variables constant. Therefore, it would be useful if the influence of a given input on the ventilatory output could be discriminated from the others without having to control inputs.

A powerful tool for the identification of the dynamic characteristics of the ventilatory system is the autoregressive model with exogenous inputs (ARX model).

1. This model estimates the unbiased frequency responses of a system with feedback loops,
2. The relative influence of each input are assessed in a multi-input system,
3. This model predicts the time course of the system for a given input, and
4. It provides an optimal input to produce and maintain a target output.

This chapter outlines several important points regarding the application of the ARX model to an analysis of the respiratory control system. The entire procedure for system identification and controller implementation has been described by Nakamura and Akaike.[3]

2. APPLICATION OF THE ARX MODEL TO PHYSIOLOGIC SYSTEMS

The autoregressive (AR) model is a linear expression applied to a time series of system variables. The general form of the autoregressive model is shown in Eq. 1.

$$x_i(s) = \sum_{m=1}^{M} \sum_{j=1}^{K} A_{ij}\, x_j(s-m) + u_i(s) \tag{1}$$

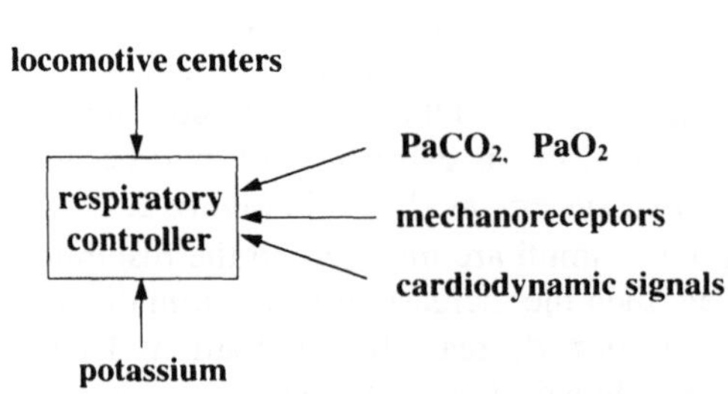

Figure 2. Factors involved in the control of exercise hyperpnea.

Each system variable x_i at time=s is expressed as the sum of the past data with a weighting factor A_{ij}. M is the order of the model, which represents the length of the history we will consider. K is the number of the variable, and $u_i(s)$ is the white noise. If the variables x_j (j=1,···,K) include an exogenous variable, then the expression is designated as the ARX model.

The respiratory control system is obviously more complex than the linear expression in Eq. 1, and is non-linear. The system described by Eq. 1 is assumed to be excited by stochastic noise (u_i), and the system is in a steady state without exogenous noise. However, recent studies have indicated that nonlinear dynamic mechanisms underlie respiratory pattern generation, and that these mechanisms can produce variable behavior that is not stochastic[4-6]. Therefore, applying the ARX model to the respiratory control system inevitably involves a certain amount of estimation error, which all models have to some extent when they are applied to physiologic systems. Nevertheless, this does not render the ARX model useless - what is important is to know the limitations of the model. Take the example of measuring the hypercapnic ventilatory responses, a very common clinical examination. The sensitivity of hypercapnic ventilatory response is expressed as the slope of the linear portion of the steady state response. Although this index is only good over the linear portion of the steady state relationship between ventilation and arterial PCO_2, it still provides important information regarding the system characteristics.

2.1. Assumption of the Stationary State

The characteristics of the respiratory system change under many conditions, for example depending on the sleep-waking cycle. Therefore, A_{ij} from Eq.1 is generally a variable. However, if we only look at a specific state, such as the awake resting state or the slow wave sleep state, then A_{ij} is relatively constant. Generally, if the system characteristics do not depend on the period being analyzed, then the system (or the signal the system delivers) is called stationary. Under the assumption of a stationary state, A_{ij} can be estimated to be a constant and the impulse and frequency responses can be calculated and thus the responses to a given stimulus can be predicted. In other words, if the signal is not stationary, then the application of the AR model with constant AR coefficients (A_{ij}) will produce erroneous results.

Even if the system is in a non-stationary state, A_{ij} could be estimated using a number of methods such as a recursive least squares method[7]. Modarreszadeh *et al.*[8] have used this technique to predict and control the end-tidal PCO_2. In this case, we can only make a one-step-ahead prediction of the output variables; A_{ij} must be renewed each time for further prediction. This is called 'adaptive' AR model, and will be discussed in detail in another chapter.

2.2. Assumption of Linearity

The ARX model provides information on linearized system behavior. If the system is non-linear, as is the case with the respiratory control system, then the estimation is only useful over a limited range, including the operating point where the linear approximation is applied. For the system depicted in Figure 1, we can describe two AR expressions: one representing the characteristics of the ventilatory controller and the other representing the characteristics of the CO_2 gas exchange system:

$$\dot{V}(s) = a \cdot PaCO_2(s-1) + b \cdot \dot{V}(s-1) + u(s) \tag{2}$$

$$PaCO_2(s) = c \cdot \dot{V}(s-1) + d \cdot PaCO_2(s-1) + v(s) \tag{3}$$

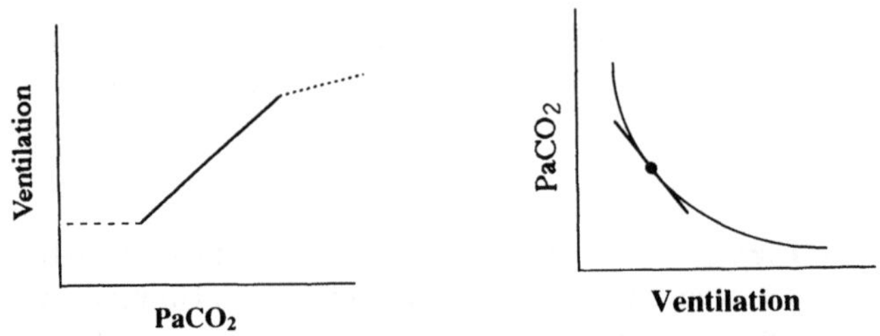

Figure 3. Schematic representation of the input-output relationships at the CO_2 ventilatory controller and the gas exchange system (lungs).

where a, b, c and d are AR coefficients, and u(s) and v(s) are white noise. These expressions can be compared with the actual input-output relationship at the controller and at the plant (gas exchange) system.

Figure 3 shows the input-output relationship at the ventilatory controller and the gas exchange system. It is intuitively obvious that the AR expression for the CO_2 gas exchange system is only valid over a range where the hyperbolic function can be approximated to its tangent at the operating point. Similarly, the relationship between ventilatory output and the $PaCO_2$ is not linear because of the 'dog-leg' region below the normocapnic point, and it may not be linear at very high CO_2 levels because of mechanical limitations. Therefore, the AR expression for the ventilatory controller can only be used over a limited range.

The ARX model also assumes that the effect of each input on the output in a multi-input system is additive, and that there are no interactions between these inputs. However, there is a multiplicative interaction between hypoxia and hypercapnia in peripheral chemoreceptors[9]. The effects of exercise and hypercapnia on ventilation are also multiplicative[10]. Therefore, using the ARX model to analyze a multi-input system that includes these elements will inevitably involve a certain amount of error.

2.3. Feedback Loops in the System

The right side of Eq. (1) not only includes previous data from the inputs ($x_j(s-m)$, $j \neq i$) but also previous data from the output itself ($x_i(s-m)$). This is important to whiten the residual $u_i(s)$ when feedback loops exist in the system. (The definition of the 'white' noise is described in ref. 19.) Without this autoregressive term, applying this model to a system with feedback loops will yield biased results[11]. For example, in Figure 1, the variation in the ventilation is not only affected by the inherent variation in the $PaCO_2$, but also by the previous value of the ventilation that was transmitted through the chemical feedback loop. Coefficients of $x_i(s-m)$ that are statistically different from zero can therefore be attributed to either the noise transmitted through the feedback loops, or to the memory of the system. Khatib *et al.*[12] applied a first-order AR model to analyze the breath-to breath fluctuations in the tidal volume and the peak of the integrated phrenic neurogram. The time series of the tidal volume in spontaneously breathing rats had a non-zero AR coefficient, whereas the AR coefficient for the time series of the peak integrated phrenic neural activity in artificially ventilated rats was not significantly

Table 1. Normalized correlation coefficients between innovations

	$\dot{V}_E$	$\dot{V}_{O_2}$	$\dot{V}_{CO_2}$	P_{ETCO_2}	HR	WR
$\dot{V}_E$	1.0	0.87	0.96	−0.27	0.25	0.002
$\dot{V}_{O_2}$		1.0	0.94	−0.06	0.42	−0.02
$\dot{V}_{CO_2}$			1.0	−0.20	0.31	−0.0006
P_{ETCO_2}				1.0	0.13	0.10
HR					1.0	0.23
WR						1.0

different from zero. Therefore, they inferred that chemical feedback loops contribute to the breath-by-breath variability in the tidal volume.

2.4. Choice of Variables

If the residuals show a strong correlation, then the simultaneous inclusion of these variables in the model must be avoided because the estimation of the AR coefficients becomes very unreliable. This situation could occur when the same noise is imposed on two or more variables. Therefore, analysis of the correlation coefficients of the residuals is useful for selecting the system variables to be used. For example, suppose that $\dot{V}_E$, $\dot{V}_{CO2}$, $\dot{V}_{O2}$, PETCO$_2$, the heart rate and the work rate were taken as system variables to estimate the system characteristics during randomly-inhaled CO_2 and random work rate forcing. Table 1 shows the crosscorrelation coefficients of the residuals normalized by autocorrelation of the residuals. Note that $\dot{V}_E$, $\dot{V}_{CO2}$, and $\dot{V}_{O2}$ show a very strong correlation - this is probably because the tidal volume fluctuation influences not only $\dot{V}_E$, but also $\dot{V}_{CO2}$, and $\dot{V}_{O2}$ Hence, this result indicates that we should choose only one of these three variables.

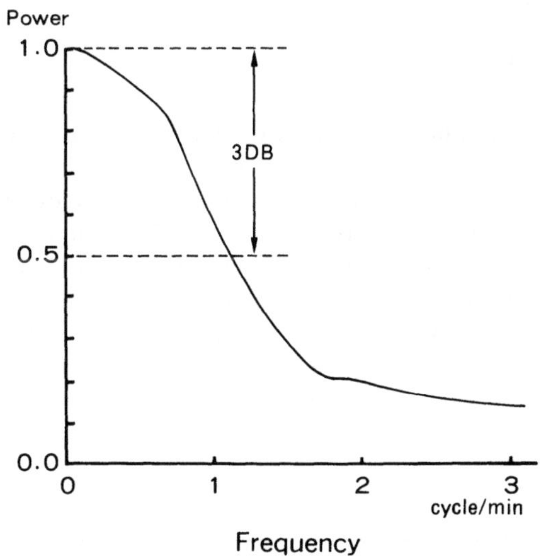

Figure 4. Normalized power spectral density for the input signal. The upper limit for the useful frequency is the frequency at which the power falls below 3 dB level (Reprinted with permission from Oku, Y., K. Chin, M. Mishima, M. Ohi, K. Kuno, and Y.H. Tamura. *Med. Biol. Eng. Comput.* 30: 51-56, 1992.).

2.5. Choice of Exogenous Inputs

To characterize the system's behavior over a certain frequency range, the inputs must have sufficient power to extend over the frequency range of interest. A random or pseudorandom binary signal is appropriate for this purpose. Many investigators have used pseudorandom binary sequences as the exogenous input that excites the system, over a wide range of frequencies[3,13-16]. Figure 4 shows a power spectral density for work rate fluctuations that consists of pseudorandom binary sequences.

The frequency at which the power drops by one half is the highest available frequency. A random signal generated by a computer algorithm is not completely white, and thus could affect the results considerably. Therefore, the assessment of test signals is necessary to ensure reliable results.

2.6 Determination of the Order of the Model

In applying this model, it is desirable to minimize the residual $u_i(s)$. This requirement increases the order of the model, and may cause overestimation. To manage this problem, several criteria listed below are used[17].

$$FPE = \left(1 + \frac{M \cdot K + 1}{N}\right)^{K} \left(1 - \frac{M \cdot K + 1}{N}\right)^{-K} |\Sigma| \tag{4}$$

$$AIC = N \cdot \log(|\Sigma|) + 2 \cdot M \cdot K \tag{5}$$

where N is the number of the data and Σ represents the prediction error covariance matrix. Eqs. (4) and (5) represent the Final Prediction Error (FPE) and Akaike's Information Criteria (AIC), respectively. The principle behind these criteria is a compromise between two opposing requirements: the minimization of the prediction error and the minimization of the number of parameters. Even if these criteria are used correctly, the model order may not be determined appropriately because of the non-linear dynamics of the system. We will discuss this issue later in the following section.

2.7 Validity of the Model

The validity of the model can be evaluated by examining whether the residual is almost white, and whether the model can predict the behavior of the system for a given input[18]. To test whether the signal is white, the autocorrelation, amplitude histogram, and power spectrum of the residual should be examined. White noise with a Gaussian distribution should meet the following criteria: a) the autocorrelation function should have a single sharp peak at lag=0, b) the amplitude histogram should have a Gaussian distribution, c) there should be a constant power spectral density function. Any failure to satisfy these criteria indicates the existence of significantly non-linear system characteristics. Figure 5 shows an example that successfully fulfills these criteria.

Figure 5. Histogram, autocorrelation, and power spectrum of the residual remaining after the ARX model was applied to the minute ventilation during simultaneous CO_2 inhalation and work rate forcing. In this example, the histogram has a single peak and the autocorrelation has a sharp peak at lag=0, which support the whiteness of the residual. However, the power spectrum is not uniform.

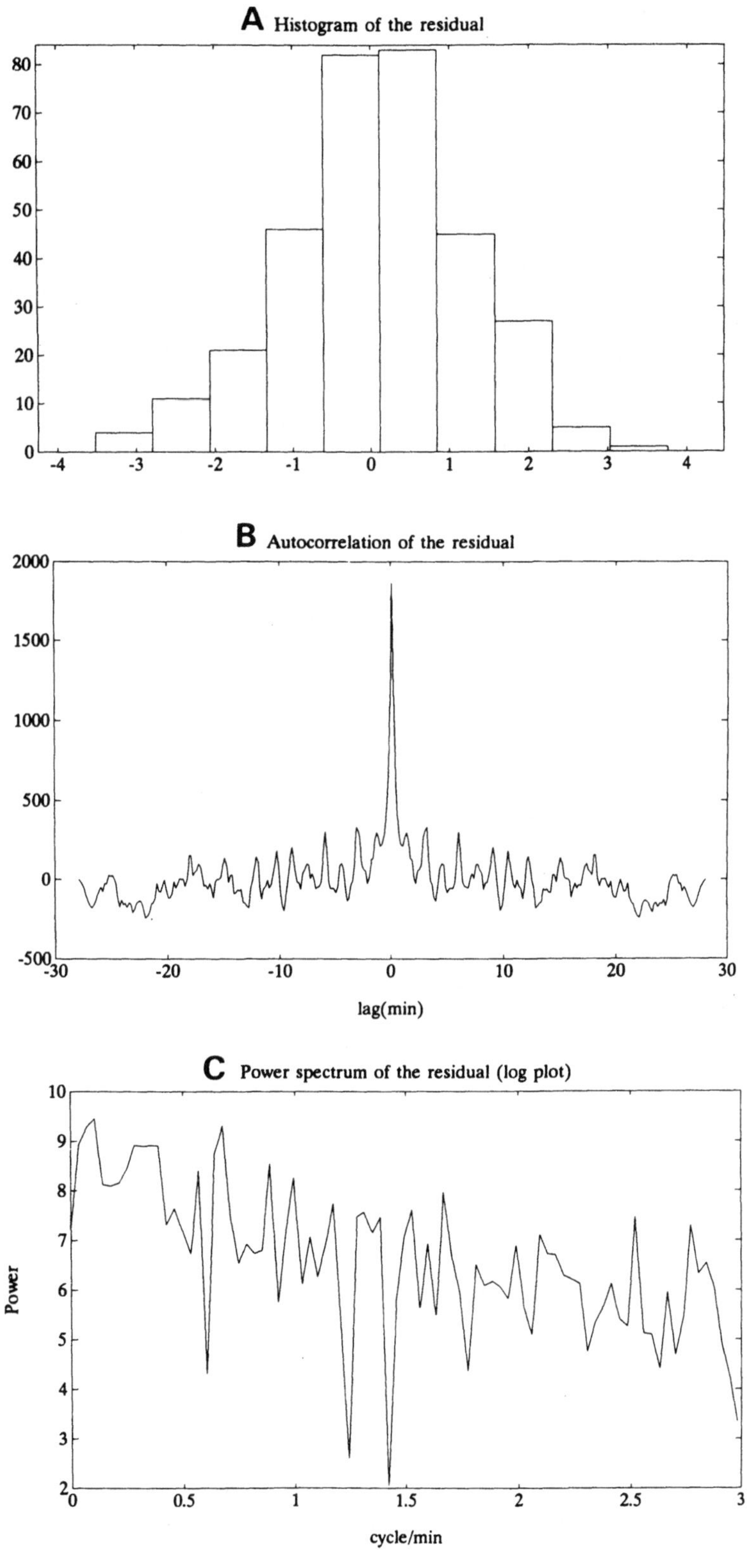
A Histogram of the residual
B Autocorrelation of the residual
lag(min)
C Power spectrum of the residual (log plot)
Power
cycle/min

The autocorrelation function has a sharp peak at lag=0, which means that the signal is not correlated. The amplitude histogram shows a single peak, which is compatible with a Gaussian distribution. If the histogram had two peaks, then we can hypothesize that the signal has two "preferable" states, or that we are recording a signal coming from two different populations[19]. The power spectrum has a relatively broad peak over the range of interest.

Figure 6 shows an example that did not fulfill the criteria for a white residual. The autocorrelation has multiple broad peaks, indicating that the residual was colored with periodic components. This suggests that the system generates discrete oscillations independent of the autoregressive structures, or that the system is receiving exogenous oscillatory inputs. In this case, it might be better to add a periodic term to the AR model, such as a sine function. Modarreszadeh *et al.*[20] analyzed the breath-to-breath fluctuations of ventilatory parameters during stage 2 sleep, and found that the power spectra for T_E had at least one peak in the high frequency band (0.2-0.5 cycles/breath). They then applied the AR model, with the exogenous variables having discrete oscillations as shown in Eq. (6).

$$x(s) = a_1 x(s-1) + \sum_{j=1}^{P} [b_{j1}\sin(2\pi f_j n) + b_{j2}\cos(2\pi f_j n)] \tag{6}$$

They determined that the AR coefficient a_1 was not significantly different from zero, and concluded that fluctuations of T_E in some subjects had discrete oscillatory components that could not be attributed to autoregressive structure.

2.8 Interpretation of the Model

Even if the model is validated, it is critical to be cautious in interpreting the results. Since the results are obtained by model fitting, and not by direct measurements, they do not necessarily reflect the real *in vivo* situation. The characteristics we can presently ascertain only deal with the input-output relationship, and the characteristics

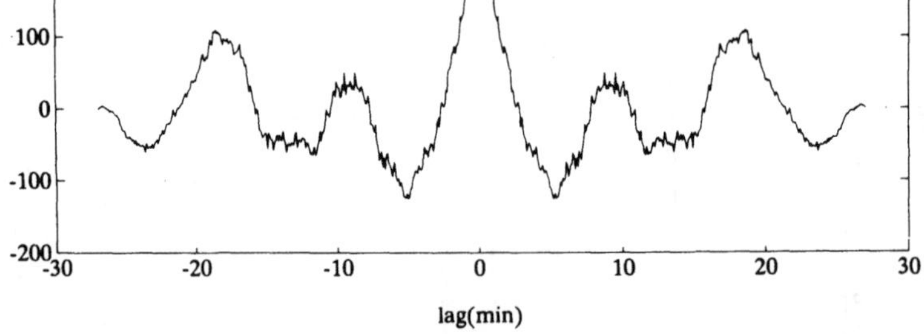

Figure 6. Autocorrelation of the residual remaining after the ARX model was applied to the minute ventilation during random CO_2 inhalation under hypoxic conditions. The symmetric broad peaks indicate oscillatory components in the residual.

of the system itself or factors mediating exogenous inputs to controller remain to be elucidated. For example, the ARX model cannot address the issue as to whether two variables interact additively or multiplicatively, because the model is by definition an additive model.

2.9 Application 1. Dynamic Ventilatory Responses during Hypoxia and Normoxia.

In this example, the dynamic ventilatory responses to hypercapnia were compared during hypoxia versus normoxia. The subjects breathed either 21% O_2 or 14% O_2, and then the CO_2 inhalation trial was performed. During the CO_2 inhalation, the concentration of the

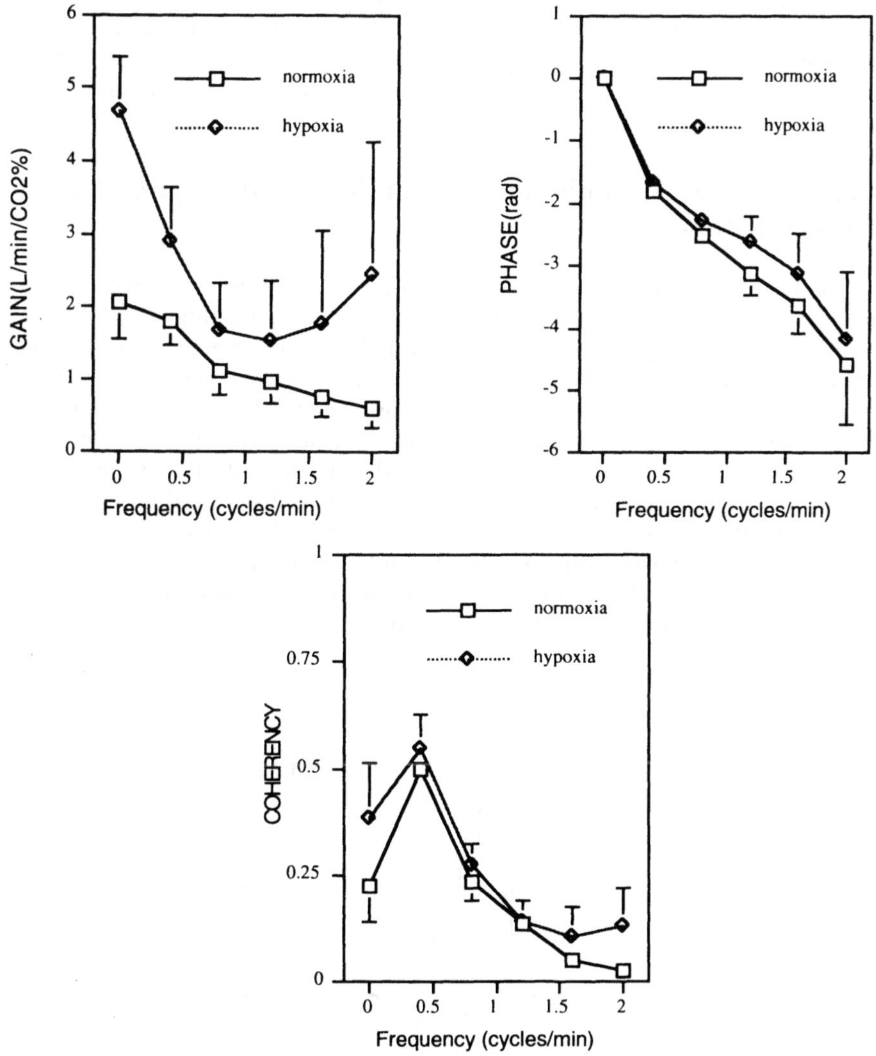

Figure 7. Weighted ensemble means of the gain (amplitude ratio of the input to the output), the phase shift and the coherency in the frequency responses of ventilation to changes in the end-tidal CO_2 (n=5). Note that hypoxia not only enhances the magnitude of the response, but also shortens the phase delay.

inspired CO_2 was varied between 0 to 7 %. The tidal volume, respiratory frequency, and end-tidal PCO_2 were measured on a breath-by-breath basis, and the minute ventilation and end-tidal PCO_2 were computed at intervals of 5 seconds using a spline function. The data were fitted to the ARX model, and the AR coefficients and subsequent frequency responses were computed.

Figure 7 shows the weighted ensemble means of the gain (amplitude ratio of the input to the output), the phase shift and the coherency in the frequency responses of ventilation to changes in the end-tidal CO_2 (n=5). The gain and phase shift plots indicate that hypoxia not only enhances the magnitude of the response, but also shortens the phase delay, probably due to a decrease in the circulatory time. These results are compatible with a steady-state observation. The coherency is an index of the linearity between input and output. Low coherency implies either a highly non-linear input-output relationship, or a high contribution of the other inputs (noises). Therefore, the coherency plot provides important information as to the degree of the linear contribution of the input to the fluctuations of the output. Figure 7C shows that coherency is greater during normoxia. This result indicates that hypoxia enhances the relative influence of chemical control, and decreases the relative influence of the other factors.

2.10 Application 2. Dynamic Influence of End-Tidal CO_2 and Exercise on Ventilation.[15]

The second application identifies a two-input respiratory control system using the multivariate ARX model. The dynamic influence of end-tidal CO_2 and exercise on ventilation were compared when the CO_2 and exercise were imposed separately, and when they were imposed simultaneously. The subjects performed three trials: random work rate forcing, random CO_2 inhalation and simultaneous loading with both. The work load was varied between 20 to 80W. The duration of each work load was varied in a pseudorandom binary sequence and between 10 and 90 seconds, at 10 second intervals. The concentration of inspired CO_2 was also varied randomly between 0 and 7 percent at random intervals of 10 to 90 seconds. The variables were measured on a breath-by-breath basis, and linearly interpolated at intervals of 10 seconds. The data were fitted to the ARX model and the AR coefficients, subsequent relative power contributions and frequency responses were computed.

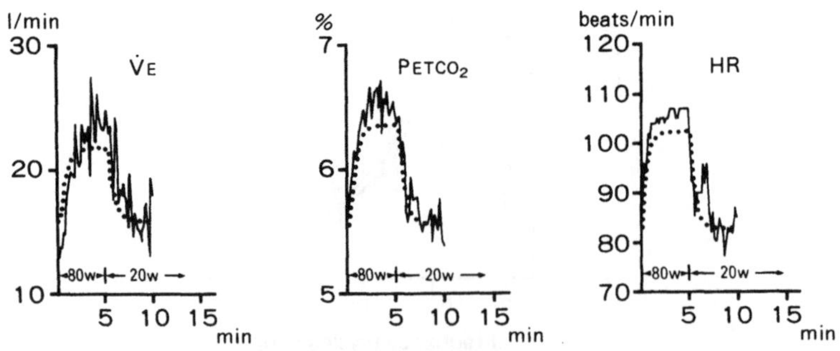

Figure 8. Comparison of the predicted (dotted line) and experimentally-obtained (solid line) ventilatory response to a step increase and decrease in work rate for *subject* N.H. The predicted response is based on the ARX model fitted to data obtained during a pseudorandom work rate forcing.

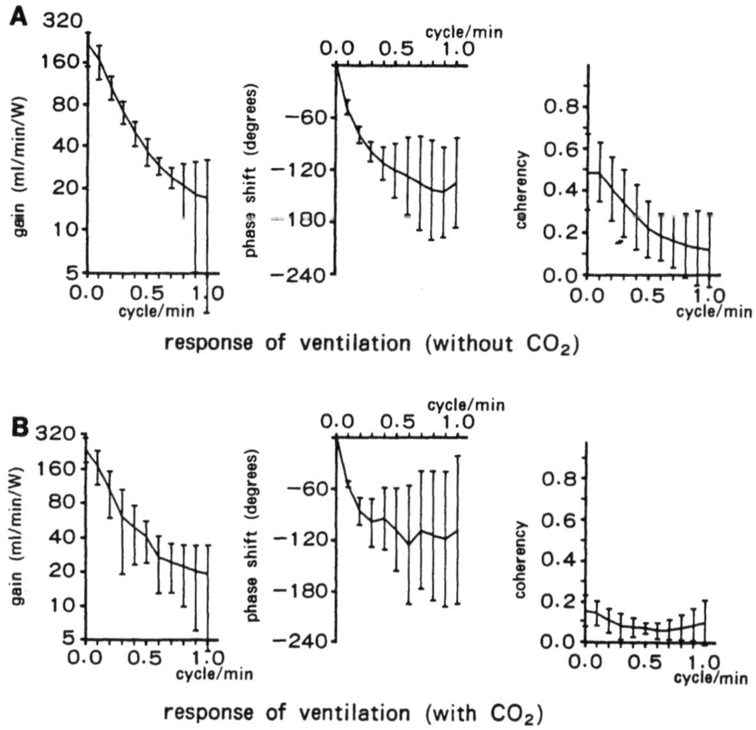

Figure 9. Weighted ensemble means of the gain, the phase shift, and the coherency in the frequency responses of ventilation to workload. The responses were normalized on a per watt basis. A: without CO_2 inhalation. B: with CO_2 inhalation. (Reprinted with permission from Oku, Y., K. Chin, M. Mishima, M. Ohi, K. Kuno, and Y.H. Tamura. *Med. Biol. Eng. Comput.* 30: 51-56, 1992.).

The model was validated by checking the whiteness of residuals. Furthermore, the predicted ventilatory response to step changes in work rate were in agreement with the experimental data, as shown in Figure 8.

The frequency responses of ventilation to work load during exercise alone and during simultaneous CO_2 plus exercise loading (Figure 9) showed similar amplitude and phase delay characteristics, although the coherency decreased greatly with CO_2 inhalation. The similarity of these response characteristics suggests that mechanisms independent of changes in the end-tidal CO_2 (and presumably the arterial CO_2) are involved in the control of ventilation during exercise. Ventilatory responses to these changes in the end-tidal CO_2 are superimposed on these mechanisms, and thus decrease the coherency of the ventilatory responses to work rate. This interpretation is compatible with the current concept regarding exercise hyperpnea[2].

If the system has two or more inputs, then the relative power contribution plot is better than the coherency plot at determining the linear contribution of each input to a given output. The relative power contribution of each variable to ventilation was given by decomposing the power spectrum of ventilation into the sum of the contributions from each variable at each frequency.

Figure 10 shows that weighted ensemble means of the relative power contribution during exercise without CO_2 inhalation, during CO_2 inhalation only, and during exercise with CO_2 inhalation. During exercise without CO_2 inhalation, work rate is the only significant contributor to the fluctuation of ventilation. However, during simultaneous CO_2 and exercise

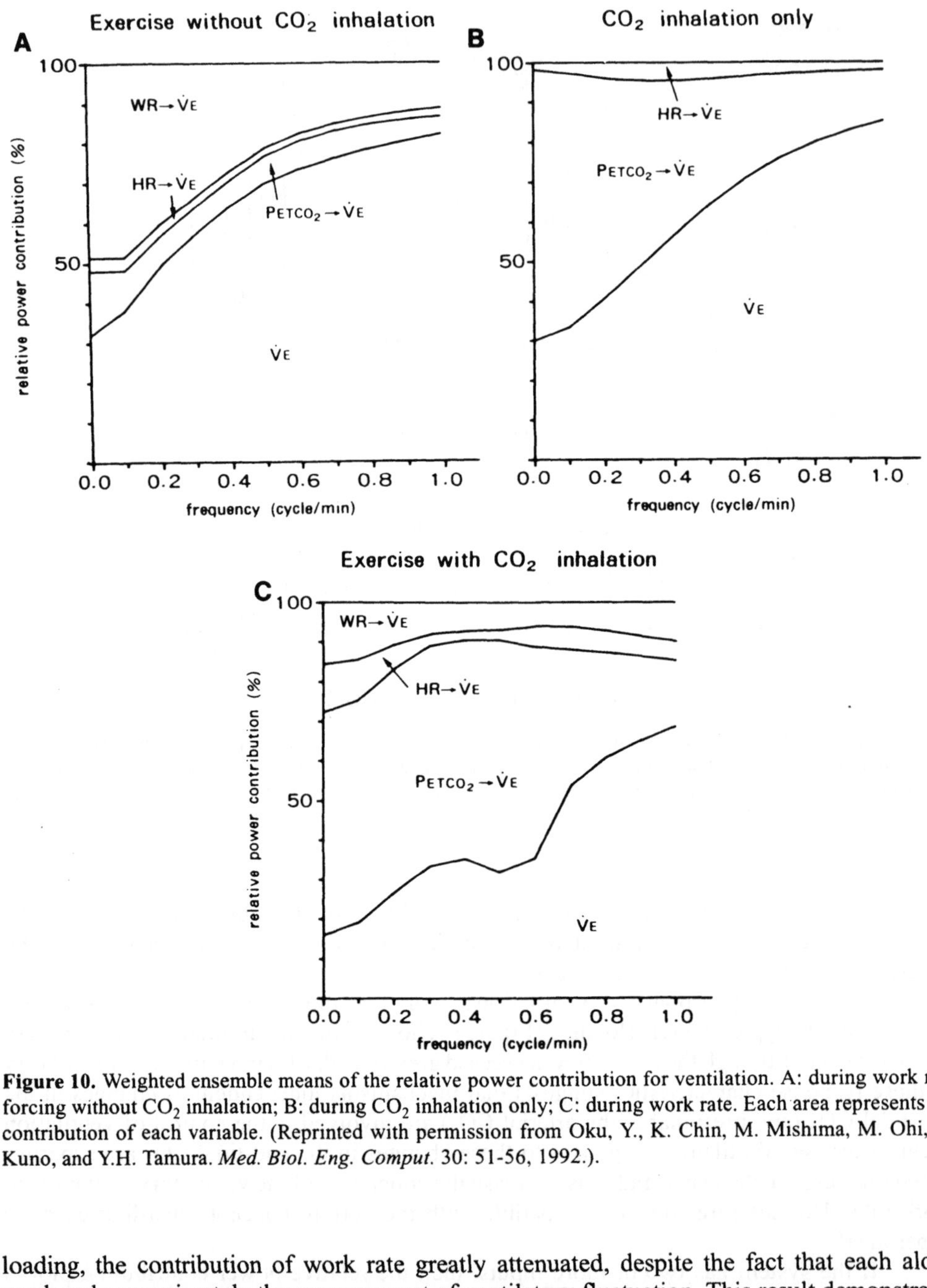

Figure 10. Weighted ensemble means of the relative power contribution for ventilation. A: during work rate forcing without CO_2 inhalation; B: during CO_2 inhalation only; C: during work rate. Each area represents the contribution of each variable. (Reprinted with permission from Oku, Y., K. Chin, M. Mishima, M. Ohi, K. Kuno, and Y.H. Tamura. *Med. Biol. Eng. Comput.* 30: 51-56, 1992.).

loading, the contribution of work rate greatly attenuated, despite the fact that each alone produced approximately the same amount of ventilatory fluctuation. This result demonstrates the non-linear system dynamics for the combination of exercise and CO_2 forcing.

3. CONCLUSION

Recently, the computing capacity of personal computers has been greatly increased, and software packages for analyzing time series data using the ARX model are now available.

Therefore, dynamic analyses are no longer exclusive to a few bioengineering labs. If we recognize the limitations discussed in this chapter, the ARX model can be one of the best methods available to estimate an individual system characteristics.

REFERENCES

1. Khoo, M.C.K., R.E. Kronauer, K.P. Strohl, and A.S. Slutsky. Factors inducing periodic breathing in humans: a general model. *J. Appl. Physiol.* 53: 644-659, 1982.
2. Dempsey, J.A., H.V. Forster, and D.M. Ainsworth. Regulation of hyperpnea, hyperventilation, and respiratory muscle recruitment during exercise. In: *Regulation of Breathing,* edited by J.A. Dempsey and A.I. Pack. New York: Mercel Dekker, 1995, p.1065-1134.
3. Nakamura, H., and H. Akaike. Statistical identification of optimal control of supercritical thermal power plants. *Automatica* 17: 143-155, 1981.
4. Bruce, E.N. and J.A. Daubenspeck. Mechanisms and analysis of ventilatory stability. In: *Regulation of Breathing,* edited by J.A. Dempsey and A.I. Pack. New York: Mercel Dekker, 1995, p. 285-314.
5. Sammon, M., J.R. Romaniuk, and E.N. Bruce. Bifurcations of the respiratory pattern associated with reduced lung volume in the rat. *J. Appl. Physiol.* 75: 887-901, 1993.
6. Sammon, M., J.R. Romaniuk, and E.N. Bruce. Bifurcations of the respiratory pattern produced with phasic vagal stimulation in the rat. *J. Appl. Physiol.* 75: 912-926, 1993.
7. Åström, K.J., and Wittenmark. *Adaptive Control.* Reading, MA: Addison-Wesley, 1989.
8. Modarreszadeh, M., K.S. Kump, H.J. Chizeck, D.W. Hudgel, and E.N. Bruce. Adaptive buffering of breath-by-breath variations of end-tidal CO_2 of humans. *J. Appl. Physiol.* 75: 2003-2012, 1993.
9. Rebuck, A.S., A.S. Slutsky, and C.K. Mahutte. A mathematical expression to describe the ventilatory response to hypoxia and hypercapnia. *Respir. Physiol.* 31: 107-116, 1977.
10. Poon, C.S., and J.G. Greene. Control of exercise hyperpnea during hypercapnia in humans. *J. Appl. Physiol.* 59: 792-797, 1985.
11. Akaike, H. Some problems in the application of the cross-spectral method. In *Spectral analysis of time series,* edited by B. Harris. New York: John Wiley, 1967, p. 81-107.
12. Khatib, M.F., Oku Y., and E.N. Bruce. Contribution of chemical feedback loops to breath-by-breath variability of tidal volume. *Respir. Physiol.* 83: 115-128, 1991.
13. Sohab, S., and S.M. Yamashiro. Pseudorandom testing of ventilatory response to inspired carbon dioxide in man. *J. Appl. Physiol.* 49: 1000-1009, 1980.
14. Bennett, F.M., P. Reischl, F.S. Grodins, S.M. Yamashiro, and W.E. Fordyce. Dynamics of ventilatory response to exercise in humans. *J. Appl. Physiol.* 51: 194-203, 1981.
15. Oku, Y., K. Chin, M. Mishima, M. Ohi, K. Kuno, and Y.H. Tamura. Dynamic control of breathing during exercise and hypercapnia. *Med. Biol. Eng. Comput.* 30: 51-56, 1992.
16. Modarreszadeh, M., and E.N. Bruce. Long-lasting ventilatory response of humans to a single breath of hypercapnia in hyperoxia. *J. Appl. Physiol.* 72: 242-250, 1992.
17. Akaike, H. A new look at the statistical model identification. *IEEE Trans. Automat. Contr.* AC-19: 716-723, 1974.
18. Wiberg, D.M. Parameter estimation for respiratory physiology. In: *Modeling and Parameter Estimation in Respiratory Control,* edited by M.C.K. Khoo. New York: Plenum, 1989, pp.71-82.
19. Marmarelis, P.A., and V.Z. Marmarelis. *Analysis of Physiological Systems, The white noise Approach.* New York: Plenum, 1978.
20. Modarreszadeh, M., E.N. Bruce, and B. Gothe. Nonrandom variability in respiratory cycle parameters of humans during stage 2 sleep. *J. Appl. Physiol.* 69: 630-639, 1990.

4

ESTIMATION OF CHANGES IN CHEMOREFLEX GAIN AND 'WAKEFULNESS DRIVE' DURING ABRUPT SLEEP-WAKE TRANSITIONS

Steve S. W. Koh, Richard B. Berry, Jang-Won Shin, and
Michael C. K. Khoo

Biomedical Engineering Department
University of Southern California
Los Angeles, California 90089
Department of Medicine
V.A. Medical Center
Long Beach, California 90822

1. INTRODUCTION

It is generally accepted that the ventilatory sensitivity to CO_2 is decreased during non-rapid eye-movement (NREM) sleep, although the degree of this reduction has varied considerably among studies[1,2]. In all these studies, the ventilatory response to hypercapnia (HCVR) was measured under steady-state conditions in quiet wakefulness and, subsequently, in one or more stable stages of NREM sleep. In sleep-disordered breathing, however, sleep is constantly punctuated by transient arousals, which are generally associated with brief periods of hyperpnea[3,4]. Thus, the standard procedures of estimating steady-state chemoresponsiveness may not produce measures that appropriately capture the changes in chemoreflex control that accompany such abrupt state changes. This may partially account for the lack of any consistent relationship between the steady-state HCVR and breathing pattern stability or instability[5].

In the present study, we propose a method for estimating chemoresponsiveness before and immediately after an abrupt change of state from sleep to wakefulness. This procedure is a simple extension of the classical Read rebreathing test[6]. The first part of the procedure is exactly analogous to the rebreathing tests that have been applied to sleeping subjects in previous studies[7,8]. In this case, however, the progressively increasing hypercapnia and hyperpnea are used to trigger an arousal. Instead of terminating the experiment at the point of arousal, as is generally done in conventional rebreathing, the hypercapnia is allowed to progress for another minute or two. By fitting a simple two-phase model to the data before and after the abrupt sleep-wake change, it is possible to estimate the chemoreflex gains

Bioengineering Approaches to Pulmonary Physiology and Medicine, edited by Khoo
Plenum Press, New York, 1996

corresponding to the two states, the P_{ACO2} threshold level for arousal, and the increase in ventilation (ie. the "wakefulness drive") immediately accompanying arousal. A practical advantage of the proposed method is that the above parameters can be estimated from a single test procedure. A preliminary account of this study has been published in abstract form[9].

2. METHODS

2.1. Experimental Setup and Protocol

We developed a novel experimental scheme which induced progressive hypercapnia without using the traditional rebreathing bag. This arrangement allowed the easy maintenance of a hyperoxic background eliminating the potential problem of oxygen desaturation that could result from prolonged rebreathing. A further advantage was the reduction in resistance in the breathing circuit. The apparatus employed is shown in Fig.1. Three electronic mass flow controllers (Model 1259, MKS Instruments, Andover, MA) were used to determine the flow rates of CO_2, O_2 and N_2 entering a gas reservoir, the exit end of which was connected to long wide-bore tubing that carried a constant high flow of the desired gas mixture to the adjacent room in which the subject slept. The inspiratory and expiratory ports of the subject's mask were each connected via a one-way valve to the tubing so that the subject would inhale from and exhale into the stream of fast-flowing gas without contamination from the ambient air. The individual mass flow controllers for CO_2, O_2 and N_2 were controlled by signals generated by a 80386-compatible computer, which sampled mask pressure, inspiratory airflow and end-tidal P_{CO2} at 20 Hz per channel.

The progressive hypercapnia of rebreathing was simulated by assuming that the subject's exhaled gas would mix with the gas in a "virtual bag" and that the resultant mixture

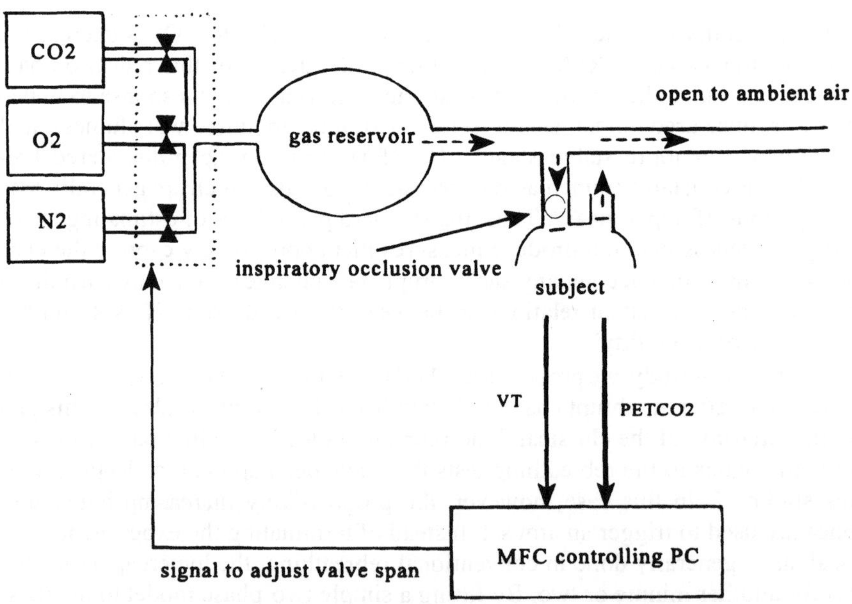

Figure 1. Experimental setup for producing progressive hypercapnia.

would be re-inhaled. Applying the principle of conservation of CO_2 to the "virtual bag" led to the following differential equation:

$$V_{bag}\frac{dF_{ICO2}}{dt} = \dot{V}_E\,(F_{ETCO2} - F_{ICO2}) \tag{1}$$

where F_{ETCO2} and F_{ICO2} represent the end-tidal and inhaled (ie. bag) CO_2 concentrations, respectively, and V_{bag} represents the volume of the "virtual bag". Assuming impulse invariance[10], the above differential equation was converted into the equivalent difference equation so that progressive hypercapnia could be simulated on a breath-by-breath basis. The resulting equation was solved in real time to predict the CO_2 concentration of the inhaled gas mixture in the next breath, $F_{ICO2}(n+1)$, with n representing the index of the current breath:

$$F_{ICO2}\,(n+1) = e^{-T_T(n-1)/\tau}F_{ICO2}(n-1) + (1 - e^{-T_T(n-1)/\tau})F_{ETCO2}\,(n-1) \tag{2}$$

where $\tau = T_T(n\text{-}1)\; V_{bag}/V_T(n\text{-}1)$. $T_T(n\text{-}1)$, $V_T(n\text{-}1)$ and $F_{ETCO2}(n\text{-}1)$ represent the breath duration, tidal volume and end-tidal CO_2 concentration of the previous breath. F_{ICO2} at breath $n+1$ was based on the parameters of the breath $n\text{-}1$, and not breath n, because F_{ETCO2} from breath $n\text{-}1$ was available only during the inspiratory phase of the current breath n due to the finite sampling delay of the CO_2 analyzer. The parameter V_{bag}, which controlled the initial rate of rise of F_{ICO2}, was assigned the value of 6 liters in all experiments.

A pneumotachograph (Model 3700, Hans Rudolph, Kansas City, MO) and two-way balloon-valve (Model 8250, Hans Rudolph) were also connected to the inspiratory port. The balloon valve was occluded during expiration and not opened until 100-150 ms after the computer detected negative mask pressure, which indicated the start of the subsequent inspiration. The occlusion pressure 100 ms after the start of inspiration, P_{100}, was taken to represent the inspiratory drive for that breath[11]. This measurement was essential for delineating the effect of state on respiratory drive from its effect on upper airway resistance. End-tidal gas was sampled near the expiratory port of the face mask for analysis of CO_2 content (Model 223, Puritan-Bennett, Los Angeles, CA or Model LB-2, Sensormedics, Anaheim, CA).

We studied a total of 7 healthy volunteers after obtaining written informed consent. Three of the subjects returned for a repeat study. However, in the second study, the measurements did not include P_{100}. Instead, we measured upper airway pressure using a saline-filled catheter that was inserted through one nostril until its tip was positioned at a supraglottic location[12]. The catheter was connected to a disposable Transpac II pressure transducer (Abbot Critical Care, Chicago, IL). Aside from the measurements mentioned above, the following variables were also monitored: arterial O2 saturation (Model 3700, Ohmeda, Boulder, CO or Model IIA, BTI Biox,), 2 channels of EEG (C3/A2 and O1/A2), electrooculogram, chin electromyogram and a single channel of electrocardiogram. All respiratory and polysomnographic signals were digitized online at 100 Hz/channel and stored in a second computer (80486-compatible) for subsequent analysis.

Following sleep onset, each subject was allowed to progress to stable Stage 2 sleep before the "rebreathing" test commenced. Using the apparatus described above, the inhaled gas increased progressively in CO_2 content at a rate determined initially by V_{bag} and subsequently by the metabolic CO_2 production rate of the individual subject. The O_2 content of the inhaled gas was maintained at 30% or higher throughout the rebreathing procedure. Following arousal, rebreathing continued for another 1-2 minutes. The subject was generally awake by the end of the procedure. Following the administration of one test, the subject was allowed to return to sleep. The entire procedure was repeated once stable Stage 2 sleep was attained. The total number of test procedures performed on each subject ranged from 2 to 4.

2.2. Data Analysis

The EEG was used to provide a quantitative description of changes in sleep-wake state. Second-by-second power spectra of the C3/A2 EEG signal between 1 and 25 Hz were generated using autoregressive modeling. This procedure is described in detail in Khoo et al.[4]. Further techical details may be found in Shiavi[10]. By computing the total area under the spectral density function between 8 and 12 Hz (alpha band) and between 14 and 25 Hz (beta band), we extracted the high frequency power from each successive 1-second spectrum. This was normalized by the total EEG power (between 1 and 25 Hz) for that corresponding 1-second duration to yield the relative high-frequency power (RHFP). The chin EMG signal was also used to help in the detection of arousal onset. The standard deviation of the raw EMG signal about the mean was computed for each second of data. In addition, P_{ETCO2}, P_{ICO2}, V_T/T_I (L/min) and P_{100} were deduced from the respiratory signals on a breath-by-breath basis. In the cases where upper airway pressure was measured, upper airway resistance (R_{UAW}) was calculated on a breath-by-breath basis by first locating the point of occurrence of peak inspiratory airflow, and then dividing the difference between upper airway pressure and mask pressure at that instant by the peak airflow.

2.3. Modeling and Parameter Estimation

In order to arrive at a useful interpretation of the test results, the following model was employed. We assumed, as in the conventional rebreathing test, that following equilibration of P_{CO2} among the "bag", lungs and body tissues compartments,

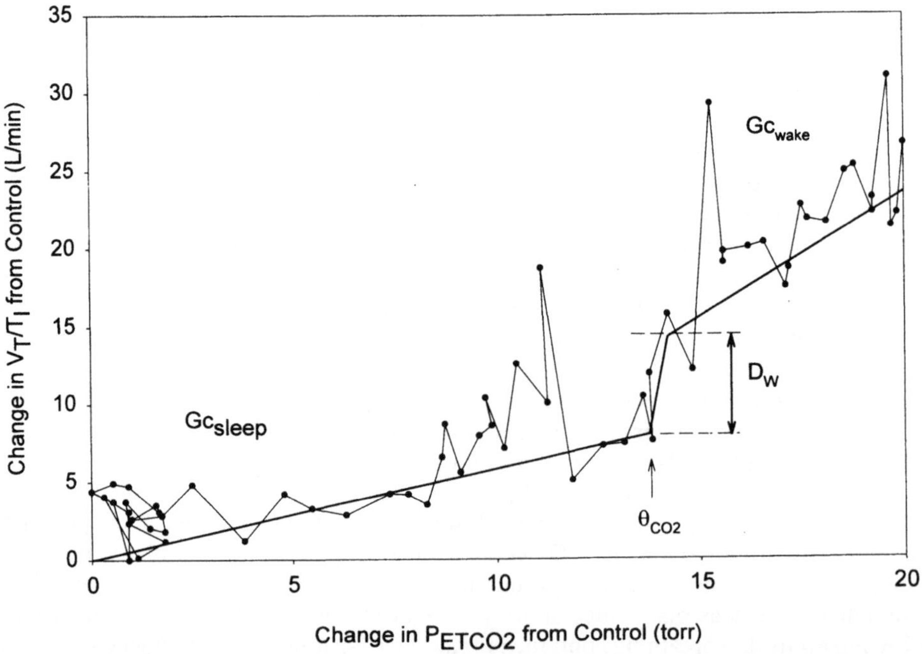

Figure 2. Illustration of the computational procedure for estimating the chemoreflex gains, during sleep (Gc_{sleep}) and during wake (Gc_{wake}), the CO_2 threshold for arousal (θ_{CO2}), and the wakefulness drive (D_W) from the test results.

P_{ETCO2} would rise linearly with time at a rate determined by the subject's metabolic CO_2 production rate[13]. During this period, with the subject still asleep, there would be a virtual "opening of the feedback loop" so that spontaneous changes in ventilation (in principle) should not affect the P_{ETCO2}. Under such conditions, ventilatory drive (D) would be a linear function of P_{ETCO2}, with the slope representing the central chemoreflex gain during sleep (Gc_{sleep}):

$$D(t) = Gc_{sleep} \cdot P_{ETCO2}(t) + D_{NC}(t) \tag{3}$$

where $D_{NC}(t)$ represents a dynamic non-chemical component of D(t). After P_{ETCO2} exceeds θ_{CO2}, the threshold for arousal, the model assumes an abrupt change of state, which is accompanied by a restoration of the wakefulness drive (D_W) and a change in chemoreflex gain to its waking value, Gc_{wake}:

$$D(t) = Gc_{wake} \cdot P_{ETCO2}(t) + D_W + D_{NC}(t) \tag{4}$$

Both V_T/T_I and P_{100} were used as indices of D.

The time-courses of RHFP derived from the central EEG channel and the second-by-second standard deviation of the chin EMG signal were used for identifying the time of the abrupt change in state. The P_{ETCO2} value corresponding to this point in time was chosen to be the best estimate of θ_{CO2}. The slope of the least-squares linear fit to the data points (V_T/T_I vs P_{ETCO2} or P_{100} vs P_{ETCO2}) from P_{CO2} equilibration to the breath immediately before the state change was taken to represent Gc_{sleep}. The slope of the corresponding straight-line fit to the data points immediately following the state change provided an estimate of Gc_{wake}. D_W was estimated from the vertical distance between the two straight-line segments at θ_{CO2}. This procedure for estimating the model parameters, in the case where V_T/T_I represents D, is illustrated in Fig.2.

3. RESULTS

Representative data from a test on Subject #2 is shown in Fig. 3. Prior to the start of the test, the subject is in Stage 2 sleep breathing air. The test begins at ~30 s as P_{ICO2} rapidly increases towards P_{ETCO2} levels. At ~t=132 s, there was an abrupt change in state to wakefulness as indicated by the shift in EEG RHFP level (Fig.3, bottom panel). This was accompanied by abrupt increases in V_T/T_I and P_{100}, followed by increases in their rates of change. Since the target P_{ICO2} always lagged P_{ETCO2} by at least two breaths, it is clear that there was incomplete equilibration between these two quantities. However, spontaneous fluctuations in V_T/T_I during the test had virtually no effect on P_{ETCO2} and P_{ICO2}, suggesting alveolar P_{CO2} and mixed venous P_{CO2} were probably equilibrated, and that the assumptions made in the model were valid.

Figure 4 illustrates the result of applying our analysis to the data displayed in Fig.3. The top panel of Fig.4 shows the superimposition of the best straight-line fits on the data, with changes in V_T/T_I plotted against changes in P_{ETCO2}. The computational procedure allows the separation of the ventilatory response into its CO_2-dependent component (2nd panel) and the wakefulness drive (3rd panel). The difference between the observed V_T/T_I and the sum of the CO_2-dependent and wakefulness drive components were taken to represent the dynamic non-chemical component, $D_{NC}(t)$ (Fig.4, 4th panel).

Gc, deduced from P_{100} measurements, increased substantially in all subjects from sleep to wakefulness (Fig.5, right panel). The average increase for the group

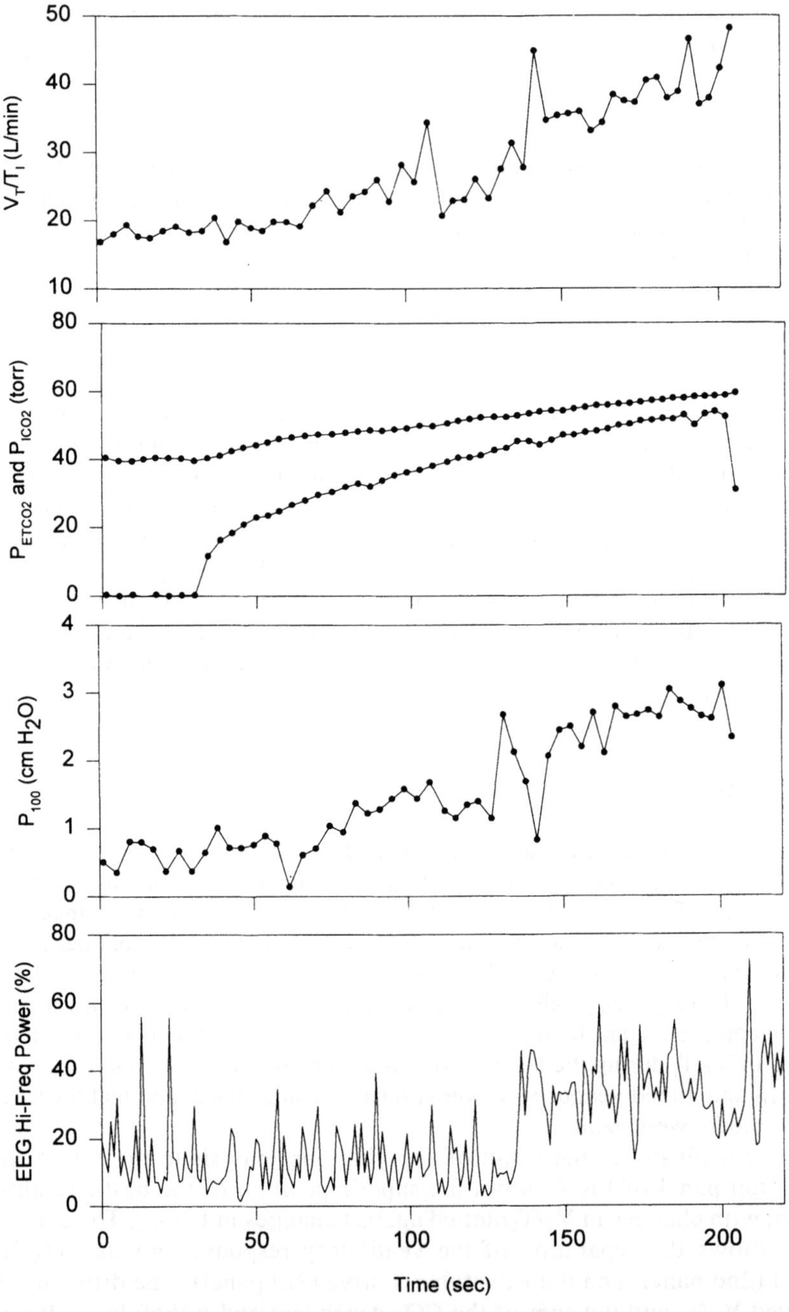

Figure 3. Effect of progressive hypercapnia (2nd panel) on mean inspiratory flow, V_T/T_I (top panel) and P_{100} (3rd panel) during Stage 2 sleep in Subject #2. The abrupt change in state at ~132 s, as indicated by the shift in EEG RHFP (bottom panel), produces a shift in V_T/T_I and P_{100} as well as an elevation in their rates of increase.

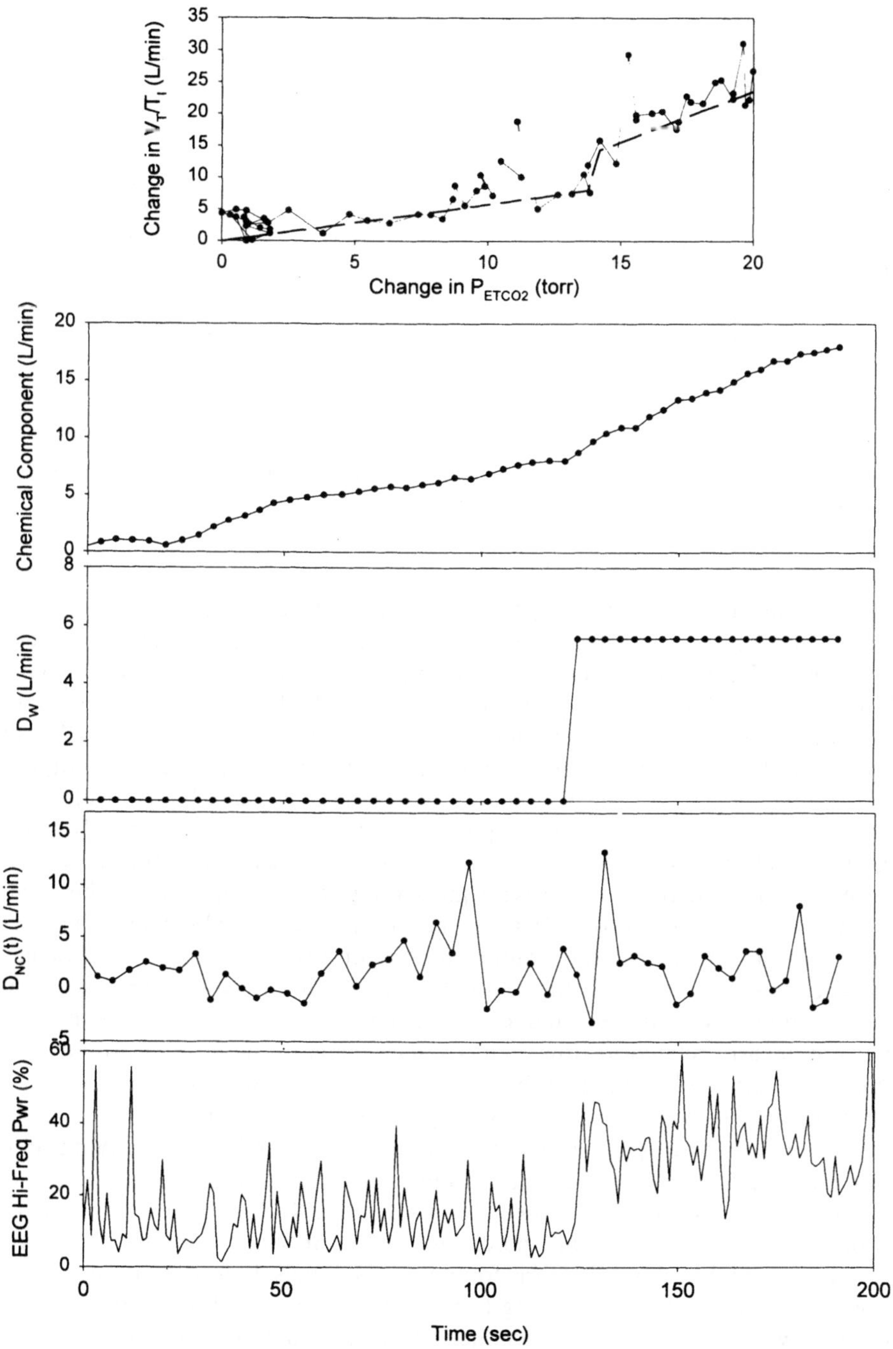

Figure 4. Decomposition of rebreathing response into its chemical (2nd panel), wakefulness drive (3rd panel) and dynamic non-chemical (4th panel) components. Top panel shows ventilatory response plotted against corresponding P_{ETCO2}, together with best model fits.

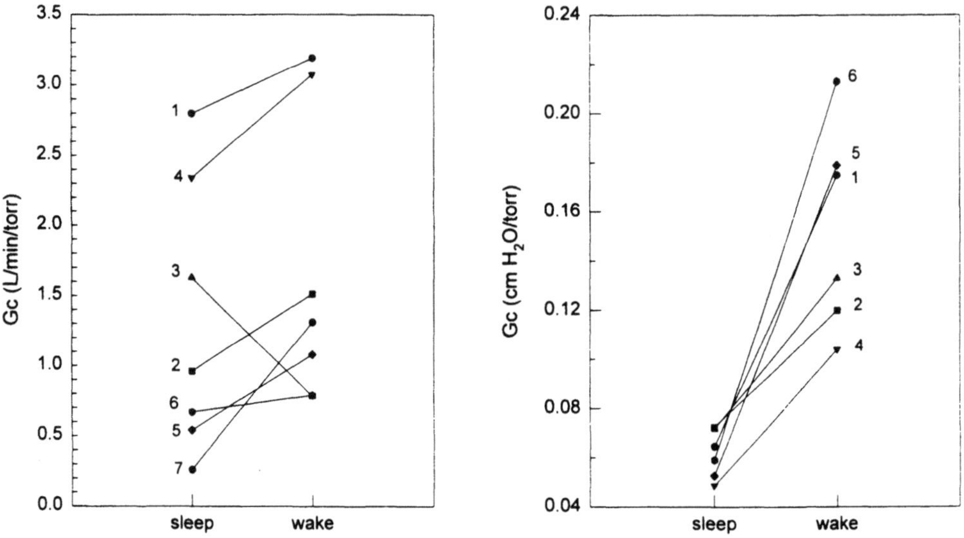

Figure 5. Changes in Gc with state, as computed from changes in V_T/T_I (left panel) and changes in P_{100} (right panel). Numbers in figure represent individual subjects.

was 150% and highly significant (P<0.005). Gc, deduced from V_T/T_I measurements, increased in 6 subjects but decreased in the remaining subject during the sleep-wake transition (Fig.5, left panel). The values of Gc_{wake} estimated in this study were generally within the ranges reported by previous workers[6,7,8,11,14], although our estimates fell more towards the low end of the spectrum. The changes here were smaller than in the case for P_{100}, averaging only +27.6% and not achieving statistical significance (P=0.16). It is unclear why the discrepancy between the P_{100} and V_T/T_I results occurred. The slope of P_{100} vs P_{ETCO2} during sleep was lower than had been previously reported[11], suggesting that our values for Gc_{sleep} (based on P_{100}) may have been underestimated - hence, the large increase in Gc with the abrupt state change. Since the face mask was not glued on to the subject's face, the possibility of leaks cannot be excluded. Another possibility is that the equipment circuit resistance may have inceased in importance towards the end of the rebreathing procedure following arousal, as airflow became progressively higher. Under turbulent flow conditions, driving pressure becomes nonlinearly related to flow, as resistance tends to increase in proportion to the magnitude of airflow[15]. This may have lowered V_T/T_I values in the awake portion of the test, leading to smaller vaues of Gc_{awake} (based on V_T/T_I).

The values of P_{ETCO2} immediately before the start of the rebreathing procedure and at the breath in which the state change was first detected (θ_{CO2}) are shown in Fig.6. The range of values obtained for θ_{CO2} (from 48.7 to 64.0 torr) is consistent with the hypercapnic arousal thresholds reported in previous studies[1,7,16]. On average, P_{ETCO2} increased by about 13 torr from control levels before leading to a clearly defined state change. It is interesting to note from Figs. 5 and 6 that the subjects who had higher waking Gc values (as deduced from P_{100} measurements) also had higher CO_2 arousal thresholds.

Values of D_W derived from the V_T/T_I measurements ranged from 3.1 to 7.2 L/min , whereas estimates of D_W computed from P_{100} measurements ranged from 0.5 to 2.8 cm H_2O. The corresponding means (± SD) were 4.74 (± 1.40) L/min and 1.77 (± 0.93) cm H_2O. In

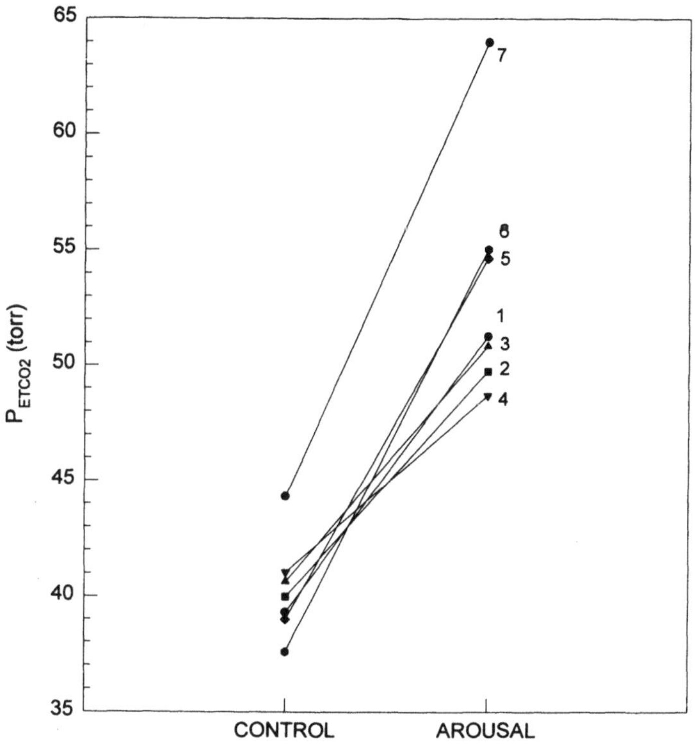

Figure 6. End-tidal P_{CO2} values in the individual subjects before the test and at the abrupt transition from sleep to wake.

the subjects where upper airway pressure was measured, R_{UAW} decreased from 5.60 (± 2.65 SD) cm H_2O min L^{-1} asleep to 4.00 (± 1.61 SD) cm H_2O min L^{-1} awake.

4. DISCUSSION

In recent years, it has become increasingly evident that transient changes in state can play a major role in producing ventilatory instability during sleep[3,4,17]. Since the interplay between the chemoreflexes and fluctuations in state is dynamic and complex, computer modeling is proving to be a useful approach for better understanding the mechanisms underlying sleep-disordered breathing[18]. However, one limitation of computer modeling is that the outcome of any given simulation depends critically on the parameter values used to produce the simulation. For the most part, population averages are used, but these produce simulation results that are representative only of the group characterized by the particular combination of parameter values. From the clinical standpoint, computer modeling would be much more useful if predictions could be made for individual subjects. In order to do this, the most important parameters for the individual in question would have to be estimated before or during the recording of the sleep study. The present work demonstrates the feasibility of estimating the model parameters most pertinent to sleep-disordered breathing from a single test procedure lasting approximately 5 minutes. Although the test procedure is a simple extension of the widely-used rebreathing method, we know of only one previous study[19] that has alluded to an experimental protocol analogous to ours. Figures 3 and 4 from

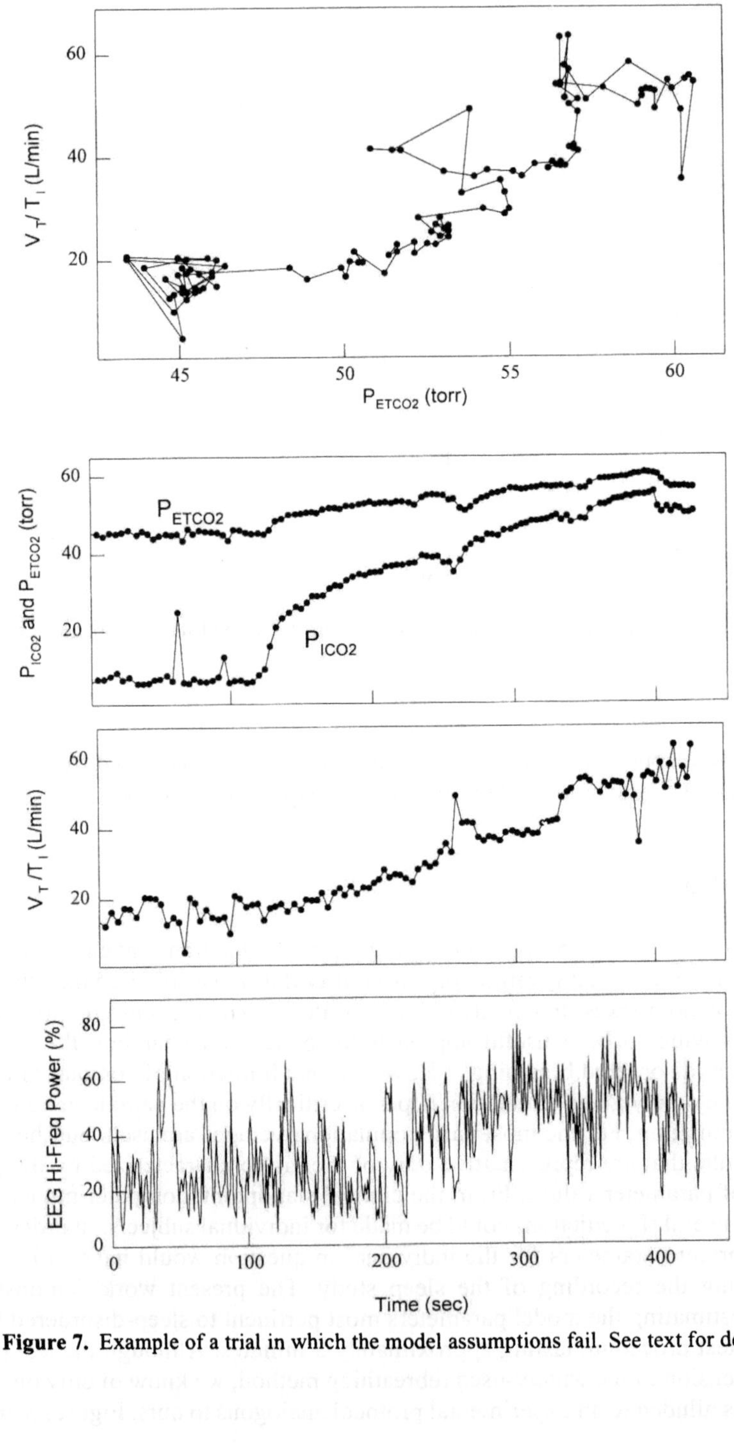

Figure 7. Example of a trial in which the model assumptions fail. See text for details.

the study of Bellville et al.[19] showed results that closely resemble our rebreathing data. However, little attention was paid to the abrupt awakening during rebreathing, while the study focussed primarily on characterizing the shift in hypercapnic response line between stable periods of wakefulness and sleep.

In order to derive estimates for Gc for sleep and wake, we assumed a simple model in which there would be an abrupt transition from one (steady) state into the other. In most cases, the EEG did show an abrupt shift in frequency at the transition point. However, in some, the "abrupt" transition point was not so easily defined. Figure 7 illustrates an example of this. In this trial, the subject shows signs of transient arousal before (at ~t=30 s) and during (at ~t=120 s) the rebreathing procedure (see bottom panel of Fig.7), as well as a more gradual change in RHFP (between t=200 and t=300 s). Furthermore, the model assumes (as in all rebreathing calculations) that following equilibration of alveolar and brain tissue P_{CO2}, the progressive rise in P_{ETCO2} should be isolated from the effects of transient fluctuations in ventilation. Again, Figure 7 illustrates that such an assumption is not true in this particular case. The transient increase in V_T/T_I at ~t=240 s produces dips in P_{ICO2} and P_{ETCO2} (2nd and 3rd panels of Fig.7). This kind of transient leads to the "looping" behavior seen in the plot of V_T/T_I versus P_{ETCO2} (top panel of Fig.7), which tends to confound the fitting of straight-line segments to the "sleep" and "wake" portions of the plot. In cases such as this, the simple abrupt state-change model would fail to produce meaningful parameter estimates. One possible means of overcoming this problem would be to incorporate dynamics into the chemoreflex responses and use the RHFP time-course as a template for the changes in state. We have explored the use of this approach, as applied to another series of sleep studies, in a companion chapter in this volume[20].

Another major limitation of our method pertains to the interpretation of the slope of hypercapnic response during rebreathing. Recent simulation and experimental studies[21,22] have demonstrated that the slope of the ventilation-P_{CO2} line during rebreathing consistently overestimates the steady-state CO_2 sensitivity by a factor close to 2. This discrepancy appears to be due primarily to changes in the cerebral blood flow following progressive increase in arterial P_{CO2}. To circumvent this difficulty, one could employ a slightly more sophisticated model of central chemoreflex dynamics, such as that proposed by Bellville et al.[23], which allows for the effect of P_{ETCO2} on the time constant associated with the central chemoreceptors.

ACKNOWLEDGMENTS

This work was supported by NIH Grant RR-01861, the American Lung Association and NHLBI Research Career Development Award HL-02536 (M.C.K. Khoo).

REFERENCES

1. Phillipson, E.A., and G. Bowes. Control of breathing during sleep. In: *Handbook of Physiology, Section 3: The Respiratory System II, Part 2*, edited by N.S. Cherniack and J.G. Widdicombe. Bethesda, MD: American Physiological Society, 1986, p.649-690.
2. Stradling, J.R. *Handbook of sleep-related breathing disorders*. Oxford:Oxford Univ. Press, 1993, p.17-19.
3. Xie, A., B. Wong, E.A. Phillipson, A.S. Slutsky, and T.D. Bradley. Interaction of hyperventilation and arousal in the pathogenesis of idiopathic central sleep apnea. *Am. J. Respir. Crit. Med.* 150:489-495, 1994.
4. Khoo, M.C.K., J.D. Anholm, S.W. Ko, R. Downey III, A.C.P. Powles, J.R. Sutton, and C.S. Houston. Dynamics of periodic breathing and arousal during sleep at extreme altitude. *Respir. Physiol.* 103(1):1-11, 1996.

5. Chapman, K.R., E.N. Bruce, B. Gothe, and N.S. Cherniack. Possible mechanisms of periodic breathing during sleep. *J. Appl. Physiol.* 64:1000-1008, 1988.
6. Read D.J.C. A clinical method for assessing the ventilatory response to carbon dioxide. *Australas Ann. Med.* 16:20-32, 1967.
7. Berthon–Jones, M and C.E. Sullivan. Ventilation and arousal response to hypercapnia in normal sleeping humans. *J. Appl. Physiol.* 57: 59-67, 1984.
8. Douglas, N.J, D.P White, J.V. Weil, Cheryl K.P., and C.W. Zwillich. Hypercapnic ventilatory response in sleeping adults. *Am. Rev. Respir. Dis.* 126:758-762, 1982.
9. Koh S., J.W. Shin, R.B. Berry and M.C.K. Khoo, Assessment of state-chemical drive interaction during abrupt sleep-to-wake transition. *Am. J. Respir. Crit. Med.* 151:A637, 1995.
10. Shiavi, R. *Introduction to Applied Statistical Signal Analysis*. Homewood, IL: Aksen, 1991.
11. White, D.P. Occlusion pressure and ventilation during sleep in normal humans. *J.Appl. Physiol.* 61:1279-1287, 1986.
12. Berry, R.B., K.G. Kouchi, J.L. Bower and R.W. Light. Effect of upper airway anesthesia on obstructive sleep apnea. *Am. J. Respir. Crit. Care Med.* 151:1857-1861, 1995.
13. Read, D.J.C., and J. Leigh. Blood-brain tissue P_{CO2} relationships and ventilation during rebreathing. *J. Appl. Physiol.* 23:53-70, 1967.
14. Hirshman, C.A., R.E. McCullough, and J.V. Weil. Normal values for hypoxic and hypercapnic drives in man. *J. Appl. Physiol.* 38:1095-1098, 1975.
15. Pedley, T.J., R.C. Schroter, and M.F. Sudlow. Gas flow and mixing in the airways. In: *Bioengineering Aspects of the Lung*, edited by J.B. West. New York: Marcel Dekker, 1977, pp.163-265.
16. Gleeson, K., C.W. Zwillich and D.P. White. The influence of increasing ventilatory effort on arousal from sleep. *Am. Rev. Resp. Dis.* 142:295-300, 1990.
17. Khoo, M.C.K., S.S.W. Koh, J.J.W. Shin, P.R. Westbrook, and R.B. Berry. Ventilatory dynamics during transient arousal from NREM sleep: Implications for respiratory control stability. *J. Appl. Physiol.* 80, 1996 80:1475-1484, 1996.
18. Khoo, M.C.K., A. Gottshalk and A.I. Pack. Sleep-induced periodic breathing and apnea: a theoretical study. *J. Appl. Physiol.* 70: 2014-2024, 1991
19. Bellville, J.W., W.S. Howland, J.C. Seed, and R.W. Houde. The effect of sleep on the respiratory response to carbon dioxide. *Anesthesiology* 20:628-634, 1959.
20. Khoo, M.C.K. Understanding the dynamics of state-respiratory interaction during sleep: A model-based approach. In: *Bioengineering Approaches to Pulmonary Physiology and Medicine*, edited by M.C.K. Khoo. New York: Plenum Press, 1996, chap.1.
21. Berkenbosch, A., J.G. Bovill, A. Dahan, J. DeGoede, and I.C.W. Olievier. The ventilatory CO_2 sensitivies from Read's rebreathing method and the steady state method are not equal in man. *J. Physiol. (Lond.)* 411:367-377, 1989.
22. Dahan, A., A. Berkenbosch, J. DeGoede, I.C.W. Olievier, and J.G. Bovill. On a pseudo-rebreathing technique to assess the ventilatory sensitivity to carbon dioxide in man. *J. Physiol. (Lond.)* 423:615-629, 1990.
23. Bellville, J.W., B.J. Whipp, R.D. Kaufmann, G.D. Swanson, K.A. Aqleh, and D.M. Wiberg. Central and peripheral chemoreflex loop gain in normal and carotid body-resected subjects. *J. Appl. Physiol.* 46:843-853, 1979.

5

REALISTIC COMPUTATIONAL MODELS OF RESPIRATORY NEURONS AND NETWORKS

Jeffrey C. Smith

Laboratory of Neural Control
National Institute of Neurological Disorders and Stroke
National Institutes of Health
Bethesda, Maryland 20892

1. INTRODUCTION

Breathing movements in mammals result from networks of neurons in the central nervous system that produce a complex spatial and temporal pattern of rhythmic neural activity. The underlying mechanisms are not fully understood, and computational modeling is becoming an essential tool for achieving a mechanistic understanding. It is now recognized that the activity of respiratory neurons results from a complex dynamic interaction of biophysical properties of individual neurons and network mechanisms that arise from the interconnections of cells. These interactions of cellular and network processes remain difficult to investigate experimentally, however, and computational approaches that permit modeling of biologically realistic neurons and networks, in particular, provide a powerful modeling approach. Methods for modeling networks of realistic neurons incorporating biophysical properties and synaptic interactions have been developed in computational neuroscience over the past several decades[1-4], and the availability of software has now made computer simulation of these realistic types of models practical[4,5]. In this chapter, I outline this approach and provide examples of our simulations with a first generation of realistic models of respiratory neurons and networks. These models are a significant departure from earlier models[6-9] of the respiratory network that have lacked many of the biophysical and synaptic properties of neurons that are required to replicate the behavior of real networks. The approach outlined here can be applied to model any aspect of respiratory neural function where neurobiological realism is sought. The models I present below are designed mainly to explore mechanisms in the brainstem involved in the generation of the rhythm and pattern of neuron activity underlying the respiratory cycle. I first consider features of the organization and properties of brainstem networks that must be incorporated in the realistic models.

2. RESPIRATORY NEURON PROPERTIES AND NETWORK ARCHITECTURE

The respiratory pattern is known to be generated by bilaterally distributed, interconnected columns of neurons in the ventral medulla, called the ventral respiratory group (VRG), and all current models that attempt to explain brainstem mechanisms of pattern generation focus on properties of VRG neurons and their interconnections. Several recent reviews[10-13] should be consulted for an overview and background of the relevant neuroanatomy and electrophysiology. Here I emphasize several main features of network activity and organization of the VRG that are considered essential. The VRG contains the main populations of interneurons that form the pattern generation network, frequently called the *central pattern generator*, which is the basic neural machinery for producing the rhythmic drives to cranial and spinal respiratory motoneurons innervating the musculature of the upper airway and respiratory pump. There are two major features of the respiratory motor pattern that must be replicated in models: the rhythm or neural oscillation underlying the respiratory cycle, and the temporal pattern of neuronal activity within the respiratory cycle which consists of three major phases: *inspiratory, stage-1 expiratory or post-inspiratory, and*

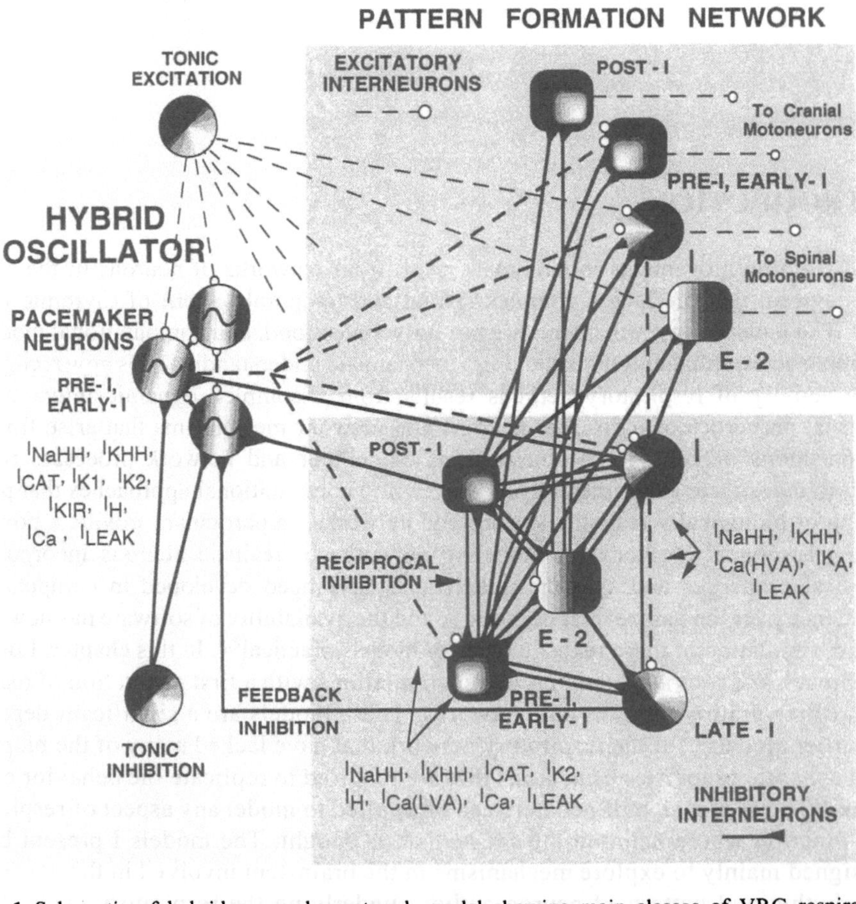

Figure 1. Schematic of hybrid pacemaker-network model showing main classes of VRG respiratory interneurons and their synaptic connections. Intrinsic membrane currents postulated for each type of neuron are indicated. See text for full explanation. Modified from ref. 14.

stage-2 expiratory phases. The network of VRG interneurons exhibits this three-phase pattern, and there are a number of different cell populations in the network, classified by their pattern of spiking in relation to these phases (e.g., inspiratory (I) phase spiking neurons (I neurons), pre-inspiratory (Pre-I) neurons that begin spiking before the I phase, post-inspiratory (Post-I) neurons, stage-2 expiratory (E-2) neurons). The different classes are indicated in Fig. 1 which summarizes what is currently known or postulated about the overall organization of the network and the architecture of synaptic connections among the main types of interneurons.

2.1. Network Components

In general, as depicted in Fig. 1, the network is postulated to be organized into two main functional components: the *oscillator*, which generates the rhythm, and the *pattern formation network* which is driven by the oscillator and generates the three phases of neural activity. The pattern formation network consists of populations of excitatory interneurons, which produce the excitatory drives to inspiratory and expiratory motoneurons, and highly interconnected populations of inhibitory neurons, which generate a pattern of synaptic inhibition that acts on the excitatory neurons to organize their activity into the three phases. Within the inhibitory and excitatory populations are the different classes of respiratory neurons, and it is postulated that most of the neuron classes have excitatory and inhibitory subpopulations (i.e., there are inhibitory and excitatory Post-I neurons, etc.). At least three of the inhibitory cell populations (Early-I, Post-I, and E-2) are connected reciprocally, and this reciprocal inhibition is a main mechanism generating the three phases of activity[11,13]. Other types of inhibitory cells are active during the transitions between I and E phases (Pre-I neurons, Late-I neurons), and function in phase termination (see further discussions in refs. 11, 13).

The neurons and mechanisms operating in the oscillator are currently controversial[11-13,16]. Neural oscillators in general range from pacemaker-driven networks, where intrinsic oscillatory properties of pacemaker neurons play a critical role, to network oscillators where the oscillation is an emergent property of synaptic interactions. There is evidence that neurons with intrinsic pacemaker-like properties are involved, particularly in the neonatal and juvenile (rodent) nervous system under some experimental conditions[15-18]. These cells receive synaptic connections from other network elements, however, and in general, network mechanisms are importantly involved in controlling the pacemaker neuron and network oscillations. The most comprehensive model therefore incorporates both pacemaker and network mechanisms, and has accordingly been named the *hybrid pacemaker-network model*[16]. This hybrid model, which is depicted in Fig. 1, has complex dynamical properties arising from the interactions of cellular and network mechanisms, and can exhibit functional states where rhythm generation is dominated by the intrinsic oscillations of the pacemaker neurons, or others where synaptic mechanisms, operating in combination with intrinsic cellular properties, have particularly critical roles in regulating the rhythm. The behavior of the respiratory oscillator changes in different brain states (e.g., sleep vs. wakefulness), during development, or between different experimental conditions (e.g., intact vs. deafferented CNS; *in vivo* vs. *in vitro* CNS preparations). The hybrid pacemaker-network model provides a framework for analyzing the contributions of cellular and network properties to such changes. Refs. 15 and 16 should be consulted for more complete discussions of this model.

In the hybrid model, the kernel of the oscillator is represented by the population of pacemaker neurons with intrinsic oscillatory properties (Fig. 1) which provide the rhythmic excitatory drive to the rest of the network during the inspiratory phase. This kernel is located in a functionally specialized subregion of the VRG, called the pre-Bötzinger Complex[16,17].

function to synchronize their activity. The membrane currents that promote the intrinsic bursting oscillations of the pacemaker cells also promote synchronized bursting within the cell population as the wave of excitation from their interconnections spreads through the population at the start of the I phase. Synchronization is not absolute, however, and there is a temporal dispersion in the onset of pacemaker neuron spiking throughout the I phase, and many of these neurons are postulated to become active before the inspiratory phase (Pre-I neurons) and during early inspiration (Early-I). These earliest spiking neurons function to initiate inspiratory phase activity in the network. The pacemaker neurons receive input through connections from a number of different neuron populations (Fig. 1), including from inhibitory and excitatory neurons with steady (tonic) patterns of spiking activity, which regulate the pacemaker neuron excitability and oscillatory behavior. Inhibitory neurons in the pattern formation network (Late-I, Post-I neurons) provide (phasic) feedback inhibition that resets and terminates the pacemaker neuron activity each cycle. The mechanisms discussed above, while consistent with present data, have not been proven, and represent our current working hypotheses about the operation of the oscillator (see ref. 13 for other models).

2.2. Electrophysiological Properties of Neurons

The different classes of respiratory neurons have characteristic patterns of spiking and time courses of the underlying changes in membrane potentials[10,11,13]. This electrophysiological behavior results from the combination of synaptic input and ionic conductance mechanisms intrinsic to the neuronal membrane. In general, the intrinsic electrophysiological properties of neurons result from a mix of ionic (Na^+, K^+, and Ca^{2+}) currents regulated by membrane conductance mechanisms that have highly nonlinear properties (below). The intrinsic conductances in the different classes of respiratory neurons have not been completely identified, but particular sets of currents can be postulated (Fig. 1) (see also refs. 11, 13) based on electrophysiological functions known to be subserved by different membrane currents in a variety of neurons (e.g., see reviews in refs. 19, 20). These include Na^+ and K^+ conductances producing action potentials (I_{NaHH}, I_{KHH} in Fig. 1), K^+ conductances modulating spiking (I_{KA}), persistent Na^+ conductances (I_{CAT}) producing long-lasting depolarizing currents, and persistent K^+ conductances (I_{K2}) that interact dynamically with I_{CAT} to regulate the duration of neuronal depolarization and bursting. Neurons exhibiting particularly complex behavior, such as pacemaker neurons with oscillatory properties, have the more elaborate set of currents (Fig. 1), but in general each of the cell types exhibit temporally complex spiking and bursting behaviors (below) that would involve a number of interacting Na^+, K^+, Ca^{2+}, and mixed cationic (Na^+, K^+) currents. Realistic neuron models are constructed by incorporating membrane conductance mechanisms producing these currents along with synaptic conductance mechanisms.

3. MODELING REALISTIC ELECTROPHYSIOLOGICAL PROPERTIES: THE HODGKIN-HUXLEY APPROACH

3.1. Equivalent Electrical Circuit Representation of the Neuronal Membrane

The methods outlined here for modeling intrinsic conductance mechanisms are based on the landmark work of Hodgkin and Huxley (HH) which provided the theoretical and experimental basis for modeling neuron excitability in terms of biophysical properties of the

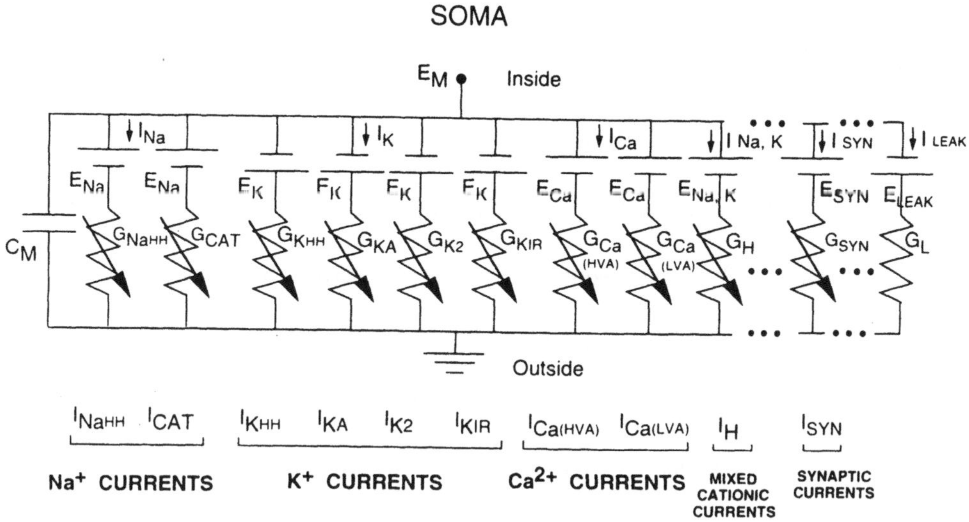

Figure 2. Equivalent electrical circuit representation of neuron membrane incorporating multiple types of currents. Subsets of the currents indicated are incorporated in single neuron models for the various types of respiratory neurons as indicated in Fig. 1. Variable conductances are indicated by arrows. See text for definitions of abbreviations and symbols.

experimental basis for modeling neuron excitability in terms of biophysical properties of the neuronal membrane. Although their work focused on explaining the electrophysiological properties of the squid axon in terms of the Na^+ and K^+ conductances specific to the axon, their approach is generally applicable for modeling a variety of neuronal conductance mechanisms[1-4, 20].

The HH modeling approach is based on the idea that the electrical properties of the neuronal membrane can be modeled by an equivalent electrical circuit (Figure 2). In the equivalent circuit, current (I) flow across the membrane has two main components- current charging the membrane capacitance (I_{C_M}) associated with the membrane dielectric properties, and the other component due to ion movement (I_{ion}) through resistive pathways. Each ionic current pathway is considered in parallel, and each type of current has a characteristic variable conductance (the inverse of electrical resistance) and a potential source. Thus each I_{ion} is determined by the conductance and its potential gradient:

$$I_{ion}(E,t) = G_{ion}(E,t)(E_M - E_{ion}) \tag{1}$$

where E_M is the membrane potential, and E_{ion} is the ionic equilibrium or reversal potential which functions like a battery. In contrast to a conductance which obeys Ohm's law, G_{ion} in general is variable and a function of both E_M and time (t). The particular voltage- and time-dependence distinguishes different conductances, and the many different conductances for the different types of K^+, Na^+ and Ca^{2+} currents that exist in neurons can be represented. In general, only one type of ionic current, called the leakage current (I_{Leak}), present in all neurons, is assumed to have a conductance that is voltage-independent and nearly Ohmic.

The total membrane current is the algebraic sum of the individual ionic currents and I_{C_M}:

$$I_M = I_{C_M} + \Sigma\, I_{ion} = I_{C_M} + \Sigma G_{ion}(E,t)(E_M - E_{ion}) + G_{Leak}(E_M - E_{leak}) \tag{2}$$

where $I_{C_M} = C_M dE/dt$.

In addition to ionic currents, synaptic currents (I_{syn}), resulting from connections that release neurotransmitters and activate synaptic conductances (G_{syn}) in the membrane, are similarly represented by parallel conductance pathways (Fig. 2):

$$I_{syn}(t) = G_{syn}(t)(E_M - E_{syn}) \tag{3}$$

where G_{syn} is also variable and time-dependent (below), and E_{syn} is the synaptic reversal potential.

The behavior of the entire equivalent circuit is described by:

$$C_M dE/dt = \Sigma G_{ion}(E,t)(E_M - E_{ion}) + G_{Leak}(E_M - E_{Leak}) + \Sigma G_{syn}(t)(E_M - E_{syn}) + I_{ext} \tag{4}$$

where I_{ext} is introduced to represent externally applied current (e.g., via a microelectrode during intracellular recording). Integration of this equation gives E_M as a function of time which can be directly compared with intracellular recordings of membrane potential.

3.2. HH Conductance Equations

The voltage- and time-dependence of the ionic conductances are described by HH style equations of the general form:

$$G(E,t) = G_{max} m^X(E,t) h^Y(E,t), \tag{5}$$

$$dm/dt = \alpha_m(1 - m) - \beta_m m, \qquad dh/dt = \alpha_h(1 - h) - \beta_h h, \tag{6}$$

$$m_\infty = \alpha_m/(\alpha_m + \beta_m), \ \tau_m = 1/(\alpha_m + \beta_m), \tag{7}$$

$$h_\infty = \alpha_h/(\alpha_h + \beta_h), \quad \tau_h = 1/(\alpha_h + \beta_h). \tag{8}$$

Concise discussions of these equations and parameter definitions, and their original derivation by Hodgkin and Huxley, can be found in refs. 20-23. These equations have been extensively and successfully employed to develop biologically realistic models of neurons (see reviews in 1-4). In essence, these equations incorporate the idea that membrane conductance and ion fluxes are regulated by separate "activation" and "inactivation" gates (associated with membrane channels) that undergo transitions between "permissive" and "non-permissive" states in a probabilistic way with voltage and time. The parameters *m* and *h* are conventionally used to represent, respectively, the activation and inactivation probabilities, which scale the maximum possible conductance G_{max} (which reflects the total density of channels of a given type in the membrane). The transitions are assumed to obey first order kinetics (Eq. 6), with kinetic parameters (rate factors) α_m, β_m, α_h, β_h that are functions of E. This voltage-dependence is usually expressed by equations of the general form:

$$\alpha_m(E) = (a + bE)/[c + \exp(d + E/(f)] \tag{9a}$$

$$\beta_m(E) = (a' + b'E)/[c' + \exp(d' + E/f')] \tag{9b}$$

$$\text{idem for } \alpha_h \text{ and } \beta_h \tag{9c, 9d}$$

where a,a′, b,b′,...f,f′ are constants with values that differ for activation and inactivation and are particular to each type of conductance. In Eqs. 7 and 8, m_∞ and h_∞ represent the steady-state values of the activation and inactivation parameters that are reached following a change in voltage to new (fixed) voltage level. The time course for approaching this

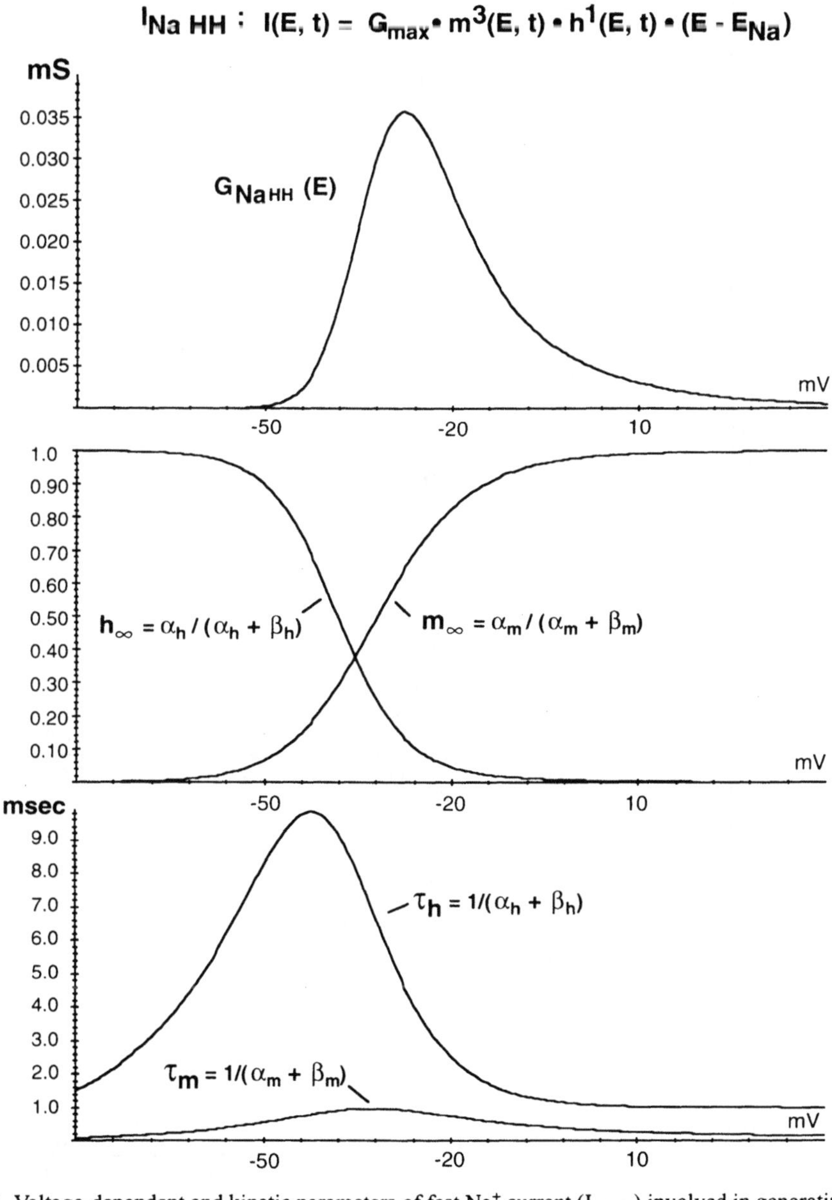

Figure 3. Voltage-dependent and kinetic parameters of fast Na^+ current (I_{NaHH}) involved in generating action potential. Form of Eq. 5 describing conductance is indicated. Current exhibits fast activation and slower inactivation in a voltage-dependent manner during membrane depolarization (i.e., to less negative values of membrane potential). Parameter *h* by convention represents the probability that the Na channel is *not* inactivated and therefore has values decreasing from 1 (i.e., the inactivation probability increases) during depolarization. Parameter values in Eqs. 5 and 7- 9 used to model this current were modified from Hodgkin and Huxley.[23]

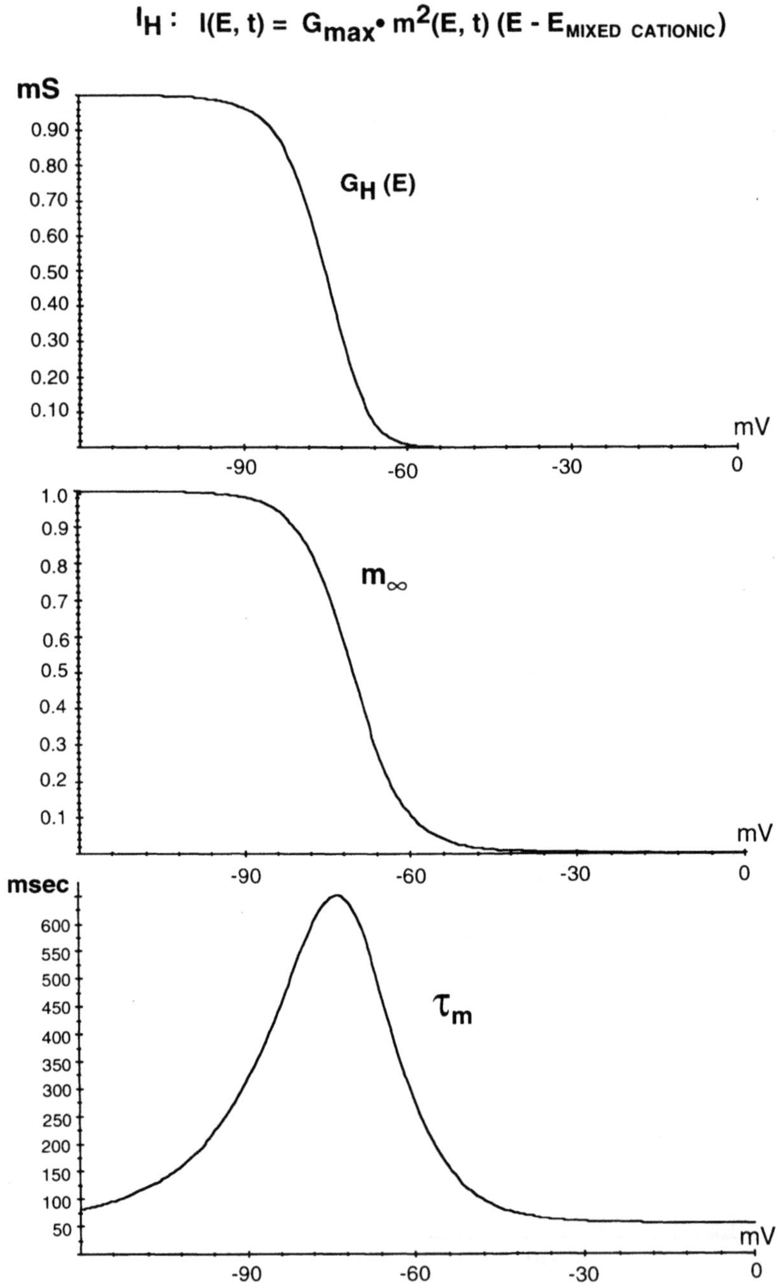

Figure 4. Voltage-dependent and kinetic properties of hyperpolarization-activated, mixed cationic (Na^+, K^+) conductance I_H. In contrast to fast Na^+ current shown in Fig. 3, I_H very slowly activates during hyperpolarization and does not exhibit inactivation. Parameter values used to model current are adapted from I_H formulation by Huguenard and McCormick.[24]

steady-state is described by a simple exponential with time constants τ_m, τ_h. The activation and inactivation properties of different types of conductances are remarkably diverse and suited to their particular electrophysiological functions[19,20]. Values of G, m_∞, h_∞, τ_m, τ_h for two of the currents used in the respiratory neuron models described below are shown in Figs. 3 and 4, which serve to illustrate some of the variations in kinetic and voltage-dependent properties of activation and inactivation of the different types of currents used in the modeling.

4. SYNAPTIC CONDUCTANCE EQUATIONS

Neurons interact in networks by synaptic connections where neurotransmitter, released due to activity in presynaptic neurons, activates transmitter-gated synaptic conductances (G_{syn}) in the postsynaptic membrane. The conductance changes and synaptic currents generated are time-dependent, reflecting the time course of transmitter release and conductance kinetics. In certain special cases, G_{syn} is also voltage-dependent, but in our modeling, only time-dependent processes are utilized. The time course of changes in G_{syn} that have been observed experimentally in neurons can be modeled by analytical functions that increase rapidly with time to a peak conductance value followed by a slower decay. Two types of analytical functions are commonly used that provide good approximations: a dual exponential function, and the *alpha function* (see discussions in ref. 25). We use the alpha function in combination with a presynaptic transmitter release function (T_s) which is initiated by presynaptic activity and activates G_{syn}:

$$G_{syn}(t) = G_{max} T_s t^{\alpha} \exp(-\alpha t/\tau) \tag{10}$$

G_{max} is the maximum conductance which depends on the density of transmitter-gated channels in the postsynaptic membrane and determines the strength of the synaptic connection. Inhibitory or excitatory synapses are represented by using the appropriate synaptic reversal potentials (E_{syn} in Eq. 3 above). For conventional fast excitatory conductances, $E_{syn} \cong 0$ mV, resulting from a mixed cationic (Na^+, K^+) conductance; for conventional (Cl^--mediated) inhibitory conductances, $E_{syn} \cong -80$ mV (i.e., the Cl^- equilibrium potential). T_s, in general, can be a constant in the simplest case, or variable, incorporating a dependence on the pattern of presynaptic activity and kinetics of transmitter release. We use a constant T_s which is triggered each time a presynaptic neuron generates an action potential (at the somal compartment, see below) which arrives at the synapse and activates G_{syn} after an assigned propagation delay (typically 5 msec). With the alpha function, $G_{syn}(t)$ is set by the two independent parameters α and τ, where α determines the initial slope and τ determines the time to peak when the function reaches Gmax. Accordingly, synaptic processes with different kinetics can be modeled by varying these parameters. In our models, kinetics for fast excitatory or inhibitory connections with $\tau \leq 3$ msec are used. Individual neurons in the model typically have multiple excitatory and inhibitory synapses producing currents that are summed according to Eq. 4.

5. REPRESENTATION OF NEURON STRUCTURE

Neurons exhibit morphological complexity, with soma, axon, and highly branched dendritic processes, each with particular ionic conductance mechanisms and electrophysiological properties. In general, compartmental models can be built that include morphological features and spatially distributed properties by incorporating multiple compartments representing different membrane regions, each represented by equivalent circuits, and linked via resistive cytoplasmic current flow pathways (see discussions in 26). Computational models for single mammalian neurons with greater than one thousand compartments, used to represent dendritic structure and emphasize the role of dendritic electrical properties, have been developed based on morphological data[28]. Development of highly realistic respiratory neuron models is currently not possible, given the lack of detailed information on somatodendritic morphologies of the different classes of neurons, the unknown heterogeneous spatial distribution of synapses and ion channels on somatodendritic membranes, and unknown membrane conductances. Furthermore, highly complex single neuron models present computational difficulties, particularly when networks of cells are to be modeled (See ref. 26 for discussions of representing complexity in single neuron models).

Our approach in this initial modeling has therefore been to use simplified models of individual respiratory neurons and their synaptic interactions in relatively small networks. We typically use one compartment models for each type of neuron, with the intrinsic and synaptic conductances lumped into the single compartment which represents the neuron soma, with a spherical geometry. We also do not include the axon and explicitly model axonal signal propagation to synapses at connections to other neurons. Action potential generation in the model neurons occurs in the somal compartment, and the activation of postsynaptic conductances is triggered by each action potential after an assumed propagation delay (above), which avoids the problem of incorporating analytical solutions for the action potential propagation. In simulating the entire network, we have used either single neurons for each neuron class with the connections indicated in Fig. 1, or small populations (10 - 20 neurons) for each class. Although highly simplified, these models contain what are currently postulated to be essential features of the synaptic and ionic conductance mechanisms in each class of neurons, with network connections that represent the postulated main types of excitatory and inhibitory interactions of the different cell populations.

6. EQUATION PARAMETERS AND SIMULATION TECHNIQUES

The conductances and kinetic parameters in the above equations that need to be specified for the different currents postulated for the various classes of respiratory neurons (see Fig. 1) have not been directly measured. The type of voltage-clamp measurements required to establish parameters of the HH equations (see discussions in refs. 20, 22) have not been conducted and are technically difficult under some experimental conditions, particularly *in vivo* conditions. *In vitro* preparations with functional respiratory networks (e.g., from the neonatal rodent nervous system 16,17,27) provide some of the required experimental conditions, but analysis of conductance mechanisms in these preparations is in its very initial stages. However, there is a substantial database in the literature and published equations for a variety of conductances which we have adapted including: generic fast Na^+ (I_{NaHH}) and delayed rectifier K^+ (I_{KHH}) currents generating somatic action potentials[23, 28]; various K^+ conductances (inward rectifier I_{KIR}[20,29], I_{KA}[28,30]); Ca^{2+} currents (low voltage-activated I_{Ca}(LVA)[24, 31]; intermediate voltage-activated I_{Ca}, and high voltage-activated I_{Ca} (HVA)[20,32]), and mixed cationic (I_H) currents[24]. The persistent Na^+ current (I_{CAT}) and the slowly activating, persistent K+ (IK2) current used were modified from formulations for other persistent Na^+ [28] and K^+ [33] currents. There is natural variability in the voltage-dependence of activation/inactivation, estimated values of Gmax, and kinetic properties for a given type of current in different neurons (e.g., see discussions in refs. 20, 28). Our starting point has therefore been to use parameter values that are generally consistent with published properties of currents, and that give simulated time courses of neuron spiking and membrane potential trajectories resembling those measured experimentally for the different respiratory neuron types. The complete set of equations will be presented elsewhere; Figs. 3 and 4 provide examples of the voltage and time-dependent behavior given by Eqs. 5,7-9 with chosen parameter values for two of the currents. Values of C_M are assumed to be 1 $\mu F/cm^2$, diameters of the spherical somal compartments were 20-25 μm, and G_{max} values are normalized to the membrane surface area. Simulations were performed with the neuron/network simulation software NODUS[34] which is designed to implement relatively small networks (currently $\leq$ 200 neurons) of compartmental models of neurons incorporating HH style conductances, and synaptic conductances specified by Eq. 10 or a number of other user-defined analytical functions. (See ref. 5 for comparisons of neuron simulation software). Integrations were performed in NODUS with a fast hybrid Euler as well as fifth order Runge-Kutta (Fehlberg) methods[34].

7. EXAMPLES OF SIMULATIONS

A number of aspects of the dynamic behavior of individual respiratory neurons and networks have been simulated and explored with earlier versions of the models [14,16,36] and are being analyzed with the present versions. Here we present two examples illustrating the types of behavior that can be simulated at the cellular and network levels.

7.1. Oscillatory Behavior of Model Pacemaker Neurons

The pacemaker neurons exhibit particularly complex dynamic behavior involving temporal interactions of a number of intrinsic membrane currents with a range of voltage-dependent activation/inactivation and kinetic properties. The pacemaker neuron models are accordingly particularly good examples of how the Hodgkin–Huxley modeling approach can represent complex properties of individual neurons. Candidate respiratory pacemaker neurons have been identified under *in vitro* experimental conditions[17,18] (i.e., in the pre-Bötzinger Complex in brainstem slice preparations from neonatal rat); the model pacemaker neurons have been constructed to replicate features of the behavior of these candidate cells. These neurons exhibit highly voltage-dependent behavior in which the oscillatory frequency increases as the baseline membrane potential is depolarized[16,17]. Simulations with the model pacemaker neurons exhibit this type of voltage-dependent oscillatory bursting behavior (Fig. 5) with the range of oscillatory frequencies, burst durations, and spiking pattern observed experimentally. The oscillation in membrane potential in the model results from the interactions of the voltage-dependent conductances activated in the subthreshold range of potentials (i.e., I_{CAT}, I_{K1}, I_{K2}, I_{KIR}, I_H, I_{Ca}). The persistent Na^+ conductance I_{CAT} provides the main depolarizing current underlying the bursting phase; the persistent K^+ conductance I_{K2}, along with the faster activating I_{K1} interacts with I_{CAT} to terminate bursting and control the burst duration. The hyperpolarization activated currents I_H and I_{KIR} importantly control the membrane potential trajectory during the interburst interval.

We postulate that this voltage-dependence of oscillatory frequency represents a basic mechanism for control of network oscillation frequency when the network is operating in the pacemaker cell-driven mode[15,16]. Depolarizing/hyperpolarizing shifts in baseline potential induced by synaptic inputs from tonic excitatory/inhibitory neurons (refer to Fig. 1) would regulate frequency and this mechanism represents one form of frequency control in the hybrid pacemaker-network model. In general, the regulation of bursting in intrinsic oscillator neurons involves a rich variety of synaptic and conductances mechanisms (see ref. 16 for further discussion), including transmitter-gated control of G_{max} and possibly the

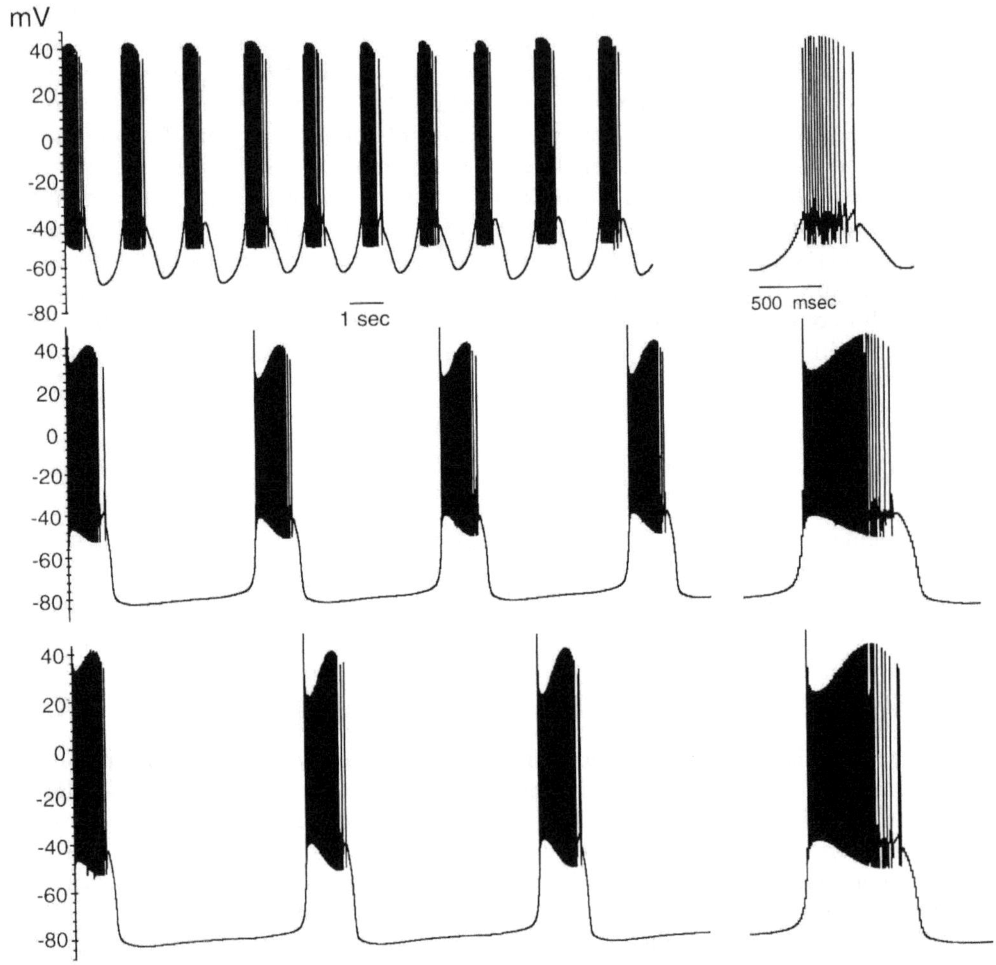

Figure 5. Example of simulated voltage-dependent oscillatory behavior of model pacemaker neuron. Model exhibits oscillatory bursting with the bursting frequency increasing as the baseline membrane potential is progressively shifted to more depolarized levels (top panels) due to injection of steady depolarizing current (i.e., I_{ext} in Eq. 4). Expanded time scale traces at right show neuron spiking profiles in more detail. The neuron oscillations only occur over a range of baseline potentials (~ -80 to -45 mV), outside of which the neuron exhibits non-oscillatory behavior. Neuron becomes quiescent with hyperpolarizing shifts to baseline potentials more negative than -80 mV, whereas steady depolarization to levels above threshold for action potential generation (-50 to -45 mV) causes neuron to undergo transition from oscillatory bursting to a mode of steady action potential generation (not shown, see ref. 16).

kinetics and voltage-dependence of activation/inactivation of subthreshold conductances. The models allow exploration of potential control mechanisms[16]. Although the pacemaker neuron model in its present form can mimic some features of the oscillatory behavior of the candidate pacemaker neurons, none of the actual currents in these cells have been identified, and the behavior of these neurons are only beginning to be explored experimentally[17,18]. The pacemaker neuron model will undoubtedly require revision as additional experimental information on conductance mechanisms becomes available.

7.2. Network Oscillations and Discharge Patterns of Respiratory Neurons

Simulations have been done for the adult and neonatal forms of the network[14,35-36] which enable comparisons to the databases of intracellular recordings of neuron spiking patterns and membrane potential trajectories during the respiratory cycle from neonatal (rat)[27,37,38] and adult mammals (cat, rat)[10,11,39-41]. Both adult and neonatal versions have been constructed to account for postulated developmental changes in strengths of synaptic interactions, and in some cases, changes in conductance properties.

Figure 6 shows an example of simulations with the adult form, illustrating the oscillatory behavior of cells from five of the neuron classes (Pre-I, Early-I, I, Post-I, and E-2). The model network exhibits a range of oscillatory frequencies characteristic of the adult with an overall three-phase pattern of neuronal discharge. The simulated time courses of spiking, membrane potential trajectories, and synaptic potentials for each neuron type resemble those measured experimentally (e.g., see refs.10, 11, 40). Some of the behavior characteristic of the different neuron types replicated by the model include: (i) Early-I and Post-I neurons exhibit a steep depolarization following release from inhibition (post-inhibitory rebound), due in part to activation of I_H during hyperpolarized phases, followed by a decrementing spiking pattern due to interactions of repolarizing inhibitory inputs and persistent Na^+ currents; (ii) I and E-2 neurons exhibit ramping membrane potential trajectories shaped by interactions of excitatory synaptic drives, inhibitory drives, and intrinsic conductances; (iii) Pre-I neurons exhibit stage-2 expiratory phase spiking and high frequency spike discharge during early I; the accompanying excitatory synaptic drive provided by these cells to the rest of the network plays a major role in initiating the inspiratory phase.

Within each class of neurons, experimentally there is substantial cycle-cycle variation in the time course of membrane potential excursions and spiking (see discussions in ref. 11). For example, there is variation in the steepness and ramping of depolarization in I neurons which accordingly can exhibit discharge patterns ranging from rapidly augmenting to constant, plateau-like spiking patterns. Pre-I neurons exhibit varying degrees of membrane depolarization and pre-inspiratory spiking during the 2- E phase[41]; Post-I neurons (also called E-Dec neurons[10]) exhibit various declining patterns of discharge and trajectories during E-2 with some neurons showing decrementing spiking throughout E-2, as in Fig. 6. These differences are presumed to reflect cycle-cycle variability in strength of inhibitory/excitatory inputs to a given neuron. Such variation is reproduced by the models by varying strengths of synaptic connections, and the model can generate the full range of behaviors found experimentally.

There are many questions regarding the interactions of cellular and synaptic properties in the control of neuronal discharge patterns and the timing of inspiratory and expiratory phases, including interactions involved in control by synaptic input from neurons transmitting afferent signals (e.g., mechano- and chemosensory afferents). Such issues can be readily addressed with the current network models. The network configuration in Fig. 1 should be viewed as the basic neural substrate for rhythm and pattern generation upon which a large number of afferent inputs involved in control of respiratory rhythm and pattern would converge. To explore hypothesis about control mechanisms, the model can be easily expanded, for example, to accommodate additional cells and circuits representing afferent pathways with connections to specific classes of neurons.

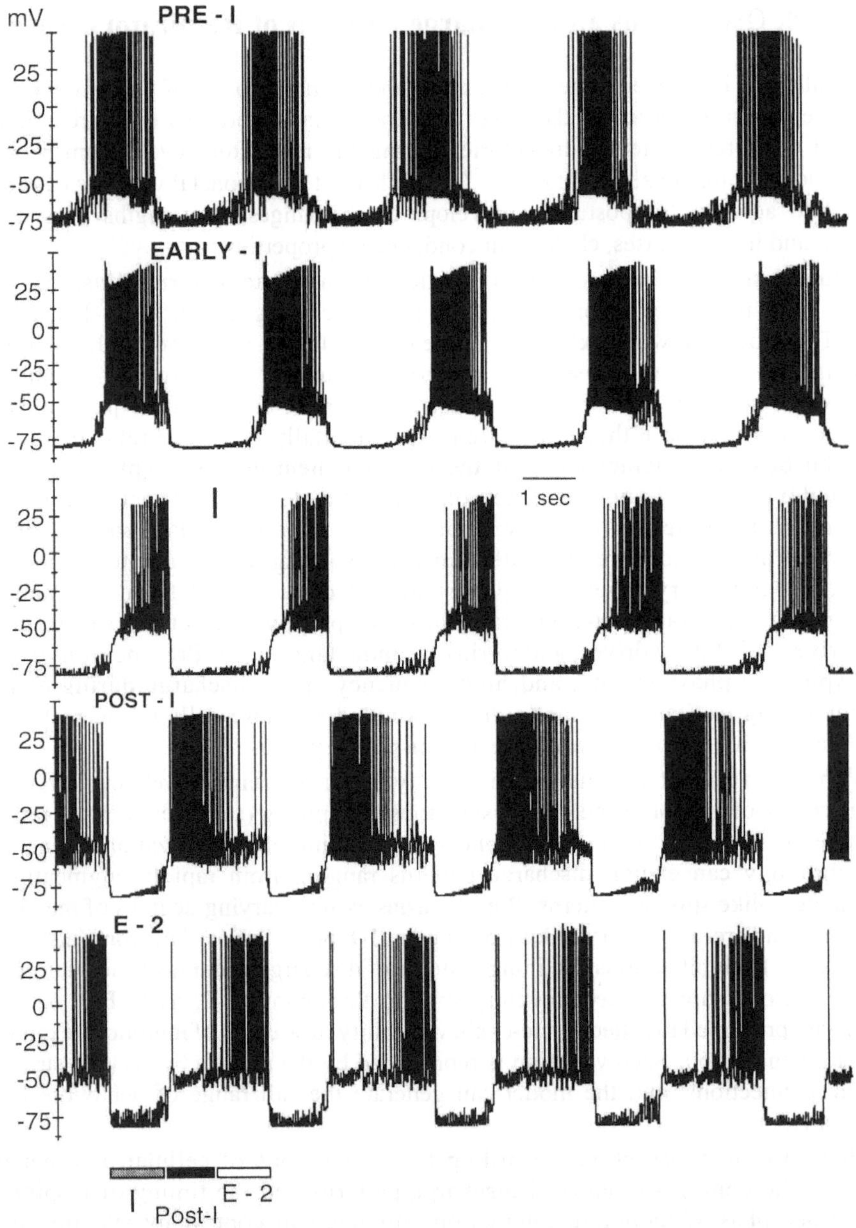

Figure 6. Simulations of membrane potential vs. time for adult form of network with neuron interconnections as in Fig. 1 showing oscillations and spiking patterns of individual neurons from several of the neuron classes. Three main phases (I, Post-I, E-2) of respiratory cycle are indicated.

8. SYNOPSIS

Computational models are important tools for understanding neural systems at all levels from membrane properties and synaptic mechanisms to complex network dynamics. The developments in computational neuroscience, beginning with the work of Hodgkin and Huxley, have resulted in methods for modeling of neurons and networks that are capable of providing neurobiologically realistic, mechanistic representations of neural function. Application of these approaches to the respiratory network is allowing construction of models which provide the type of synthesis of biophysical processes at cellular and network levels that is required to understand how the respiratory network operates. The models presented here are evolving and there are obvious deficiencies of the modeling in its present form, given the paucity of detailed information on cellular and network parameters of the real system. Nevertheless, this modeling approach allows a close interaction between experimental and computational studies, and will enable simulations with increasing neurobiological realism as additional experimental data becomes available. The first-generation models presented here represent the initial stages of establishing a computational framework for exploring complex dynamical properties in ways not possible experimentally.

REFERENCES

1. Koch, C. and I. Segev. *Methods in Neuronal Modeling. From Synapses to Networks*, MIT Press, Cambridge, MA, 1989.
2. McKenna,T., Davis, J., and S.F. Zornetzer. *Single Neuron Computation*, Academic Press, San Diego, CA, 1992.
3. Bower, J.M. (ed). Modeling the Nervous System. *Trends Neurosci.*, 1992, vol. 15.
4. Bower, J. M. and D. Beeman. *The Book of Genesis. Exploring Realistic Neural Models with the GEneral NEural SImulation System*. Springer-Verlag, New York and TELOS, Santa Clara, CA, 1995.
5. De Shutter, E. A consumer guide to neuronal modeling software. *Trends Neurosci.* 15: 462-464, 1992.
6. Botros, S.M. and E.N. Bruce. Neural network implementation of the three-phase model of respiratory rhythm generation. *Biol. Cybern.* 63: 143-153, 1990.
7. Duffin, J. A model of respiratory rhythm generation. *Neuroreport* 2: 623-626, 1991.
8. Ogilvie, M.D., Gottschalk, A., Anders, K., Richter, D.W., and A.I. Pack. A network model for respiratory rhythmogenesis. *Am. J. Physiol.* 32: R962-R975, 1992.
9. Balis, U.J., Morris, K.F., Koleski, J., and B.G. Lindsey. Simulations of a ventrolateral medullary neural network for respiratory rhythmogenesis inferred from spike train cross-correlation. *Biol. Cybern.* 70:311-327, 1994.
10. Ezure K. Synaptic connections between medullary respiratory neurons and considerations on the genesis of respiratory rhythm. *Prog. Neurobiol.* 35: 429-450, 1990.
11. Bianchi, A. L., Denavit-Saubie, M., and J. Champagnat. Centrol control of breathing in mammals: Neuronal circuitry, membrane properties, and neurotransmitters. *Physiological Rev.* 75:1-45, 1995.
12. Feldman, J.L. and J.C. Smith. Neural control of respiratory pattern in mammals: An overview. In: *Regulation of Breathing*, edited by J.A. Dempsey and A.I. Pack. Marcel Dekker, Inc, New York, 1995, p. 39-69.
13. Richter, D.W., Ballanyi, K., and S. Schwarzacher. Mechanisms of respiratory rhythm generation. *Curr. Opin. Neurobiol.* 2: 788-793, 1992.
14. Smith, J.C. New computational models of the respiratory oscillator in mammals. In: *Modeling and Control of Ventilation*, edited by Semple, S., L. Adams, and B.J. Whipp, Plenum Press, New York.
15. Smith, J.C. Integration of cellular and network mechanisms in mammalian oscillatory motor circuits. Insights from the respiratory oscillator. In: *Neurons, Networks, and Motor Behavior*, edited by Stein, P., Grillner, S., Selverston, A.I., and D.G. Stuart. MIT Press, Cambridge, MA, (in press).
16. Smith, J.C., Funk, G.D., Johnson, S.M., and J.L. Feldman. Cellular and synaptic mechanisms generating respiratory rhythm: Insights from *in vitro* and computational studies. In: *Ventral Brainstem Mechanisms and Control of Respiration and Blood Pressure*, edited by O. Trouth, R. Millis, H. Kiwull-Schone, M. Schlafke. Marcel Dekker, Inc, New York, 1995, p. 463-496.

17. Smith, J.C., Ellenberger, H.H., Ballanyi, K., Richter, D.W., and J.L. Feldman. Pre-Bötzinger Complex: A brainstem region that may generate respiratory rhythm in mammals. *Science* 254: 726-729, 1991.
18. Johnson, S.M., Smith, J.C., Funk, G.D., and J.L. Feldman. Pacemaker behavior of respiratory neurons in medullary slices from neonatal rat. *J. Neurophysiol.* 72: 2598-2608, 1994.
19. Llinas, R. The intrinsic electrophysiological properties of mammalian neurons: Insights into central nervous system function. *Science* 242: 1654-1664, 1988.
20. Hille, B. *Ionic Channels of Excitable Membranes*. 2nd ed.,Sinauer, Sunderland, MA, 1992.
21. Cronin, J. *Mathematical Aspects of Hodgkin-Huxley Neural Theory*. Cambridge Studies in Mathematical Biology, Vol. 7, Cambridge Univ. Press, Cambridge, Eng., 1987.
22. Nelson, M. and J. Rinzel. The Hodgkin–Huxley Model. In: *The Book of Genesis. Exploring Realistic Neural Models with the GEneral NEural SImulation System*, edited by J. M.Bower, and D. Beeman, Springer-Verlag, New York and TELOS, Santa Clara, CA, 1995, p. 29-51.
23. Hodgkin, A.L. and A.F. Huxley. A quantitative description of membrane current and its application to conduction and excitation in nerve. *J. Physiol. Lond.* 117: 500-544, 1952.
24. Huguenard, J. R. and D.A. McCormick. Simulation of the currents involved in rhythmic oscillations in thalamic relay neurons. *J. Neurophysiol.* 68: 1373-383, 1992.
25. Segev, I. Temporal interactions between post-synaptic potentials. In: *The Book of Genesis. Exploring Realistic Neural Models with the GEneral NEural SImulation System*, edited by J. M. Bower, and D. Beeman. Springer–Verlag, New York and TELOS, Santa Clara, CA, 1995, pp. 83-101.
26. Segev, I. Single neurone models: oversimple, complex, and reduced. *Trends Neurosci.* 15: 414-421, 1992.
27. Smith, J.C., Greer, J., Liu, G., and J.L. Feldman. Neural mechanisms generating respiratory pattern in mammalian brain stem-spinal cord *in vitro*. I. Spatiotemporal patterns of motor and medullary neuron activity. *J. Neurophysiol.* 64: 1149-1169, 1990.
28. De Schutter, E. and J.M. Bower. An active membrane model of the cerebellar purkinje cell I. Simulation of current clamps in slice. *J. Neurophysiol.* 71: 375-400, 1994.
29. Wilson, C.J. Dendritic morphology, inward rectification, and the functional properties of neostriatal neurons. In: *Single Neuron Computation*, edited by T. McKenna, J. Davis, and S.F. Zornetzer, Academic Press, San Diego, CA, 1992, p. 141-172.
30. De Schutter, E. Alternative equations for the molluscan ion currents described by Connor and Stevens. *Brain Res.* 382:134-138, 1986.
31. Wang, X.-J., Rinzel, J., and M.A. Rogawski. A model of the T-type calcium current and the low-threshold spike in thalamic neurons. *J. Neurophysiol.* 66: 839-850, 1991.
32. Traub, R.D, Wong, R.K.S., Miles, R., and H. Michelson. A model of a CA3 hippocampal pyramidal neuron incorporating voltage-clamp data of intrinsic conductances. *J. Neurophysiol.* 66: 635-650, 1991.
33. Yamada, W.M., Koch,C., and P.R. Adams. Multiple channels and calcium dynamics. In: *Methods in Neuronal Modeling. From Synapses to Networks*, edited by C Koch,and I. Segev, MIT Press, Cambridge, MA, 1989, p. 97-133.
34. De Schutter, E. Computer software for development and simulation of compartmental models of neurons. *Comput. Biol. Med.* 19: 71-78, 1989.
35. Smith, J.C. Computational models of the respiratory oscillatory in mammals. *Soc. Neurosci. Abstr.* 20: 1202, 1994.
36. Smith JC. A model for developmental transformations of the respiratory oscillator in mammals (Abstr.) *FASEB J.*: 8:A394, 1994.
37. Smith J.C., Ballanyi, K., and D.W. Richter. Whole-cell patch-clamp recordings from respiratory neurons in neonatal rat brainstem *in vitro*. *Neurosci. Lett.*, 1992; 134: 153-156.
38. Smith, J.C. The network generating respiratory rhythm and pattern in medullary slices *in vitro*. Experimental analysis and computational model. Soc. Neurosci. Abs. 21:689, 1995.
39. Richter, D.W. and K.M. Spyer. Cardiorespiratory control. In: *Central Regulation of Autonomic Functions*, edited by A.D. Lowey, and K.M. Spyer, Oxford Univ. Press, New York, 1990, p. 189-207.
40. Schwarzacher, S.W., Wilhelm, Z., Anders, K., and D.W. Richter. The medullary respiratory network in the rat. *J. Physiol. London* 435: 631-644, 1991.
41. Schwarzacher, S.W., Smith, J.C., and D.W. Richter. Pre-Bötzinger Complex in the cat. *J. Neurophysiol.* 73:1452-161, 1995.

6

SYNAPTIC PLASTICITY AND RESPIRATORY CONTROL

Chi-Sang Poon

Harvard-MIT Division of Health Sciences and Technology
Massachusetts Institute of Technology
Cambridge, Massechusetts 02139

1. INTRODUCTION

In the past two decades there have been considerable advances in the understanding of the neural mechanisms of learning and memory in the mammalian higher brain. A preeminent view based upon the classical model of Hebb[27] holds that learning and memory may result from activity-dependent modifications of neural transmission at certain chemical synaptic junctions. Generally referred to as *synaptic plasticity*, such neuronal modifications are widely believed to occur continually in infancy and, to some extent, throughout adulthood. One of the best known examples of such synaptic modification is hippocampal long-term potentiation (LTP) of neural transmission which can be robustly induced by a brief period of tetanic (high-frequency) afferent stimulation, both *in vivo* and *in vitro*[3,8,9]. Recently, many other forms of synaptic plasticity have been identified in the hippocampus and other brain structures[4,33,37,38]. It thus appears that synaptic plasticity is probably a generic property of many types of neurons which may be expressed throughout the mammalian central nervous system and may subserve a wide variety of neural functions.

In contrast, little is known about the expression and possible role of synaptic plasticity in the brainstem. Indeed, the brainstem is probably one of the few brain structures that would be deemed by many investigators as highly unlikely to express synaptic plasticity. For one thing, exquisite cognitive behaviors such as learning and memory are generally thought to belong exclusively in the higher brain. By contrast, physiological functions such as respiratory and circulatory control-which are primarily governed by the brainstem - are traditionally considered as rudimentary processes that consist in nothing more than simple reflex behavior.

Recently, however, there has been increasing recognition that many forms of physiological regulation cannot be explained solely on the basis of simple feedback or feedforward reflex control[61]. Rather, there is growing speculation that these enigmatic regulatory functions may reflect sophisticated, hitherto unknown computational structures within the brainstem. In particular, it has been suggested that homeostatic regulation may represent an adaptive neural behavior involving some form of learning and memory similar to those found in the higher brain[63]. This is consonant with recent revelations that certain physiological systems such as respiratory control may exhibit highly complex computational characteristics which are compatible with an adaptive behavior in the respiratory controller[47].

The postulation of synaptic plasticity in the brainstem opens a new and exciting paradigm for experimental and modeling investigations of physiological control mechanisms. To put this new paradigm in perspective, we begin in this Chapter with a brief overview of the various forms of synaptic plasticity found in the mammalian and invertebrate nervous systems. This is followed by an introduction to various forms of synaptic plasticity that have recently been discovered in the mammalian brainstem. Finally, an attempt is made to formalize specific hypotheses as to how various forms of synaptic plasticity might contribute to respiratory control in varying physiological states such as chemoreflex, periodic breathing, central hypoxic depression, behavioral conditioning of respiration, and exercise hyperpnea.

2. SYNAPTIC PLASTICITY: A MODEL OF LEARNING AND MEMORY

Learning is the acquisition of behavioral changes by experience, and memory is the retention of the modified behaviors over time. Many different forms of learning and memory have been identified in invertebrate and mammalian nervous systems. It is now generally believed that synaptic plasticity is the neuronal basis for most forms of learning and memory[2,8,12,26] although the mechanisms underlying many forms of synaptic plasticity are not yet fully understood. A common property of all forms of synaptic plasticity is that they are *activity dependent*, i.e., their expression is critically dependent on the modes of pre- and/or post-synaptic activity. This property allows the separation of synaptic plasticity into several classes.

2.1 Non-Associative Learning

In non-associative learning synaptic modification is dependent solely on the stimulus. Suppose two neurons are connected together via a chemical synapse (Fig. 1a). Let W be the synaptic strength and x be the stimulus intensity in the presynaptic neuron. Then a general adaptation rule for non-associative learning is of the form:

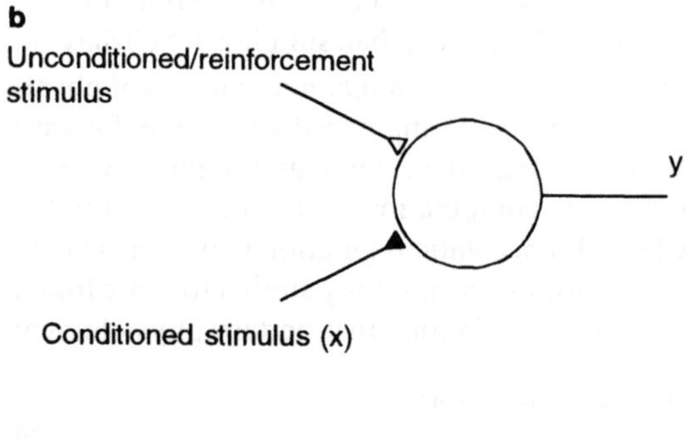

Figure 1. (a) In homosynaptic plasticity synaptic strength is modified by activity within the same neural pathway. The plasticity is non-associative if it is dependent only on the presynaptic activity (x), and is associative if it is dependent on the pairing of pre- and postsynaptic (y) activity. (b) In heterosynaptic (associative) plasticity the pairing of pre- and postsynaptic activity is influenced by a second input which modulates postsynaptic activity. The modulatory input may be in the form of an unconditioned stimulus (classical conditioning) or a reinforcement signal (operant conditioning). Modifiable and (possibly) non-modifiable synapses are indicated by filled and open triangles, respectively.

$$\delta W = k f(x) \tag{1}$$

where δW is the change in synaptic strength; k is the adaptation rate; and f(x) is a monotone non-decreasing function of x. The neuronal adaptation is non-associative because synaptic modification is strictly a function of stimulus intensity only. Two types of non-associative learning may be distinguished.

Habituation: synaptic strength decreases with repeated presentation of a stimulus. This may be represented by Eq. 1 with a negative adaptation rate ($k < 0$). With habituation, an animal learns to recognize and then gradually ignore the continued presence of an innocuous stimulus (e.g. a special odor).

Sensitization: synaptic strength increases with repeated presentation of a (usually noxious) stimulus (as in hyperalgesia). This is described by Eq. 1 with $k > 0$.

Non-associative learning has been studied most extensively in invertebrates. A similar form of non-associative synaptic plasticity is found in the CA_3 region of the hippocampus (see Hawkins[26]).

2.2 Associative Learning

Associative synaptic plasticity is a more complex form of learning in that its expression is not only activity dependent but also *pairing specific*, i.e., it requires a particular pairing of pre- and postsynaptic activities. A conjunctive form of associative synaptic plasticity first proposed by Hebb[27] is represented by the following associative adaptation rule:

$$\delta W = k f(xy) \tag{2}$$

where the term xy is the product of pre- and postsynaptic activity. In this event the adaptation may be either homosynaptic or heterosynaptic, respectively, depending on whether the changes in postsynaptic activity are induced by the same or other presynaptic inputs (Fig. 1a,b). There are two main classes of associative synaptic plasticity[11].

Hebbian learning: In agreement with Hebb's original conjecture[27], coactivity of pre- and postsynaptic neurons often results in synaptic long-term potentiation (LTP). This is obtained in Eq. 2 if the adaptation rate is positive ($k > 0$). Hebbian synapses are commonly found in the mammalian hippocampus and neocortex[11,32]. The pairing-specific property is due to the activation of postsynaptic NMDA receptors - a process that is both ligand (presynaptic) and voltage (postsynaptic) dependent. In invertebrates, similar pairing specific effects are achieved by presynaptic facilitation[26].

Anti-Hebbian learning: This is represented by Eq. 2 with $k < 0$. A well-known example is the heterosynaptic long-term depression in cerebellar Purkinje cells[28]. In this system, synaptic transmission via the parallel fibers is depressed by coactivity of climbing fibers which depolarizes the Purkinje cells.

Hebbian (or anti-Hebbian) synaptic plasticity may provide a cellular account for some forms of associative learning at the behavioral level. Two important forms of associative learning are classical conditioning (Pavlovian conditioning) and operant conditioning (also known as instrumental conditioning or reinforcement learning).

Classical conditioning: In Pavlov's classic experiment[43] a physiological response (salivation) was shown to be provoked by a normally neutral conditioned stimulus (CS) such as the ringing of a bell, after the CS was repeatedly paired with a powerful unconditioned stimulus (US) such as feeding. This behavior is mimicked (to the first order) by heterosynaptic Hebbian long-term plasticity of the CS pathway after pairing it with an US which causes strong depolarization in the postsynaptic neuron (Fig. 1b).

Operant conditioning: In this mode the animal learns a favorable behavior by responding to positive and/or negative reinforcement feedback signals derived from the environment with which it interacts. This may be modeled by heterosynaptic long-term plasticity in a similar fashion as in classical conditioning except that a reinforcement signal becomes the US (Fig. 1b).

It has been suggested that Hebbian learning may be a useful mechanism of supervised learning and adaptive control in the brain[50].

2.3 Short-Term Synaptic Plasticity

In contrast to long-term memory, which may last for hours or days (or even longer), short-term synaptic plasticity has a much shorter time span[36] and may be accounted for by the residual activity of presynaptic Ca^{2+}.[30] Some common forms of short-term plasticity (with their time spans in parenthesis) include: facilitation (milliseconds), augmentation (seconds), and short-term potentiation (minutes). A simple adaptation rule corresponding to the associative form of short-term synaptic plasticity is[49]:

$$\delta W = k f(xy) - W g(y) \tag{3}$$

where g(y), the forgetting factor, is a positive-definite function of y. The rate of recovery of synaptic modification is determined by the magnitude of g(y). A non-associative form of short-term synaptic plasticity is obtained by substituting the term f(x) for f(xy) in Eq. 3.

Similarly, as will be shown in Section 3, short-term decreases (with $k < 0$) in synaptic strength may occur in some neurons. We refer to such short-term decreases in synaptic strength as synaptic *accommodation* (seconds) and *short-term depression* (minutes), depending on the recovery rate g(y).

2.4 Hebbian Covariance Learning

One drawback of the classical Hebbian model of synaptic plasticity is that synaptic modification is unidirectional (potentiation or depression). Thus, synaptic changes are irreversible, saturable, and sensitive to extraneous perturbations. These difficulties are obviated in a modified Hebbian adaptation rule in which synaptic modifications are dependent on the *covariance* instead of coactivation of pre- and postsynaptic activity[2]. That is,

$$\delta W = k f(\delta x \delta y) \tag{4}$$

where $\delta x \delta y$ is the covariance of x and y over a specific time period. Thus, the synapse is strengthened if pre- and postsynaptic activities are positively correlated over time and is weakened otherwise (the opposite is true for anti-Hebbian covariance learning). Thus, both up- and down-regulation of synaptic strength is possible depending on the correlation of the input and output activities. This adaptation rule appears to satisfactorily describe the induction of synaptic LTP and LTD in many types of neurons[2].

Similarly, short-term plasticity of the Hebbian covariance form may be described by the following equation:

$$\delta W = k f(\delta x \delta y) - W g(\delta y) \tag{5}$$

In Section 4.4 we will show that Hebbian covariance learning may provide a useful neuronal substrate for complex optimization computations.

3. SYNAPTIC PLASTICITY IN THE BRAINSTEM

The existence of memory systems in the respiratory controller has been recognized for more than half a century[24,25], even before the postulation of synaptic plasticity by Hebb[27] and the discovery of hippocampal long-term potentiation by Bliss and Lømo[9]. Since that time, many forms of respiratory memory of both long and short durations have been found (for a review see Eldridge and Millhorn[17]). The memory effect can be robustly elicited in a variety of ways including electrical stimulation of peripheral nerves, mechanical stimulation of skeletal muscles, chemical stimulation by inhalation of hypoxic or hypercapnic gas mixtures, voluntary hyperventilation, or muscular exercise. Furthermore, there is evidence that respiratory memory may prevail even during eupneic (unstimulated) breathing and may play an important role in respiratory control in the quiescent state[5,57].

For a long time, the memory effect in the respiratory system has been ascribed to the existence of some reverberating neuronal networks in the brainstem, perhaps within the reticular system[17]. Another possibility is that the long-lasting memory effect may be mediated by certain endogenously produced neurotransmitters or neuromodulators - the accumulation (or depletion) of which may serve to maintain neuronal activity.

More recently, it has been suggested that respiratory memory may be a manifestation of synaptic plasticity, much like long-term potentiation and depression in the higher brain[47]. However, unlike the hippocampus, long-term potentiation of respiratory activity induced by carotid nerve stimulation is mediated by serotonergic (instead of glutamatergic) neural transmission[17,20], perhaps in the mid-line neurons of the raphe nucleus[41].

One form of respiratory memory that has received a great deal of attention recently is respiratory short-term potentiation (STP) which is elicited by sustained afferent stimulation[19,66]. Upon cessation of the primary stimulus, an "afterdischarge" of respiratory activity may persist for several minutes before decaying to the baseline level. This behavior is probably analogous to the short-term memory described in earlier studies under similar experimental conditions[17], which is now being referred to as STP in keeping with the notion of short-term synaptic plasticity found in other types of neurons[30,36]. However, experimental studies of respiratory STP in the whole animal *in vivo* cannot distinguish whether the memory effect is due to synaptic potentiation or other factors.

Recently, two experimental studies in our laboratory provided strong evidence at the cellular and molecular level that synaptic plasticity may play an important role in respiratory control. A genetically engineered strain of mutant mice was recently created[35] in which the NMDA receptors in the brain were rendered malfunctional by targeted deletion of a key subunit of the NMDA receptor using embryonic stem cell technology. Since activation of NMDA receptors is a key step in many forms of synaptic plasticity[38], the NMDA receptor deficient mutant is an excellent animal model of impaired learning and memory in the brain. However, the neonates homozygous for the mutation were found to suffer cyanotic respiratory failure and died within 20 hours after birth even though they appeared normal and well oxygenated at birth[35]. Respiratory measurements in these mutant animals[56] revealed the development of severe respiratory depression and irregularities of respiratory movements including recurrent apneic episodes well before the onset of cyanotic respiratory failure. The marked respiratory abnormalities in the NMDA mutant mice raise the specter that disturbance of the processes involved in synaptic plasticity might lead to respiratory depression, respiratory failure and eventual death.

Direct evidence that synaptic connections in brainstem neurons are indeed modifiable (i.e., plastic) was recently heralded by Zhou, Champagnat and Poon[70,71]. Using an *in vitro* brainstem slice preparation (400 μm) in newborn (1-3 weeks) rats, these investigators studied the synaptic transmissibility in the nucleus tractus solitarius (NTS) of the brainstem, a region that is richly innervated by primary chemoreceptor, baroreceptor, vagal, and glossopharyngeal afferent fibers in the tractus solitarius (TS). Synaptic strength of the TS-NTS junction was measured by the amplitude of the monosynaptic excitatory postsynaptic potential, EPSP (recorded by means of a whole-cell patch electrode), in NTS neurons evoked by an electrical impulse delivered to the contralateral TS fibers. The synaptic response was mediated mainly by glutamate, an excitatory amino acid neurotransmitter which activates both NMDA and non-NMDA receptors on the membrane surface to allow a transient influx of cations through the corresponding ligand-gated ion channels, thereby increasing the membrane potential. Inhibitory postsynaptic responses (via inhibitory interneurons) were blocked by adding bicuculline (an antagonist for $GABA_A$ receptors) to the perfusate of the brainstem slice.

By using a conventional protocol of sustained low-frequency stimulation (5 Hz, 3-5 min) of afferent fibers in the TS, Zhou et al.[70,71] showed that neurons in the medial and commissural regions of NTS exhibited a tri-phasic response during and after the stimulation protocol (Fig. 2):

1. In nearly all neurons, low-frequency TS stimulation resulted in an *accommodation* response characterized by a progressive decrease in synaptic efficacy. Typically, the amplitude of evoked EPSPs decreased to < 50% of control in 3-5 min and remained depressed with continued TS stimulation.
2. In one group of neurons (Type I) the synaptic strength recovered rapidly following the cessation of low-frequency stimulation, with the EPSP amplitude returning to the baseline level in ~2 min. In another group of neurons (Type II), the synaptic strength only partially recovered during the post-stimulation period with the EPSP amplitude approaching 70-80% of control value after ~5 min.
3. In Type II neurons, EPSP amplitude remained depressed (after the initial partial recovery) for > 30 min suggesting the presence of *long-term depression* (LTD) in the TS-NTS synaptic junction.

Furthermore, the LTD phase of the response seemed to be dependent on the level of intracellular calcium ions in the post-synaptic neuron. This was demonstrated by the application of APV, an antagonist of NMDA receptors which mediate the influx of calcium ions, or the use of EGTA (a chelator of calcium ions) in the recording micropipette thereby buffering the level

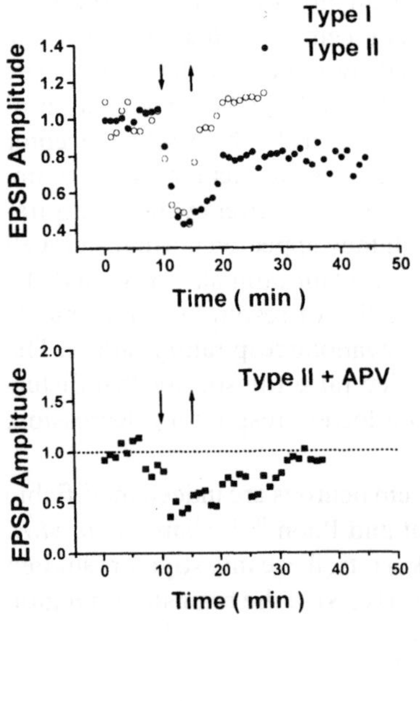

Figure 2. Synaptic plasticity in nucleus tractus solitarius region of rat brainstem. Arrows indicate the beginning and end of low-frequency stimulation (LFS; 5 Hz) of afferent fibers in the tractus solitarius. Synaptic efficacy is measured by the amplitude of excitatory postsynaptic potential (EPSP) evoked by an electrical impulse delivered to the tractus solitarius. All EPSPs are normalized with respect to the mean value in the control period prior to LFS. *Upper panel* shows typical responses of Type I and Type II neurons. *Lower panel* shows the response of a Type II neuron treated with APV which blocked NMDA receptors. See text for explanation.

of intracellular calcium ions in the recorded cell. In both cases, the LTD phase of the response was abolished and the EPSP amplitude gradually returned to the baseline level in some 20-30 min, resulting in a synaptic *short-term depression* (STD; Fig. 2).

The synaptic long-term depression in NTS neurons is, in certain ways, similar to the homosynaptic LTD found in hippocampal and neocortical neurons. For example, the procedure of low-frequency stimulation for the induction of LTD in the NTS is common to the induction of LTD in many neuronal types in the higher brain[33]. Also, activation of NMDA receptors and the level of intracellular calcium ions in the post-synaptic cell, which are important for many forms of synaptic plasticity in the higher brain[14], seem to play a role in the development of the LTD in the NTS as well.

This finding is important because it is one of the first demonstrations at the cellular level that synaptic plasticity is not an exclusive attribute of neurons in the higher brain but may also be intrinsic to brainstem neurons. An interesting implication is that cognitive processes such as learning and memory, which have been generally thought to be synonymous with conscious behavior in the higher brain, may as well be an integral part of physiological regulation which occurs at a subconscious level in brainstem nuclei.

However, there are also some important differences in the LTD of NTS neurons compared to other types of neurons. First, although NMDA receptors and intracellular calcium ions contributed to the maintenance of the LTD, they did not seem to be important for its induction in NTS neurons: blocking the NMDA receptors or the buildup of Ca^{2+} in the post-synaptic cell converted the LTD into STD but did not entirely abolish the synaptic depression. This is in contradistinction to the key role played by calcium flow through NMDA receptors in the induction of both LTP and LTD in hippocampal and neocortical neurons[14,38]. Second, the development of LTD (or STD) in NTS cells was always preceded by an accommodation response during the low-frequency stimulation period. Since the accommodation effect was expressed in both Type I and Type II cells while LTD was expressed only in Type II cells, it appears that accommodation and LTD represent two distinct forms of synaptic plasticity that are simultaneously elicited by low-frequency stimulation in the NTS. The accommodation response to low-frequency stimulation is a unique feature of NTS cells that is not commonly found in the mammalian higher brain. Third, in contrast to the low-frequency stimulation protocol, high-frequency (50-100 Hz) stimulation of the TS did not elicit LTP which is commonly found in other neurons. Instead, a short-term post-tetanic potentiation is obtained in some NTS neurons which typically decays completely within 1-2 min[18].

The accommodation and LTD responses in NTS neurons represent two distinct forms of synaptic plasticity that are unique to the brainstem. The distinctive expressions of synaptic plasticity in the brainstem should not be surprising since the NTS is involved in very different physiological functions than those corresponding to higher brain centers. In the following section we explore the possible roles of different forms of brainstem synaptic plasticity in explaining several important behaviors of the respiratory controller under varying physiological conditions.

4. SYNAPTIC PLASTICITY: A NEW PARADIGM IN RESPIRATORY CONTROL

Classical models of respiratory control are derived from the notion of a long-loop reflex, i.e., respiratory output is assumed to be reflexly driven by phasic chemoafferent and mechanoafferent inputs via the brainstem controller. According to this view, the respiratory controller is nothing more than a hard-wired neural network governed by unique stimulus-response relationships. Consequently, much of the research efforts in the past have focused

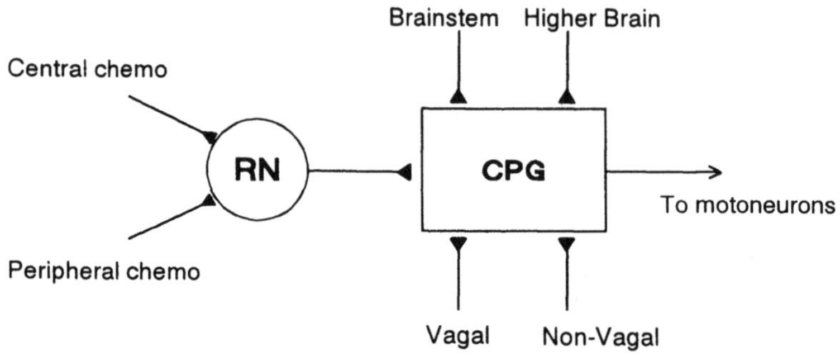

Figure 3. Schematic diagram showing possible roles of synaptic plasticity in modulating various ascending and descending inputs to the respiratory controller. RN, respiratory neuron; CPG, central pattern generator. Adapted from Poon.[47]

on defining the characteristics of afferent processes as primary explanations for integrative respiratory behaviors.

Although this conceptual framework seems to fit certain experimental observations (e.g., chemoreflex response), it fails to explain many critical phenomena where a simple stimulus-response relationship cannot be verified. A long-loop reflex model of respiratory control may be over-simplified in light of the recently discovered forms of synaptic plasticity in the brainstem. Furthermore, it is possible that the brainstem respiratory controller may be influenced by other forms of synaptic plasticity that remain to be elucidated.

We surmise that various forms of brainstem synaptic plasticity may contribute importantly to distinct aspects of respiratory control. This hypothesis is formalized in the model shown in Fig. 3 in which the respiratory controller is assumed to be comprised of a respiratory central pattern generator (RCPG) network driven by various chemoreceptor, vagal and non-vagal afferent inputs as well as descending inputs from the pons and higher brain. A possible locus of the RCPG is the rostral ventrolateral medulla where the respiratory rhythm is generated as a result of reciprocal inhibition between various neuronal groups[15].

It is assumed that synaptic changes occur primarily in the central pathways mediating the afferent and cortical inputs to the RCPG (Fig. 3). This assumption is based upon several considerations. First, *in vitro* studies (Fig. 2) have identified the existence of several forms of synaptic plasticity in the excitatory pathways in the NTS which is the gateway for various primary afferent inputs to the respiratory controller. Second, *in vivo* studies have shown that respiratory STP may be induced even if a stimulus is delivered during the expiratory cycle when the primary stimulatory effect is gated off by the RCPG[17]. Finally, since synaptic plasticity is found mostly in excitatory synapses, it is unlikely to be expressed in reciprocally inhibitory networks in the RCPG.

The proposed model provides an integrative description as to how the respiratory responses to various inputs might be influenced and modulated by synaptic changes in the corresponding neural pathways.

4.1 Short-Term Potentiation

Respiratory STP is one of the most well studied forms of neural memory in the respiratory system. The physiological significance of STP in respiratory control, however, remains poorly understood and highly controversial. Several important implications of STP in the measurement and analysis of the chemoreflex response warrant special consideration.

4.1.1 Chemoreflex Response. One immediate implication of STP is its effect on the measurement and interpretation of ventilatory CO_2 sensitivity. Traditionally, the hypercapnic ventilatory response is taken as a measurement of the sensitivity of the central and peripheral CO_2 chemoreceptors. A more accurate interpretation should include the effect of STP of the respiratory controller. This may be of physiological and clinical significance since changes in the magnitude of respiratory STP due to pharmacological intervention may influence the measurement of the hypercapnic ventilatory response. Similar considerations also apply to the hypoxic ventilatory response.

Another effect of STP that is less well appreciated is that it may affect the dynamics of the chemoreflex response. The on- and off-transients of the STP response introduce a low-pass filtering effect in addition to the primary effect of the CO_2 stimulus. An idealized, linear first-order dynamical model (Fig. 4) of STP is described by the following frequency-domain transfer function:

$$H(s) \doteq \frac{y(s)}{x(s)} = 1 + \frac{k}{s+b} \tag{6a}$$

or

$$H(s) = \frac{s+a}{s+b} \tag{6b}$$

where y(s) and x(s) denote presynaptic (input) and postsynaptic (output) activity, respectively; s is the complex frequency variable; k and b are respectively the gain factor and cut-off frequency of the low-pass filter representing the STP component of the synaptic response. At high input frequencies ($s \to \infty$) or with zero input ($x = 0$) the throughput gain of the synapse is normalized to unity (unpotentiated), whereas with a constant input ($s = 0$) the total throughput gain of the synapse is equal to $a/b = 1 + k/b$ which is > 1, indicating a potentiation effect. Upon removal of the input, the synaptic gain decays to unity with a time constant of $1/b$. The transfer function of the model is characteristic of a phase-lag controller[34]. In this simplified model the nonlinearity and difference in the speed of the on- and off-transients of the STP response are neglected.

The dynamic behavior of respiratory STP is important in experimental studies involving time-varying inputs[13,64]. Thus, the low-pass filtering and gain magnification effects of respiratory STP may affect the accuracy of any parameter estimation of the dynamical behavior of the respiratory controller. In particular, estimation of the CO_2 sensitivity using dynamical procedures such as CO_2 rebreathing[58] may be problematic if the variable effect of respiratory STP is not taken into account. Indeed, the accuracy of CO_2 rebreathing in measuring the hypercapnic ventilatory response has recently been called into question particularly regarding its use in studies of drug effects[6,10].

4.1.2 Respiratory Instability. There exists some controversy regarding the effect of STP on periodic breathing and the stability of the respiratory system. It has been argued by

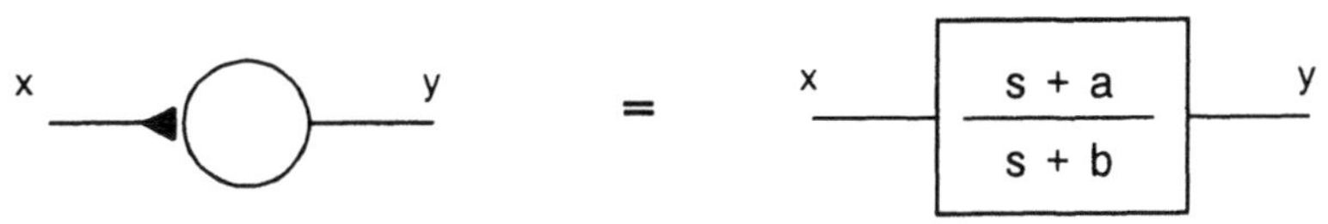

Figure 4. Transfer function model of synaptic short-term potentiation. See text.

some authors[69] that the characteristic afterdischarge of respiratory activity may have a stabilization effect on respiration since it may prevent the onset of apnea following voluntary or induced hyperventilation.

Although this conjecture seems to be supported by simulation studies which showed that STP abolished both posthyperventilation apnea and Cheyne-Stokes-like periodic breathing in a model of the respiratory system[16], such argument may be oversimplified in light of the complex dynamical behavior of the respiratory system. In general, stability of the respiratory system decreases with increasing loop gain and/or phase lag[31]. Although the low-pass filtering property of afterdischarge (Eq. 6) may tend to dampen any spontaneous oscillation, the STP may itself introduce a destabilization effect because:

1. the throughput gain of the controller is increased;
2. the total phase lag of the system is increased as a result of the slow dynamics of the afterdischarge.

It is important to note that the increase in phase lag alone does not destabilize the system (and, indeed, may enhance stability owing to the low-pass filtering effect); it is the combination of increased phase lag *and* increased loop gain that may result in instability. As is well-known in control theory[34], a phase-lag controller could increase or decrease system stability depending on the relationship between its dynamical behavior and that of the system being controlled. Thus, stability obtained under one set of gain and phase lag conditions does not ensure similar stability under other general conditions.

Consequently, in assessing the stabilization effect of STP it is imperative to take into account the intrinsic dynamical behavior of the respiratory system including the phase lags caused by the CO_2 and O_2 stores in the lung, brain and other body tissues, transport delays in the systemic circulation, as well as the response time constants of the peripheral and central chemoreceptors. Any stability analysis must include the effects of the increases in overall loop gain and phase lag resulting from STP rather than focusing on its low-pass filtering effect alone.

Nevertheless, it is worthwhile to note that certain characteristics of STP may indeed have important stabilization effects that have not been well appreciated (Table 1). Of particular relevance is the fact that the on- and off-transients of STP generally have different time constants: the development of STP typically occurs more quickly than its decay[66]. A hysteresis in response speed is an effective means of stabilizing the system toward the preferred direction of response. Another important property of STP is that it is saturable - the gain potentiation is limited to a maximum value. The saturation effect may help to curtail any continual gain potentiation that would lead to runaway instability.

Table 1. Effects of short-term potentiation on respiratory instability

Property of STP	Effect on stability	Remark
Gain magnification	↓	Decrease in gain margin
Phase lag	↑ or ↓	1) Low-pass filtering (↑ stability) 2) Decrease in phase margin (↓ stability)
Hysteresis	↑	Halts bidirectional oscillation
Saturation	↑	Halts runaway instability

In summary, STP has complex and multifarious effects on respiratory instability that cannot be described by simple consideration of the slow decay of the afterdischarge alone. A comprehensive study must include all relevant factors (Table 1) that may contribute importantly to the stability of the respiratory system under varying physiologic conditions.

4.2 Central Hypoxic Depression

In humans and cats, the respiratory response to sustained moderate hypoxia is biphasic: an initial rapid increase in ventilation is followed by a slower, gradual decline or roll-off of the ventilation. The first phase of the response is ascribable to peripheral chemoreceptor stimulation. The second phase, known as central hypoxic depression (or decline), is generally thought to be of central origin[7] although a peripheral chemoreceptor mediation cannot be totally ruled out especially in the awake state[60].

The mechanism of central hypoxic depression is unclear. During hypoxia, the ability of medullary respiratory related neurons to fire action potentials remains intact even though their activities are decreased along with a decrease in phrenic nerve activity[59]. Thus, the depression of neural activity is not caused by neuronal failure due to O_2 deprivation. Instead, attention has recently been focused on various inhibitory neurotransmitters or neuromodulators which may be elicited by hypoxic stimulation. Among them are adenosine, GABA, dopamine, serotonin, and endogenous opioids[7] all of which have been shown to have some influence on the development of central hypoxic depression. However, none of these neurochemicals can completely account for central hypoxic depression since application of the corresponding antagonists does not block its induction.

In contrast to previous studies, an alternative hypothesis for the neurogenesis of central hypoxic depression is suggested by the recent discovery that NTS neurons exhibit synaptic plasticity in the form of accommodation and long-term depression (Fig. 2). One possible scenario is that synaptic transmission of excitatory inputs from the peripheral chemoreceptors or medullary neurons sensitive to the hypoxic stimulus[40] may be progressively attenuated by accommodation and long-term depression of NTS or other respiratory related neurons. This contention is supported in part by consideration of the time course of the depression response (Table 2). Specifically, the development of central hypoxic depression typically occurs slowly and continually over the course of $>$ 30 min. The slow development of central hypoxic depression cannot be explained by the direct effect of inhibitory neurotransmitters which is expected to occur much more rapidly following afferent stimulation. It is possible that certain neurochemicals may accumulate slowly over time to produce a roll-off effect. This may also account for the observed persistence of the depression effect after hypoxic stimulation[7], since any neurochemicals accumulated during hypoxic stimulation are unlikely to be completely washed out immediately upon removal of the hypoxia. However, this mechanism of ventilatory depression may be non-specific to the

Table 2. Mechanisms of central hypoxic depression: synaptic depression vs. neural inhibition

Characteristic of hypoxic depression	Mechanism	
	Synaptic depression	Neural inhibition
Slow onset	yes	probably yes
Slow offset	yes	probably yes
Specific to input pathway	yes	no

hypoxic stimulus since an excess of neurochemicals induced by hypoxia is likely to influence other neural pathways due to molecular diffusion in the extracellular space. This is in contradistinction to the observation that the hypercapnic ventilatory response remained unchanged during hypoxic ventilatory depression[23].

In contrast, activity dependent synaptic plasticity is highly specific to the afferent stimulation. The accommodation response of NTS neurons has a slow onset (Fig. 2). In type II cells the depression of synaptic transmission persists for a long period of time after afferent stimulation and slowly recovers only if the calcium influx is blocked (Fig. 2). These characteristic features of synaptic plasticity in NTS neurons bear close resembance to those of central hypoxic depression (Table 2).

However, several important issues deserve further consideration. For example, the accommodation and LTD responses in NTS were elicited only by low-frequency stimulation; the synaptic strength was not changed and might even be potentiated by high-frequency afferent stimulation. The frequency dependent behavior of brainstem synaptic plasticity may explain why respiratory depression is provoked by hypoxia (a weak respiratory stimulant) and not by hypercapnia (a relatively strong stimulant) although the correspondence of the frequency of stimulation *in vitro* to the strength of a stimulus *in vivo* is not clear. This and many other issues must be addressed in elucidating a possible synaptic origin of central hypoxic depression.

4.3 Behavioral Conditioning

The respiratory rhythm is modulated continuously by cortical inputs. In the awake state, voluntary and psychosomatic behaviors such as speech, diving, and fright and anger can completely override the automatic control of respiration by the RCPG. During sleep, the respiratory rhythm exhibits varying characteristics in different sleep stages and arousal states. These direct effects of cortical influence on respiration, although well recognized, have nonetheless often been neglected because they are generally thought to be transitory and random in nature without any long-lasting impact on the automatic control of respiration.

Recent studies, however, have suggested that there are at least two ways in which cortical inputs may exert long-term influence on ventilatory pattern. First, volitional modulation of the respiratory rhythm during visually guided feedback training may elicit a learning effect similar to sensorimotor learning in the musculoskeletal system[21]. The learned ventilatory pattern, which may be retained for many hours or days, cannot be explained in terms of chemical control of breathing. Second, there is increasing evidence that somatosensory inputs may be classically conditioned (by repeated pairing with respiratory stimuli) to exert long-term modulation of ventilatory pattern in humans[22] and neonatal rats[65].

Synaptic plasticity is a possible neuronal mechanism of classical conditioning[12] and sensorimotor learning[28,52]. Presently, it is not clear whether such synaptic modifications occur in the cerebromedullary pathway or within higher brain structures. Regardless of its origin, synaptic plasticity is likely to play an important role in behavioral conditioning of ventilatory pattern which could begin as early as the embryonic or neonatal stage and may continue throughout adulthood[63].

4.4 Exercise Hyperpnea

4.4.1 Central and Peripheral Mechanisms. Despite numerous research efforts continuing for over a century, the mechanism of exercise hyperpnea has remained one of the greatest unsolved mysteries in contemporary respiratory physiology. Following the classic reflex model of respiratory control, most previous studies have approached this problem by searching for a putative "exercise stimulus" that would account for the isocapnic exercise hyperpnea response. Unfortunately, although many afferent mechanisms have been found to exert some influence on exercise hyperpnea under specific conditions, the protracted search of the exercise stimulus has continued to trail over the years to no avail[67].

Experimental investigations of the exercise stimulus ignore the possibility that the exercise hyperpnea response may be modulated or even engendered entirely by mechanisms within the respiratory controller, such as synaptic plasticity. For example, it has been suggested that the development of STP may be important for the rising phase (phase 2) of the exercise hyperpnea response[66]. In goats, the exercise ventilatory response remained elevated for several days after pairing it with increased respiratory dead space, suggesting that part of the exercise ventilatory response might be learnt[39].

4.4.2 Optimization Model of Respiratory Control. The possibility that the exercise hyperpnea response may have a central instead of peripheral origin is suggested by an optimization model of the respiratory controller proposed by Poon[44,45]. The model is based upon a simple proposition: the exercise hyperpnea response may be generated without the need for any explicit exercise related stimulus provided the respiratory controller is capable of certain computational abilities that allow for the optimization of some implicit objective of respiration. In this model the respiratory controller does not simply act as a feedback/feed-forward controller. Rather, it adaptively adjusts the feedback and/or feedforward gains in order to minimize the total cost of respiration under varying physiologic conditions. A particular cost function relevant to the control of exercise hyperpnea is of the form:

$$J \begin{cases} = J_c + J_m \\ = \alpha^2 (P_{aCO_2} - \beta)^2 + \ln \dot{V}_E^2 \end{cases} \tag{7}$$

where J_c and J_m are respectively the chemical and mechanical costs of respiration; P_aCO_2 is arterial PCO_2; α, β are parameters corresponding to the sensitivity and threshold of chemoreception, respectively; and $\dot{V}_E$ is total ventilation. The term J_c reflects a departure of P_aCO_2 from the desired level whereas the term J_m is a penalty function corresponding to the rate of work done or energy expenditure of breathing expressed as a logarithm function of $\dot{V}_E$ The relationship between P_aCO_2 and $\dot{V}_E$ is given by the pulmonary gas exchange equation:

$$P_{aCO_2} = P_{ICO_2} + \frac{863 \dot{V}_{CO_2}}{\dot{V}_E (1 - V_D/V_T)} \tag{8}$$

where P_ICO_2 is the PCO_2 in the inspired air; $\dot{V}CO_2$ is metabolic CO_2 production rate; and V_D/V_T is ratio of respiratory dead space to tidal volume. It can be easily verified[44,45] that the minimization of J subject to the operating constraint imposed by Eq. 8 results in the following optimal steady-state characteristic for the ventilatory response:

$$\dot{V}_E = 863 \dot{V}_{CO_2}\ \alpha^2 (P_{aCO_2} - \beta)/(1 - V_D/V_T) \tag{9}$$

Implicit in this characteristic equation are the following model predictions of respiratory response behaviors:

- *Hypercapnic ventilatory response*: The model predicts a linear ventilatory CO_2 response curve at any constant metabolic level.

- *Exercise hyperpnea*: The model predicts an isocapnic behavior with a proportional relationship between exercise ventilatory response and metabolic CO_2 output when breathing room air (0% CO_2). An explicit exercise stimulus is not required.
- *Hypercapnia-exercise interaction*: The model predicts that hypercapnia and exercise have a synergistic effect on the ventilatory output.
- *Ventilatory response to increased dead space load*: The model predicts a potentiation of the exercise ventilatory response in the presence of increased respiratory dead space.

All predicted response behaviors have been experimentally corroborated. In particular, the experimentally observed effects of CO_2-exercise interaction[54] and dead space-exercise interaction[48,62] are unique to the optimization model and are not directly predicted by the conventional reflex model. Furthermore, with some elaboration of the respiratory cost function (Eq. 7) the model has been shown to be compatible with other critical behaviors such as ventilatory load compensation, respiratory pattern generation, and vagal modulation of motor trajectory[55].

4.4.3 Optimization as a Form of Learning. The optimization model suggests a possible control law whereby isocapnic exercise hyperpnea could be produced without the need for any feedforward drive from an explicit exercise stimulus. The model is also in agreement with a number of distinctive response behaviors that cannot be predicted by the reflex model (Table 3). However, the optimization model raises an intriguing question: What is the neural *mechanism* underlying such sophisticated optimization behavior in the respiratory controller?

Before tackling this important question, we first note that optimization is a complex computational process that requires iterative adjustment of the control variables until an optimal solution is obtained. Computationally, this may be achieved by a systematic search (e.g. by gradient descent) or by random trial-and-error. In both cases, some form of memory is probably necessary in order to accumulate past experience while searching for the best solution. On the digital computer, this amounts to storing in the computer memory the results from past iterative steps for the computation of future iterative steps. Could such iterative computation occur in the brainstem?

This is a difficult question because there has been little evidence that such complex optimization computation could occur even in the higher brain at all. To address this issue, we performed an experiment in which the experimental subjects were asked to perform a visuomotor task that mimicked the optimization of the respiratory system[46]. The experiment was designed in such a way that a subject's visuomotor cortex (and other related brain structures) served as a surrogate for the respiratory controller in optimizing the control cost on the job. It was found that with some trial and error, the subjects were able to reproduce

Table 3. Comparison of optimization model and reflex model

Response behavior of respiratory controller	Prediction by optimization model	Prediction by reflex model
Chemoreflex	Yes	Yes
Exercise hyperpnea	Yes	Yes
CO_2-exercise interaction	Yes	Possibly
Dead space effect	Yes	No
Ventilatory load compensation	Yes	No
Respiratory pattern generation	Yes	No

the optimal response characteristics given by Eq. 9 even though no explicit feedback signals corresponding to an exercise stimulus was made available to them. The results suggested that optimization computations similar to those found in the respiratory system were indeed feasible in the higher brain, and that some forms of learning and memory were probably involved in such computations.

If neural optimization indeed amounts to learning and memory, there would be little doubt that the visuomotor system is capable of such computations. After all, the higher brain is known to be richly endowed with many forms of synaptic plasticity which is the neural basis of learning. This observation, however, led to two fundamental questions: 1) What forms of learning and memory may achieve optimization computations in the brain? 2) Is the brainstem capable of such learning and memory as well?

The recent discovery of synaptic plasticity in the NTS (Fig. 2) suggests that the brainstem respiratory controller may be capable of learning and memory, much like the higher brain. Furthermore, the LTD induced by low-frequency stimulation in NTS neurons is compatible with the Hebbian covariance model of synaptic modification which is commonly found in the hippocampus and neocortex[11,3,33]. In the following we present a mathematical model of synaptic plasticity[49,51,53] which demonstrates that Hebbian covariance learning could lead to respiratory optimization.

4.4.4 Self-Tuning Learning Models of Respiratory Optimization. Optimization of the control cost function J (Eq. 7) under differing physiological conditions requires adaptive changes in the respiratory controller in response to changes in the environment. Thus, respiratory optimization may be considered as a form of self-tuning adaptive control[1] whereby certain control objective functions are optimized by adaptive adjustment (self-tuning) of the controller. A necessary condition for such adaptive control is that the adaptation rule must be stable[42]. We describe in the following a particular adaptation rule[49,53] which is: 1) stable; 2) realizable by Hebbian synaptic plasticity; and 3) compatible with respiratory optimization.

Suppose the controller has an adjustable (tunable) gain W in the chemoafferent path (Fig. 3) so that the hypercapnic ventilatory response is given by:

$$\dot{V}_E = W \cdot \alpha(P_{aCO_2} - \beta) \tag{10}$$

To obtain a stable adaptation rule we first derive the time derivative of Eq. 7 by using the chain rule of differentiation as follows:

$$\frac{dJ}{dt} = 2\alpha^2(P_{aCO_2} - \beta)\frac{dP_{aCO_2}}{d\dot{V}_E} + \frac{1}{\dot{V}_E} \tag{11}$$

From Lyapunov theory[42], an adaptation rule for the gain W is stable provided dJ/dt ≤ 0 at all time. From Eq. 11 and noting that $d\dot{V}_E/dW \geq 0$, the above condition is satisfied provided:

$$\frac{dW}{dt} = -k \cdot \alpha^2(P_{aCO_2} - \beta) \cdot \frac{dP_{aCO_2}}{d\dot{V}_E} + \frac{1}{\dot{V}_E} \tag{12}$$

where the term k may be any constant or function that satisfies $k > 0$.

Equation 12 is a stable adaptation rule that would cause the function J to approach the minimum value under all physiological conditions. In order to compute this adaptation rule the controller must evaluate the partial derivative $\delta P_aCO_2/\delta \dot{V}_E$ We postulate that Hebbian learning may be useful for such computation. Thus, multiplying Eq. 12 by the factor $W.\delta^2\dot{V}_E/\dot{V}_E > 0$ and rearranging we obtain:

$$\frac{dW}{dt} = -k\alpha \cdot \delta P_{aCO_2}\delta \dot{V}_E + W(\delta \dot{V}_E/\dot{V}_E)^2 \qquad (13)$$

The right hand side of Eq. 13 is comprised of two terms. The first term is the covariance of afferent and efferent activity. Thus the adaptation rule is in the form of Hebbian covariance learning. The negative sign on the adaptation rate (-k) suggests that the adaptation rule is anti-Hebbian instead of Hebbian. Hence, the controller gain is potentiated whenever a given increase in $\dot{V}_E$ causes a large decrease in P_aCO_2. The second term on the right hand side of Eq. 13 is a decay (forgetting) term that is proportional to the auto-covariance of efferent activity. In the steady state where dW/dt = 0, the sum of these two terms is zero giving an optimal controller gain

$$W = \alpha \dot{V}_E^2 \frac{\delta P_{aCO_2}}{\delta \dot{V}_E} = \frac{863\alpha \dot{V}_{CO2}}{(1 - V_d/V_T)} \qquad (14)$$

Substitution of Eq. 14 into Eq. 10 yields the optimal ventilatory response characteristic given in Eq. 9.

Hebbian covariance learning is a possible substrate of neural optimization because it provides a useful means of computing the gradient of changes in afferent activity with respect to changes in efferent activity. Thus, during muscular exercise the controller gain is potentiated because a given increase in $\dot{V}_E$ produces a large decrease in P_aCO_2, whereas during CO_2 inhalation the controller gain is not increased because the covariance of $\dot{V}_E$ and P_aCO_2 is relatively small and is totally negated by the auto-covariance term $W(\delta\dot{V}_E/\dot{V}_E)^2$ (Fig. 5).

The Hebbian covariance model suggests the feasibility of synaptic plasticity as a mechanism of respiratory optimization. However, several important issues remain. For example, a necessary condition for the covariance rule is that the afferent and efferent activities must constantly fluctuate in order to produce the variational terms $\delta\dot{V}_E$ and δP_aCO_2. This is probably a reasonable assumption since neural activities are known to fluctuate spontaneously in a random or chaotic fashion. It is also possible that fluctuations in $\dot{V}_E$ and P_aCO_2 may result from the tidal nature of respiration. It has been proposed that the oscillations of the P_aCO_2 signal caused by tidal volume changes may induce an exercise stimulus for ventilatory output[68,47]. The Hebbian covariance model presently proposed suggests a possible neuronal mechanism for such an oscillation induced ventilatory effect. Furthermore, it suggests that the ventilatory response to increases in arterial blood gas oscillation during exercise may be caused by a potentiation of controller gain instead of the elicitation of an oscillation sensitive exercise stimulus.

One critical issue not considered in the above model is that the temporal relationship of the afferent and efferent activities is not instantaneous but is modulated by the slow dynamics and time delays of signal transduction in the cardiopulmonary system and chemoreceptors. This is important because any delay in the response of the controller may result in instability of the adaptive control system. This difficulty may be obviated if one assumes that the respiratory controller is responsive directly to the efferent output $\dot{V}_E$ and chemoafferent input P_c which is sensed by the chemoreceptors, where P_c is the effective chemoafferent input as seen by the controller. It has been shown[53] that minimization of the

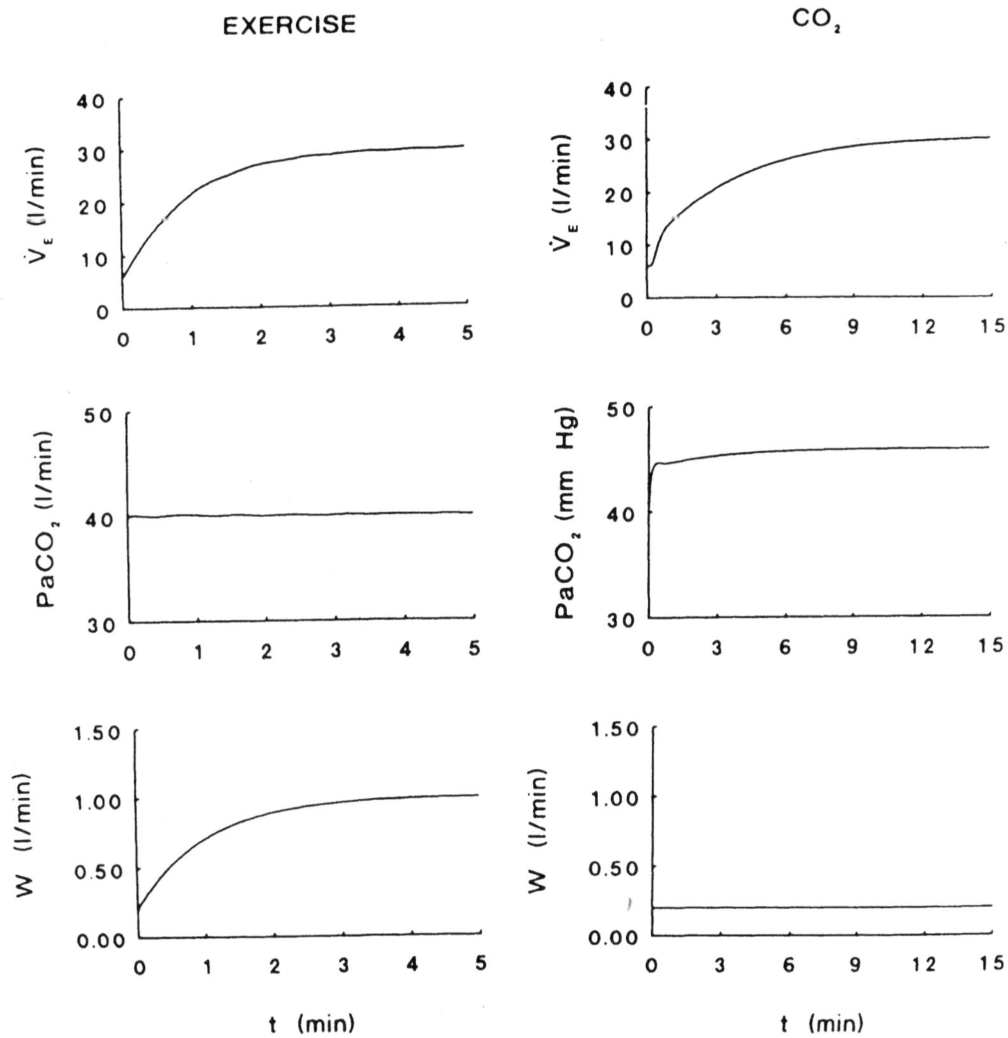

Figure 5. Hebbian covariance learning model simulation of isocapnic exercise hyperpnea (*left*) and hypercapnic ventilatory response (*right*). $\dot{V}_E$, P_aCO_2, and W are total ventilation, arterial CO_2 tension, and synaptic strength (or gain) of respiratory controller, respectively. Included in the model are dynamical equations for pulmonary, brain tissue and whole-body tissue compartments, central and peripheral chemoreceptors, circulatory delays, and a Hebbian covariance adaption rule for gain adjustment in the respiratory controller. Note the differential effects of self-tuning gain adaptation in exercise but not CO_2 inhalation. Adapted from Poon[49].

physiologic objective function J (Eq. 7) of the respiratory system is equivalent to the minimization of a neural objective function of the form:

$$Q = P_c^2(n + 2) + \lambda \dot{V}_E(n) \tag{15}$$

where n is a time index (which may be conveniently chosen as multiples of the respiratory cycle) and λ is a constant. The function Q is a neural objective function as seen by the respiratory controller. Notice that the mechanical cost term is given by the ventilatory output at the present time n whereas the chemical cost term is given by the arterial chemical level at a future time (n+2). Also, the mechanical cost to the controller is a linear (instead of logarithmic) function of $\dot{V}_E$ The differences between the neural and physiological cost functions effectively compensate for the dynamics of pulmonary gas exchange and circulatory transport.

5. CONCLUSSION

Synaptic plasticity is an important neuronal property that may be integral to normal physiological functions in the brainstem as it is in the higher brain. Until recently, the critical role of synaptic plasticity in respiratory control has not been well appreciated. This has led to a great deal of confusion in the understanding of respiratory control mechanisms. The recent discovery of various forms of synaptic plasticity in the brainstem opened a new horizon for experimental elucidation and mathematical modeling of respiratory control which may have wide ranging implications in other physiological control systems in general.

Presently, the mechanism and physiological roles of brainstem synaptic plasticity are only beginning to be appreciated and many aspects of this neuronal process remain to be elucidated. The lack of understanding of this critical phenomenon presents new and challenging opportunities for experimental investigation. Similarly, the complexity inherent in the nonlinear dynamical behavior of brainstem synaptic plasticity calls for rigorous modeling and simulation studies to elaborate its full physiological consequences. Our initial experimental and modeling results seemed to support the hypothesis that the respiratory control system is not merely a feedback/feedforward reflex system but may be more appropriately modeled as a self-tuning adaptive optimal control system. If true, this model may provide a general, unified framework for the understanding of a wide variety of respiratory control behaviors.

ACKNOWLEDGMENT

The author's research work was supported by National Science Foundation (grant BES-9216419), Office of Naval Research (grant N00014-95-0414), and National Heart, Lung and Blood Institute (grants HL45261 and HL50641).

REFERENCES

1. Åstrom, K.J. and Wittenmark. *On self-tuning regulators*. Automatica 9:185-199, 1973.
2. Barnes, C.A., Bindman, L.J., Dudai, Y., Frégnac, Y., Ito, M., Knöpfel, T., Lisberger, S.G., Morris, R.G.M., Moulins, M., Movshon, J.A., Singer, W., and Squire, L.R. Group Report: Relating activity-dependent modifications of neuronal function to changes in neural systems and behavior. In A.I. Selverston and P. Ascher (Eds.), *Cellular and Molecular Mechanisms underlying Higher Neural Functions*, New York: Wiley, 1994, pp. 81-110.
3. Baudry, M. and J.L. Davis. *Long-term Potentiation: A Debate of Current Issues*. Cambridge, MA: MIT Press, 1990.
4. Baudry, M., Thompson, R.F. and Davis, J.L. *Synaptic Plasticity: Molecular, Cellular, and Functional Aspects*. Cambridge, MA: MIT Press, 1993.
5. Benchetrit, G. and F. Bertrand. A short-term memory in the respiratory centers: statistical analysis. *Respir. Physiol.* 23:147-158, 1975.

6. Berkenbosch, A., J.G. Bovill, A. Dahan, J. DeGoede, and I.C.W. Olievier. The ventilatory CO_2 sensitivities from Read's rebreathing method and the stead-state method are not equal in man. *J. Physiol. (London)* 411:367-377, 1989.
7. Bisgard, G.E. and J.A. Neubauer. Peripheral and central effects of hypoxia. In: J.A. Dempsey and A.I. Pack (Eds.), *Regulation of Breathing*, 2nd ed., Lung Biology in Health and Disease, Vol. 79, 1995, pp. 617-668.
8. Bliss, T.V.P. and G.L. Collingridge. A synaptic model of memory: long-term potentiation in the hippocampus. *Nature* 361:31-39, 1993.
9. Bliss, T.V.P. and T. Lømo. Long lasting potentiation of synaptic transmission in the dentate area of the anaesthetized rabbit following stimulation of the perforant path. *J. Physiol. (London)* 232:331-356, 1973.
10. Bourke, D.L. and A. Warley. The steady-state and rebreathing methods compared during morphine administration in humans. *J. Physiol. (London)* 419:509-517, 1989.
11. Brown, T.H., E.W. Kairiss, and C.L. Keenan. Hebbian synapses: biophysical mechanisms and algorithms. *Ann. Rev. Neurosci.* 13:475-511, 1990.
12. Byrne, J.H. Cellular analysis of associative learning. *Physiol. Rev.* 67:329-439, 1987.
13. Casaburi, R., B.J. Whipp, K. Wasserman, and R.W. Stremel. Ventilatory control characteristics of the exercise hyperpnea as discerned from dynamic forcing techniques. *Chest* 73:, Suppl.:280S-283S, 1978.
14. Debanne, D. and S.M. Thompson. Calcium: a trigger for long-term depression and potentiationin the hippocampus. *News in Physiol. Sci.* 9:256-260, 1994.
15. Duffin, J., K. Ezure, and J. Lipski. Breathing rhythm generation: focus on the rostral ventrolateral medulla. *News in Physiol. Sci.* 10:133-140, 1995.
16. Eldridge, F.L. The North Carolina respiratory model: A multipurpose model for studying the control of breathing. (*This volume*).
17. Eldridge, F.L. and Millhorn, D.E. Oscillation, gating, and memory in the respiratory control system. In N.S. Cherniack and J.G. Widdicombe (Eds.), *Handbook of Physiology, The Respiratory System. Control of Breathing*. Bethesda, MD: American Physiological Society, 1986, sect. 3, Vol. II, part I, p. 93-114.
18. Fortin, G., J.C. Velluti, M. Denavit-Saubié, and J. Champagnat. Responses to repetititive afferent activity of rat solitary complex neurons isolated in brainstem slices. *Neurosci. Lett.* 147:89-92, 1992.
19. Fregosi, R.F. Short-term potentiation of breathing in humans. *J. Appl. Physiol.* 71, 892-899, 1991.
20. Fregosi, R.FF. and G.S. Mitchell. Long-term facilitation of inspiratory intercostal nerve activity following carotid sinus nerve stimulation in cats. *J. Physiol. (London)* 477:469-479, 1994.
21. Gallego, J., J. Ankaoua, M. Lethielleux, B. Chambille, G. Vardon, and C. Jacquemin. Retention of ventilatory pattern learning in normal subjects. *J. Appl. Physiol.* 61:1-6, 1986.
22. Gallego, J. and P. Perruchet. Classical conditioning of ventilatory responses in humans. *J. Appl. Physiol.* 70:676-682, 1991.
23. Georgopoulos, D., S. Walker, and N.R. Anthonisen. Effect of sustained hypoxia on ventilatory response to CO_2 in normal adults. *J. Appl. Physiol.* 68:891-896, 1990.
24. Gesell, R., Brassfield, C.R., and Hamilton, M.A. An acid-neurohumoral mechanism of nerve cell activation. *Am. J. Physiol.* 136:604-608, 1942.
25. Gesell, R. and Hamilton, M.A. Reflexogenic components of breathing. *Am. J. Physiol.* 133:694-719, 1941.
26. Hawkins, R.D., E.R. Kandel and S.A. Siegelbaum. Learning to modulate transmitter release: Themes and variations in synaptic plasticity. *Annu. Rev. Neurosci.* 16:625-665, 1993.
27. Hebb, D.O. *The Organization of Behavior*. New York: Wiley, 1949.
28. Ito, M. *The Cerebellum and Neural Control*. NY: Raven Press, 1984.
29. Ito, M., M. Sakurai, and P. Tongroach. Climbing fiber induced depression of both mossy fiber responsiveness and glutamate sensitivity of cerebellar Purkinje cells. *J. Physiol. (London)* 324:113-134, 1982.
30. Kamlya, H. and R.S. Zucker. Residual Ca^{2+} and short-term synaptic plasticity. *Nature* 371:603-606, 1994.
31. Khoo, M.C.K., R.E. Kronauer, K.P. Strohl, and A.S. Slutsky. Factors induing periodic breathing: a general model. *J. Appl. Physiol.* 53:644-659, 1982.
32. Kirkwood, A. and M.F. Bear. Hebbian synapses in visual cortex. *J. Neurosci.* 14:1634-1645, 1994.
33. Kirkwood, A., Dudek, S.M., Gold, J.T., Aizenman, C.D., and Bear, M.F. Common forms of synaptic plasticity in hippocampus and neocortex in vitro. *Science* 260, 1518-1521, 1993.
34. Kuo, B.C. *Automatic Control Systems*. 2nd ed., Englewood Cliffs, NJ: Prentice-Hall, 1972.
35. Li, Y., Erzurumlu, R.S., Chen, C., Jhaveri, S., and Tonegawa, S. Whisker-related neuronal patterns fail to develop in the trigeminal brainstem nuclei of NMDAR1 knockout mice. (1994). *Cell* 76, 427-437, 1994.
36. Magleby, K.L. Synaptic transmission, facilitation, augmentation, potentiation, depression. In: G. Edelman (Ed.), *Encyclopedia of Neuroscience*, Vol. 2., pp. 1170-1174, Boston: Biekhauser, 1987.
37. Malenka, R.C. Synaptic plasticity in the hippocampus: LTP and LTD. *Cell* 78:535-538, 1994.

38. Malenka, R.C. and R.A. Nicoll. NMDA-receptor-dependent synaptic plasticity: multiple forms and mechanisms. *Trends in Neural Sci.* 16:521-527, 1993.
39. Martin, P.A. and G.S. Mitchell. Long-term modulation of the exercise ventilatory response in goats. *J. Physiol. (London)* 470:601-617, 1993.
40. Mazza, E., N.H. Edelman and J.A. Neubauer. Intrinsic effects on membrane potential and input resistance of chemical hypoxia on cultured neurons from the rostral ventral lateral medulla (RVLM). *Soc. Neurosci. Abstr.* 21:1881, 1995.
41. Millhorn, D.E. Stimulation of raphe (obscurus) nucleus causes long-term potentiation of phrenic nerve activity in cat. *J. Physiol. (London)* 381:169-179, 1986.
42. Narendra, K.S. and Annaswamy, A.M. (1989). *Stable Adaptive Systems*. Englewood Cliffs: Prentice Hall.
43. Pavlov, I.P. *Lectures on Conditioned Reflexes - Twenty-Five Years of Objective Study of the Higher Nervous Activity (Behavior) of Animals, translated by W.H. Gantt.* New York: International Publishers, 1928.
44. Poon, C.-S. Optimal control of ventilation in hypoxia, hypercapnia and exercise. In B.J. Whipp and D.M. Wiberg (Eds.), *Modelling and Control of Breathing*. New York: Elsevier. 1983, pp. 189-196.
45. Poon, C.-S. Ventilatory control in hypercapnia and exercise: optimization hypothesis. *J. Appl. Physiol.* 62, 2447-2459, 1987.
46. Poon, C.-S. Optimization character of brainstem respiratory neurons: a cerebral neural network model. *Biol. Cybern.* 66, 9-17, 1991.
47. Poon, C.-S. Introduction: Optimization hypothesis in the control of breathing. In Y. Honda, Y. Miyamoto, K. Konno, and J.G. Widdicombe (Eds.). *Control of Breathing and its Modeling Perspective*. New York: Plenum, 1992, pp. 371-384.
48. Poon, C.-S. Potentiation of exercise ventilatory response by CO_2 and dead space loading. *J. Appl. Physiol.* 73, 591-595, 1992.
49. Poon, C.-S. Adaptive neural network that subserves optimal homeostatic control of breathing. *Annals of Biomed. Engr.* 21, 501-508, 1993.
50. Poon, C.-S. Hebbian synaptic plasticity: a neural mechanism of supervised learning and adaptive control. In B.W. Patterson (Ed.), *Modeling and Control in Biomedical Systems*, Madison, WI: Omnipress, 1994, pp. 497-500.
51. Poon, C.-S. Respiratory models and control. In J.D. Bronzino (Ed.), *Biomedical Engineering Handbook*, Boca Raton, Fl.: CRC Press, 1995, pp. 2404-2421.
52. Poon, C.-S. Learning to optimize performance: Lessons from a neural control system. *Preprints of the 6th IFAC/IFIP/IFORS/IEA Symposium on Analysis, Design and Evaluation of Man-Machine Systems*, Cambridge, MA, 1995, pp. 499-504.
53. Poon, C.-S. Self-tuning optimal regulation of respiratory motor output by Hebbian covariance learning. *Neural Networks*, 1996 (acceped for publication in a special issue on "Four major hypotheses in neuroscience").
54. Poon, C.-S. and J.G. Grenne. Control of exercise hyperpnea during hypercapnia in humans. *J. Appl. Physiol.* 59:792-797, 1985.
55. Poon, C.-S., Lin, S.L., and Knudson, O.B. Optimization character of inspiraotry neural drive. *J. Appl. Physiol.* 72:2005-2017, 1992.
56. Poon, C.-S., Li, Y., Li, S.X, and Tonegawa, S. Respiratory rhythm is altered in neonatal mice with malfunctional NMDA receptors. *FASEB J.* 8, A389, 1994.
57. Priban, I.P. An analysis of some short-term patterns of breathing in man at rest. *J. Physiol. (London)* 166:425-434, 1963.
58. Read, D.J.C. A clinical method for assessing the ventilatory response to carbon dioxide. *Australasian Annals Med.*, 16:20-32, 1967.
59. Richter, D.W., A. Bischoff, K. Anders, M. Bellingham, and U. Windhorst. Response of the medullary respiratory network of the rat to hypoxia. *J. Physiol. (London)* 470:23-33, 1993.
60. Robbins, P.A. Hypoxic ventilatory decline: site of action. *J. Appl. Physiol.* 79:373-374, 1995.
61. Schmidt-Nielsen, K. How are control systems controlled? *Am. Scientist.* 82, 38-44, 1994.
62. Sidney, D.A. and Poon, C.-S. Ventilatory responses to dead space and CO_2 breathing under inspiratory resistive load. *J. Appl. Physiol.* 78, 555-561, 1995.
63. Somjen, G.G. The missing error signal - regulation beyond negative feedback. *News in Physiol. Sci.* 7, 184-185, 1992.
64. Swanson, G.D. and J.W. Bellville. Step changes in end-tidal CO_2: methods and implications. *J. Appl. Physiol.* 39:377-385, 1975.
65. Thomas, A.J., L. Friedman, C.N. MacKenzie, and K.P. Strohl. Modification of conditioned apneas in rats: evidence for cortical involvement. *J. Appl. Physiol.* 78:1215-1218, 1995.

66. Wagner, P.G. and Eldridge, F.L. Development of short-term potentiation of respiration. *Respirat. Physiol.* 83:129-140, 1991.
67. Wasserman, K., Whipp, B.J. and Casaburi, R. Respiratory control during exercise. In N.S. Cherniack and J.G. Widdicombe (Eds.), *Handbook of Physiology, The Respiratory System. Control of Breathing.* Bethesda, MD: American Physiological Society. 1986, sect. 3, Vol. II, part II p. 595-620.
68. Yamamoto, W.S. Mathematical analysis of the time course of alveolar CO_2. *J. Appl. Physiol.* 15, 215-219, 1960.
69. Younes, M. The physiological basis of central apnea and periodic breathing. *Curr. Pulmonol.* 10:265-326, 1989.
70. Zhou, Z., Champagnat, J. and Poon, C.-S. Synaptic short-term depression in nucleus tractus solitarius (NTS) of rat brain stem in vitro. *FASEB J.* 9, A3283, 1995.
71. Zhou, Z., Champagnat, J. and Poon, C.-S. Intracellular calcium is required for the maintenance but not induction of long-term depression in nucleus tractus solitarius. *Soc. Neurosci. Abstr.* 21:263, 1995.

7

DYSRHYTHMIAS OF THE RESPIRATORY OSCILLATOR

David Paydarfar and Daniel M. Buerkel

Departments of Medicine and Biomedical Research
St. Elizabeth's Medical Center of Boston and
Tufts University School of Medicine
Boston, Massechusetts 02135

1. INTRODUCTION

Breathing is regulated by a central neural oscillator that produces rhythmic output to the respiratory muscles. Pathological disturbances in the rhythm of breathing can lead to prolonged apnea and severe hypoxemia. Respiratory recordings[1] of these potentially fatal episodes have shown that in some cases, most commonly in sleeping infants, the first sign of apnea is cessation of rhythmic contraction of the diaphragm without airway obstruction, suggesting that the disturbance is due to loss of rhythmicity of the brainstem respiratory oscillator. These dysrhythmias often arise unexpectedly, i.e., they are immediately preceded by a normal respiratory pattern and a normal metabolic profile. In some instances prolonged apneic pauses are associated with or triggered by brief physiological stimuli such as exposure to sound[2,3], stimulation of the face[4,5], oropharynx or larynx[6], and swallowing[7]. All infants are frequently exposed to these brief physiological stimuli, so the mechanism by which generally benign perturbations could on rare occasion cause prolonged apnea has remained a mystery.

Figure 1 shows an example[1] of prolonged non-obstructive expiratory apnea during sleep in a 30 month-old child. The respiratory pattern and oxygen saturation are normal during the first 40 seconds of the tracing followed by loss of chest wall movement and oxygen desaturation for nearly 1 min. During the apnea, there are no respiratory-related esophageal and gastric pressure changes, indicating lack of inspiratory effort, which would have been seen if the apnea was caused by acute airway obstruction. The large esophageal positive pressure wave near the onset and offset of the apnea was likely due to peristaltic activation of the esophagus several seconds after swallowing. The incidence of swallowing at the transitions between normal and abnormal respiratory rhythm has not been systematically examined. Otherwise, these tracings typify non-obstructive infant apnea syndrome.

A broad class of excitable systems can exhibit dysrhythmic behavior in response to perturbations that impinge upon the oscillation with specific combinations of strength and timing (critical stimuli)[8]. A clock-pendulum system helps to illustrate the phenomenon:

Bioengineering Approaches to Pulmonary Physiology and Medicine, edited by Khoo
Plenum Press, New York, 1996

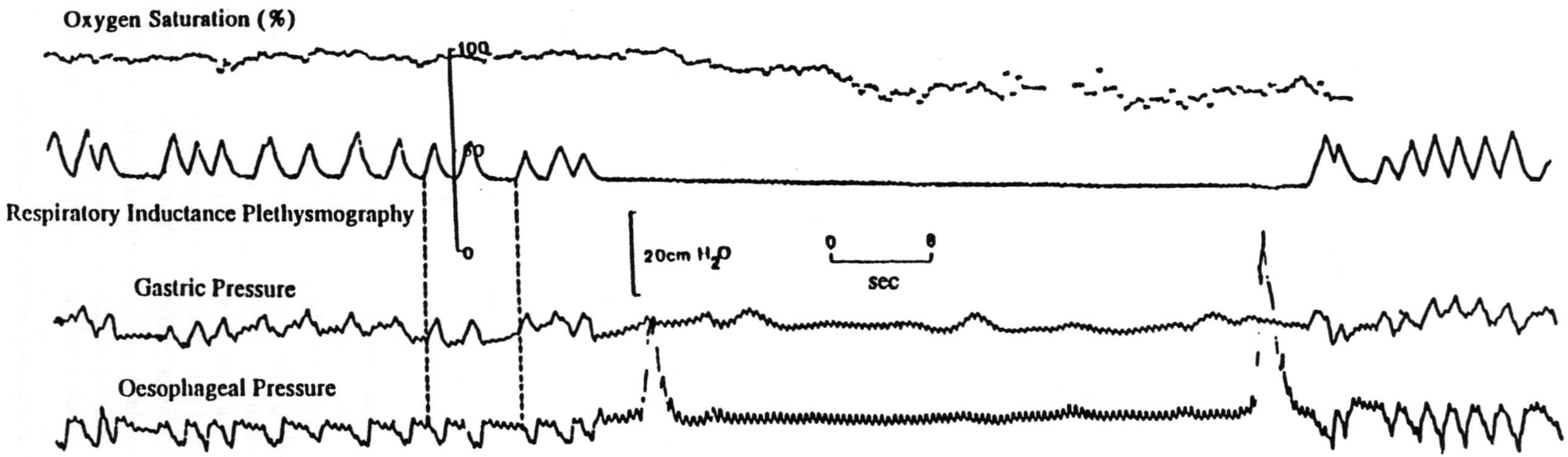

Figure 1. Example of non-obstructive sleep apnea in a 30 month-old child. The normal respiratory pattern is interrupted by progressive oxygen desaturation and apnea lasting approximately 1 min. Absence of negative esophageal pressure waves during theapnea suggests lack of inspiratory effort. Note the large esophageal pressure waves at beginning and end of apnea, likely due to peristaltic esophageal activation related to swallowing. Adapted from S.P. Southall and D.G. Talbert[1].

counter-striking the pendulum just as it passes to the bottom of its arc can stop the clock only if the strength of the impulse exactly opposes the momentum of the pendulum at that time. Other perturbations having strength or time of impact different from the critical impulse cause shifts in the phase the pendulum's rhythm of movement but do not stop the clock. Once stopped by a critical impulse, striking the pendulum with a second impulse, if strong enough, will reinitiate the normal operation of the clock.

In this chapter, we review experimental work in animals[9-11] and humans[12] that characterizes the phase resetting properties of the mammalian respiratory oscillator. In adult animal experiments, respiratory dysrhythmias can be induced by stimuli having specific combinations of strength and timing. Newborn animals readily exhibit spontaneous dysrhythmias which become more prominent at lower respiratory drives. Computational studies of the Bonhoeffer-van der Pol (BvP) equations, whose qualitative behavior is representative of many excitable systems, support a unified interpretation of these experimental findings. Rhythmicity is observed when the BvP model exhibits recurrent periods of excitation alternating with refractory periods. The same system can be perturbed to a state in which amplitude of oscillation is attenuated or abolished. Critical perturbations induce transitions between these two states, giving rise to patterns of activity that are similar to those seen in the experiments. We illustrate the importance of noise in initiation and termination of rhythm, comparable to normal respiratory rhythm intermixed with spontaneous dysrhythmias. Our hypothesis on the dynamics of the respiratory oscillator can be further evaluated in neurophysiological experiments, and may have therapeutic implications for the treatment of respiratory dysrhythmias.

2. METHODS

2.1. Animal Studies

These experiments in cats allowed for recording the rhythm of the respiratory oscillator in the absence of respiratory feedback mechanisms and influences from higher brain. Animals were anesthetized, mechanically ventilated through a tracheostomy, and paralyzed by neuromuscular blockade. In order to ablate feedback from lung and carotid body receptors, the vagosympathetic trunks and carotid sinus nerves were cut. A servo-ventilator was used to hold end-tidal PCO_2 within a narrow range ($\pm$ 0.5 Torr) around any desired level[13]. Body core temperature was monitored and servo-controlled at 37-38°C by an electronic circuit and DC heating pad. Neural respiratory rhythm was recorded from the central end of a phrenic nerve; the electrical activity was half-wave rectified and integrated for each 100 msec period by means of an integrating digital voltmeter[14]. The onset of inspiration was designated by computer as the time when integrated phrenic activity increased to a level that was twice the base-line noise level. In order to study influences of neural maturation on dysrhythmias of the respiratory oscillator, experiments were performed in six newborn kittens (age 1-10 days), using a similar preparation. Anesthesia was achieved with intraperitoneal injection of sodium pentobarbital (30mg/kg) in 2 kittens, a mixture of chloralose (40 mg/kg) and urethan (250 mg/kg) in 2 kittens. Because of the possibility that there are differential effects of anesthesia on the newborn vs. adult respiratory oscillator, therefore leading to inconclusive results on the role of maturation, intercollicular decerebration followed by removal of inhaled ether was performed in 4 kittens. Decerebration in two kittens led to severe hypotension and no phrenic rhythm. In the remaining two, the preparation was viable throughout the study. There were no qualitative differences in the findings between anesthetized and decerebrate kittens.

In the adult animal studies, two series of experiments on respiratory phase resetting were performed, distinguished by the neural structure that was stimulated: superior laryngeal nerve (SLN)[9] or midbrain reticular formation[10,11]. Each stimulus throughout an experiment was preceded by 10 control breaths, during which end-tidal PCO_2 and neural respiratory rhythm remained constant (range: ±0.5 Torr end-tidal PCO_2; ±5% change in respiratory period). After each stimulus, an additional 4-15 breaths were recorded. Initially, a specific stimulus strength was selected and given at various times in the respiratory cycle. An attempt was made to give at least one stimulus in each 5% of the respiratory cycle. On completion of this set of runs, the stimulus strength was increased or decreased and the protocol was repeated. Stimulus strength was varied by changing the duration of the stimulus train, or by changing the frequency or intensity (current) of stimulation during a fixed interval.

2.2. Human Studies

Relationships between the timing of respiration and swallowing (deglutition) were studied in 30 healthy subjects at rest[12]. Swallowing was evaluated by monitoring electrical activity of the submental muscles, pressure within the pharynx, and fluoroscopically detected movement of a swallowed bolus of barium and associated movements of the hyoid bone and larynx. Respiration was recorded by measurement of oro-nasal airflow through a pneumotachometer attached to a face mask, and chest wall movement using pressure tubing placed around the chest. Three types of deglutition were studied: injected bolus swallows, spontaneous swallows, and visually cued swallows of boluses previously placed in the mouth.

2.3. Computational Studies

Fitzhugh[15] presented a generalization of the classic van der Pol "relaxation oscillator" equation which he called the Bonhoeffer–van der Pol (BvP) model[16]:

$$dx/dt = c(y + x - x^3/3 + z)$$
$$dy/dt = -(x - a + by)/c$$

where $1 - 2b/3 < a < 1$; $0 < b < 1$; and $b < c^2$

The model is a simple representative of a broad class of non-linear systems that exhibit excitability and oscillation. The phase plane of the BvP model, depicting the two variables of state (x,y), illustrates the various states of excitability to form a "physiological state diagram" which aids in illustrating the dynamics of excitability in familiar biological terms: resting, active, and refractory phases. We implemented the BvP model as follows. The solutions of the differential equations were estimated by computer, using the Runge-Kutta difference equation method (Δt=0.01). Pseudorandom noise was incorporated into each iteration of x and y with a stochastic amplitude η. For example η of 0.01 means that a pseudorandom number between -0.005 and +0.005 was added to each iterative calculation of x and y. The noise amplitudes for x and y were independently generated. The effects of discrete perturbations on the model were studied by adding a fixed amount S_x and S_y to x and y, respectively, for a finite number of iterations. The strength of the perturbation is $|S| = (S_x^2 + S_y^2)^{1/2}$ and the direction of perturbation is $\tan^{-1} S_y/S_x$. In the present study, z was systematically varied, and a=0.7, b=0.8, c=3.0.

Excitability of the BvP system was altered by changing the z value, which allowed for study of a broad range behavior[15]. In the noiseless Runge–Kutta approximations (Δt=0.01) of the BvP equations, z>-.3356 or z<-1.4118 results in a stable singular point and no spontaneous oscillation. A discrete perturbation can lead to a single regenerative cycle (action potential) but always results in trajectories that eventually lead back to the singular point. For -0.3356>z>-0.3452 or -1.4023>z>-1.4118, the singular point is stable and trajectories within a circumscribed region (basin) converge to the singular point. However, trajectories outside the basin lead to sustained oscillatory behavior (attractor-cycles). Therefore, the approximated BvP equations can exhibit hard excitation[17]. For -1.4023 <z<-0.3452, the singular point is unstable and all trajectories eventually converge to the attractor-cycle.

The cubic term of the BvP equation of x was scaled to alter the numerical amplitude of oscillation[18] by replacing the $x^3/3$ term with $x^3/3d$; increasing d results in increase in amplitude. Unless stated otherwise, d=1. In the non-oscillatory BvP (all trajectories converging to a stable singular point), increasing d can result in spontaneous oscillation (stable attractor-cycle with a stable or unstable singular point), similar to the effects of changing z. Changing z in the oscillatory BvP, however, can produce large shifts in the position of the singular point relative to the attractor-cycle, with little change in amplitude of oscillation.

The location of the singular point of the BvP equations was determined in two ways. First, we followed the classical approach of determining the singular point by solving for the intersection of the x and y nullclines (dx/dt=0, dy/dt=0). The second approach was by Runge–Kutta approximation of the BvP equations. For non-oscillatory solutions, the computations for any x,y initial condition eventually led to constant x and y values corresponding to the coordinate of the stable singular point. This approach was also applicable to the oscillatory BvP system with stable singular points by starting with x and y values located within the singular point's basin of attraction. We found that the two methods (analytic vs computational) resulted in very close concordance; the range of differences in coordinants for the singular point was within 0.001 for x and 0.015 for y. Localization of the singular point then allowed us to determine the distance from the singular point to the approximated attractor-cycle of the BvP equations, defining D_{min} as the distance from the singular point to the most proximate portion of the attractor-cycle. If the singular point was repelling, accurate localization of the singular point by Runge-Kutta computation was difficult. As an approximation, we used the topological method of Winfree[8] to find a critical perturbation of the approximated BvP oscillation. The x,y coordinants of the singular point was determined as the x,y value at which a phase singularity was observed, i.e., a singularity in the cophase vs old phase curves of phase resetting. The accuracy of this method is limited by the resolution of old phase, perturbation, and by the errors of Runge-Kutta approximation. Increased resolution can be achieved by using smaller Δt increments.

2.4. Definitions

2.4.1. Respiratory Dysrhythmias. Expiratory apnea is identified when the chest wall and diaphragm are in the expiratory position longer than a normal expiratory period. Inspiratory apnea, also termed apneusis, is defined as tonic inspiratory activity, giving the appearance of "breath-holding". Respiratory rhythms are never exactly regular so designation of an expiratory or inspiratory period as prolonged should reflect normal variability of the respiratory cycle. In some respiratory recordings, there are low amplitude fluctuations in inspiratory or expiratory activity that are not normal and not completely arrhythmic. The distinction between normal rhythm and dysrhythmia is generally a clinical determination; the low amplitude fluctuations on the inspiratory or the expiratory side is still called apnea or apneusis because a clinically important parameter such as tissue oxygenation is compromised when the system develops such patterns.

Dysrhythmias were studied in the BvP equations with analogy to respiration as follows. BvP activity in $y(t)$ was depicted, and an activity threshold ($y=y_t$) was set which divided the observed phases into inspiratory-like activity and expiratory-like activity. Apneic-like dysrhythmia occurs if the activity remained below the activity threshold longer than the period of time expected for the unperturbed attractor-cycle below this threshold. Apneustic-like dysrhythmia was defined similarly for prolonged activity above the activity threshold. The time below the activity threshold is designated as T_b and time above as T_a, analogous to the expiratory and inspiratory periods, respectively.

In the noiseless BvP system, a state of permanent dysrhythmia is achieved when there is a stable singular point and *x*,*y* values are within the singular point's basin of attraction. We varied the position and size of the basin relative to the attractor-cycle by changing the equation parameters *z* and *d*. Transient dysrhythmia is observed if the *x*,*y* values are outside the basin, and divergence away from the vicinity of the basin to the attractor is slower than the expected period of the cycle. In the BvP system with an unstable ("repelling") singular point, the duration of dysrhythmia is similarly dependent upon the rate of divergence. This rate, and thence the duration of dysrhythmia, is influenced by all equation parameters.

2.4.2. Phases. The time from the onset of inspiration to the onset of a stimulus is defined as old phase (ϕ). Cophase (θ) is the time from the stimulus to the subsequent onsets of inspiration. The resulting times are expressed as normalized fractions, with the value of one being the average period of three control breaths prior to stimulation. Similar definitions are applied to phases of BvP oscillation, with old phase and cophase measured relative to onset of activity above y_t.

3. RESULTS

3.1. *In vivo* Studies

3.1.1. Examples of Respiratory Perturbations and Phase Resetting of Rhythm. Figure 2 illustrates different respiratory perturbations that cause transient alteration and phase resetting of respiratory rhythm. Figure 2A shows the effects of a 500 msec electrical stimulus (1 volt, 0.1 msec pulses given at 100 Hz) of the superior laryngeal nerve (SLN)[9] in an anesthetized cat. SLN stimulation during the inspiratory phase inhibits inspiratory activity and shortens the duration of the inspiratory phase of rhythm. Expiratory stimuli cause prolongation of the expiratory phase, delaying the onsets of subsequent inspirations. Figure 2B shows the effects of stimulation (1 sec duration, 10 Hz pulse frequency, 1 msec pulse duration, 600 μA mean current) of the midbrain reticular formation[10], which causes marked inspiratory facilitation. These examples show that SLN and midbrain reticular stimuli have quite different immediate effects on respiratory rhythm. However, both cause a characteristic type of phase resetting of rhythm: the latency from the stimulus to onsets of subsequent breaths (cophases) are independent of the respiratory phase of stimulus onset (old phase).

The tracings shown in Fig. 2A & B are examples of strong perturbation of respiratory rhythm in the absence of other respiratory feedback mechanisms and influences from higher brain. We also studied respiratory perturbations by swallowing in awake healthy humans[12], shown in Fig. 2C. Onset of swallowing, identified as the onset of submental electromyographic activity, is associated with inhibition of inspiratory airflow to a period of apnea, followed by expiration. The duration of expiration after swallowing is not the same in all runs: it is longest for swallows initiated at the transition between inspiration and expiration (I-E) and shortest for swallows initiated at the expiratory-inspiratory (E-I) transition. This pattern was found in all 30 subjects studied.

We have characterized the topological type of respiratory phase resetting by strong SLN and midbrain reticular stimuli, and by swallowing. Plots of cophase against old phase

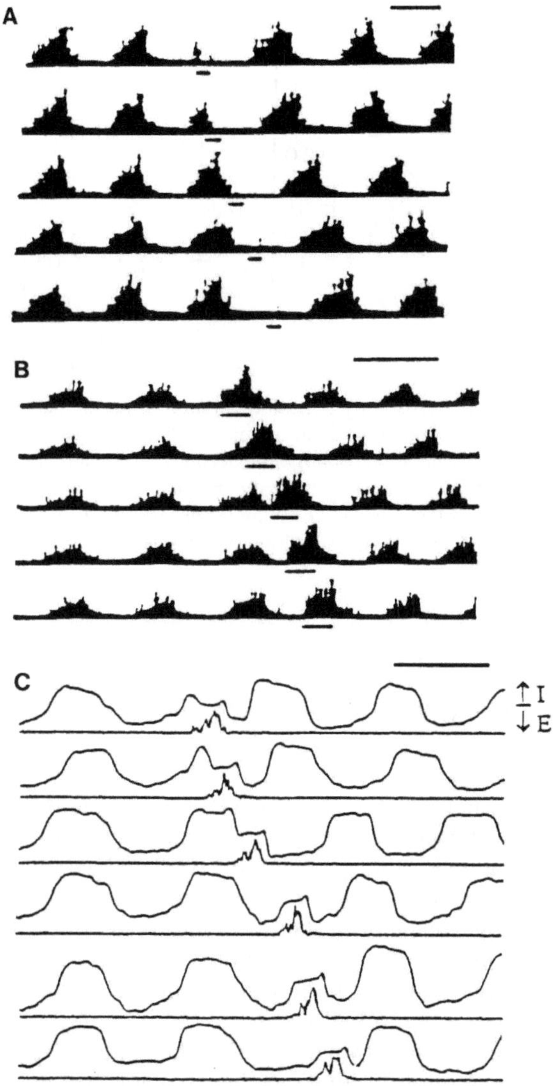

Figure 2. Phase resetting of respiratory rhythm by strong superior laryngeal nerve (*A*, stim marker: 500 ms) and midbrain reticular (*B*, stim marker: 1 s) stimulations in adult cats, and by swallowing in humans (*C*, time bar: 3 sec). In A & B, rhythm is half-wave rectified phrenic nerve activity. In *C* respiratory airflow is shown (↑ Inspiration; ↓ Expiration). Adapted from Refs. 9, 10, and 12, with permission from Am. J. Physiol. and J. Physiol. (London).

are suggestive of type 0 resetting, i.e., there is a net change of zero cycles of cophase as old phase varies through one full cycle[9-12].

3.1.2. Stimulus-Evoked Dysrhythmias of the Respiratory Oscillator. In the animal experiments, we found that stimuli having strength less than that required to produce type 0 resetting, if properly timed, could cause dysrhythmic responses[10]. Such critical stimuli were identified by giving brief stimuli of constant strength at different times in the respiratory cycle as in Fig. 2, then reducing the strength of stimulation and repeating the protocol. We

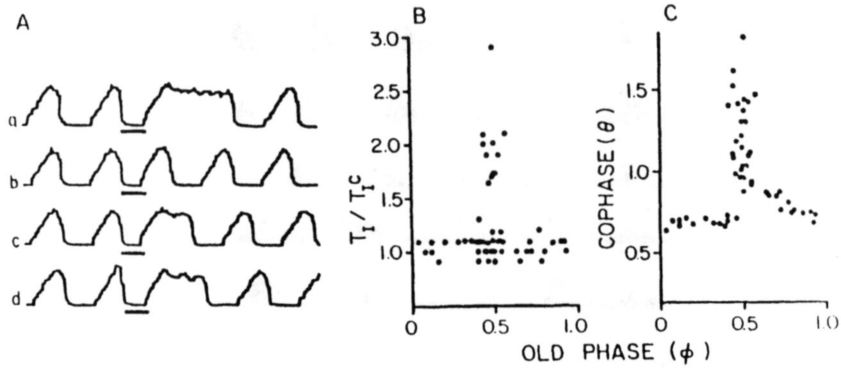

Figure 3. Respiratory dysrhythmia evoked by critical midbrain reticular stimulus in a cat anesthetized with sodium pentobarbital. *A*. The same stimulus given at approximately the same old phase (0.48 + 0.04 (SD)) had no effect on subsequent inspiratory duration (*run b*) or induced variable inspiratory prolongation (*runs a,c,d*). *B*. Plot of inspiratory duration of first breath after stimulation normalized to control inspiratory durations before stimulation, vs old phase, showing the induced dysrhythmia appears only if the stimulus is given at a critical old phase. Reprinted from Ref. 10, with permission from Am. J. Physiol.

applied the topological search methodology of Winfree[8] to identify critical stimuli that cause the rhythm to be reset in a highly irregular fashion, thereby characterizing the respiratory oscillator's phase singularity. SLN stimuli of appropriate strength caused irregular phase resetting only if given at the E-I transition, whereas midbrain stimuli caused similar irregularities only if given at the I-E transition. In the majority of experiments, "highly irregular" meant that the normal rhythm resumed after an unpredictable latency after the stimulus, but never greater than a few seconds (i.e., one normal cycle period). However, in three cats studied with midbrain stimuli and anesthetized with pentobarbital sodium (instead of chloralose-urethan which was the usual anesthetic) critically timed midbrain reticular stimuli resulted in prolonged inspiratory activity of the breath following the stimulus. Figure 3A shows an example in one cat. The same stimulus (1 sec duration, 20 Hz pulse frequency, 200 μA, 1 msec pulse duration) given at the same old phase had no effect on the subsequent inspiratory duration (*run b*) or resulted in variable inspiratory prolongation (*runs a,c,d*), lasting up to three times the inspiratory duration of the control breaths. Figure 3B (middle panel) shows a plot of the inspiratory duration of the first breath after stimulation, normalized to the mean inspiratory duration of five control breaths, versus old phase. In twelve of the 55 runs in this cat, there was significant ($P < 0.01$, unpaired Student's t test) prolongation of the inspiratory duration of the first breath after the stimulus. This prolonged response occurred only when the stimulus was given at an old phase of 0.48 + 0.04 (SD), which corresponds to stimuli initiated at the offset of inspiration. This plot therefore shows that post-stimulus prolongation of inspiratory activity occurred only when the stimulus was given at a specific time in the respiratory cycle. When the stimulus strength was increased to 50 Hz (other parameters unchanged) phase specific inspiratory prolongation did not occur, and strong (type 0) resetting similar to Fig. 2B was found.

3.1.3. Spontaneous Dysrhythmias of the Respiratory Oscillator. In experiments in newborn kittens there were episodes of variable phrenic rhythm due to prolonged expiratory apnea. Figure 4 shows examples in two kittens, one anesthetized with chloralose and urethan (Fig. 4A), the other with sodium pentobarbital (Fig 4B). Initially, the end-tidal PCO_2 was lowered to a level just below threshold of phrenic nerve activity, the apneic threshold. The

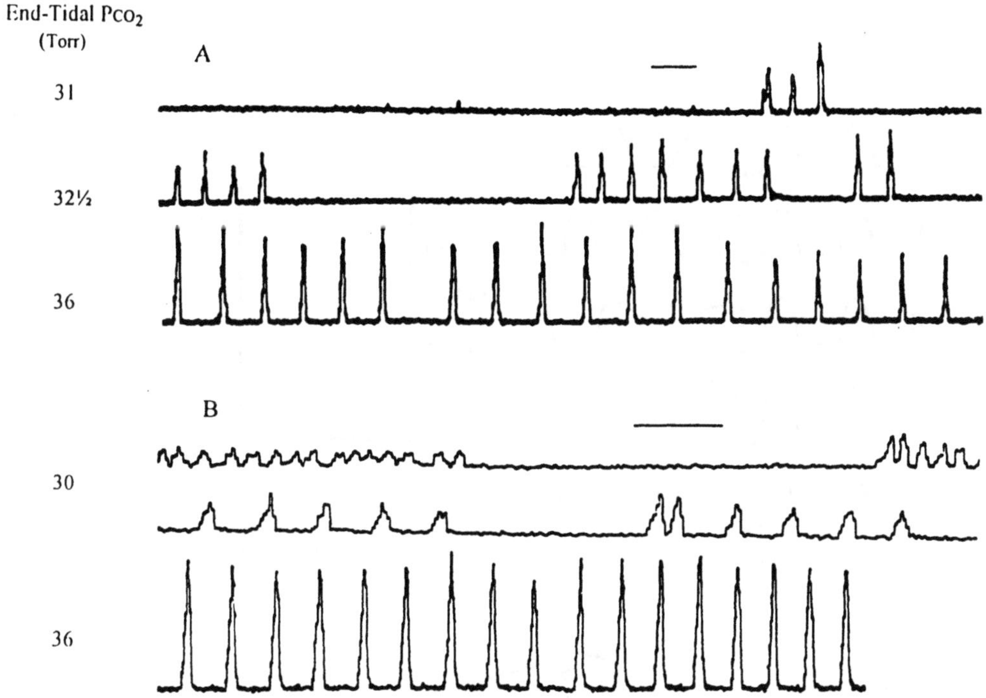

Figure 4. Spontaneous dysrhythmias of newborn respiratory pattern in two kittens. Tracings are of integrated phrenic nerve activity at different levels of constant end-tidal PCO_2. *A.* 4-day old kitten anesthetized with sodium pentobarbital. Time bar: 3 sec. *B.* 2-day old kitten anesthetized with chloralose and urethan. Time bar: 3 sec for top two tracings, 6 sec for bottom tracing.

ventilator's rate was then slowed to cause step increases in end-tidal PCO_2; no further changes were made for at least 10 minutes, allowing for near-steady state phrenic recordings.

The two examples show that spontaneous apneas were more likely to appear, and were of greater duration, at respiratory drives that were near the threshold for rhythmicity compared to higher drives, readily demonstrated by changing end-tidal PCO_2 in a step-wise manner. Apnea was often immediately preceded and followed by apparently normal breaths. Figure 4B shows an example of low amplitude rapid phrenic rhythms that were sometimes seen just above apneic threshold.

2.2. Computational Studies

2.2.1. Dysrhythmic Behavior Induced by Critical Perturbations. The approximated BvP equations exhibit sustained oscillation over a range z values ($-0.3356>z>-1.4118$). In order to find specific combinations of stimulus strength and stimulus timing that induce dysrhythmia, we applied the same topological search method that was used to find critical stimuli in the animal studies. We depict rhythm in $y(t)$ with activity threshold at $y_t=1.7$, and set $z=-1.34$ which places the singular point just below the maximum y value for the attractor. The system was perturbed with a critical stimulus ($|S|=0.06$, direction 2.94 radians) lasting a finite fraction of the cycle (0.25). The direction of perturbation is oriented to facilitate y activity, analogous to the inspiratory-facilitatory effects of reticular stimuli. We found that strong perturbations cause patterns of resetting similar to those shown in Figure 2. By systematically varying strength and timing of the stimulus, we found a unique stimulus which

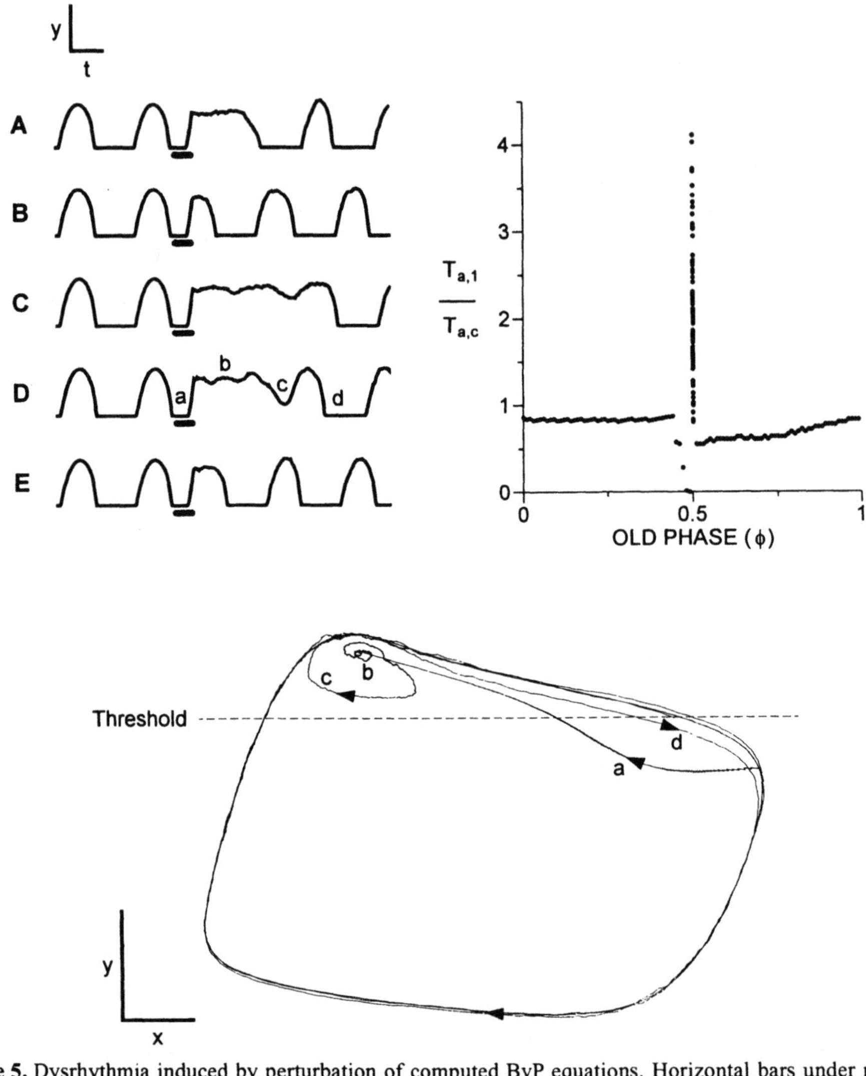

Figure 5. Dysrhythmia induced by perturbation of computed BvP equations. Horizontal bars under rhythm tracings represent interval of perturbation. *A-E:* Perturbations ($|S|$=0.06, direction 2.94 radians, duration 0.25) given at the same old phase (ϕ=0.5) cause highly variable resetting, sometimes delaying offset of T_a. Noise (η=0.05) is implemented at end of stimulation. Unit bars: y=1, t=5. Graph (*top right*) of first T_a after onset of stim, normalized to control T_a, vs old phase showing the dysrhythmic response only for perturbation at old phase of 0.5. Trajectories (*bottom*) before, during and after perturbation for run *D*. Perturbation (*a*) caused displacement of the trajectory to a locus near the singular point (*b*), followed by spiral orbits (*b*, *c*) away from singular point toward attractor-cycle (*d*). Horizontal threshold: y_t=1.7. Unit bars: x=0.5, y=0.5.

caused highly irregular resetting of the oscillator. Figure 5 shows six runs (A-E) of the simulated equations, with one full cycle before five perturbations having the same strength and old phase. These perturbations cause an unpredictable pattern of resetting, with some perturbations causing prolonged cessation of rhythmic activity, up to 4 times the duration of the positive phase of the control cycles. Figure 5 (*top right*) shows a plot of first T_a after onset of stimulation, normalized to control T_a, versus old phase. The dysrhythmic response occurs only after perturbations given at an old phase of 0.5. In this example, the BvP equations were computed with pseudorandom noise added to each calculation. Without noise, each perturbation given at ϕ=0.5 would have resulted in the same response. The addition of noise leads to highly variable responses to this stimulus which perturbs the oscillator to a locus very close to its singular point. Figure 5 (bottom) shows the trajectories of the oscillator before, during and after perturbation for the fourth run (D). The large closed-loop trajectory represents the stable attractor-cycle prior to perturbation. Perturbation (a) causes displacement of the attractor to a locus near the singular point (b). There are subsequent spiral orbits (b,c) away from the singular point leading back (d) to the stable attractor-cycle. In the example shown, the *z*-value (=-1.34) of the BvP equation was chosen to place the singular point on the "inspiratory" side of the attractor relative to the threshold $y_t = 1.7$. The singular point in this example is unstable, i.e. trajectories arbitrarily close to the singular point diverge back towards the attractor-cycle. These critical stimuli cause responses of the model oscillator that are similar to the stimulus-evoked apneusis shown in Fig. 3. Fine adjustments of the stimulus direction was required to mimic the experimental results of dysrhythmia evoked at stimulus old phase of 0.5. However, the finding of a specific combination of stimulus strength and timing which induces dysrhythmic behavior is a general feature of attractor-cycle systems[8]. Changing the equation parameters and direction of perturbation alters the combination of stimulus strength and timing needed to induce dysrhythmia, and the duration of dysrhythmia (see Sec.2.4).

2.2.2. Spontaneous Dysrhythmias. Figure 6 illustrates this effect in the noisy BvP equations (z=-0.34, η=0.02), depicting $y(t)$ and x,y plots. The activity threshold (y_t) was set below the minimum $y(t)$ so that the entire range of cycling is seen. The conversion by increasing d of steady state to oscillation is not all-or-none. Figure 6B & 6C shows intermediate states in which variable periods of rhythmic activity are interrupted by variable periods of very low amplitude fluctuations. Figure 6D shows that increasing d further leads to relatively regular rhythm. The x,y trajectories in B and C show that the periods of dysrhythmia are associated with low amplitude orbits near the attractor-cycle orbit. These smaller orbits circumnavigate around the singular point of the system and at times there is true arrhythmia when the trajectory is so close to the singular point that the observed fluctuations are indistinguishable from noise. In the BvP system, patterns similar to Fig. 6B & 6C are readily generated if the singular point is close to the attractor-cycle and sufficient noise is added to the system. Figure 6 (bottom graph) shows that increasing d above 1 results in increases in D_{min} (calculated from the noiseless BvP system, see Methods). Below d=1 the singularity point is stable and there is no oscillation. One explanation for the pro-rhythmic effects of increasing d is that by increasing the distance from the singular point to the attractor, the noise is less likely to displace the trajectory off the attractor-cycle towards a dysrhythmic locus around or at the singular point. Another consideration is the stability of the singular point. Dysrhythmia is favored if there exists a basin within which trajectories converge to singular point, allowing perturbations of sufficient strength to displaces the trajectory off the attractor-cycle to this basin. On the other hand, dysrhythmia would be less prominent if the singular point is unstable. In this situation displacement off the attractor-cycle toward the singular point would always leads to trajectories that diverge away from the singular point back toward the attractor-cycle. However, the existence of a repelling (unstable) singular point does not preclude dysrhythmic behavior because trajectories near the singular point might take a long time (e.g. more than a "normal" cycle length) to return to the attractor-cycle. Noise or other perturbations can be prorhythmic. For example if the system is at the stable singular point, noise of sufficient magnitude can cause displacements outside the basin. In order to further characterize these relationships, we studied the effects of noise on BvP activities for different levels of excitability (z).

Figure 7 illustrates the remarkable influences of noise (η) on rhythmicity of the BvP model. Noise can convert a non-oscillatory system to one that exhibits spontaneous cycles.

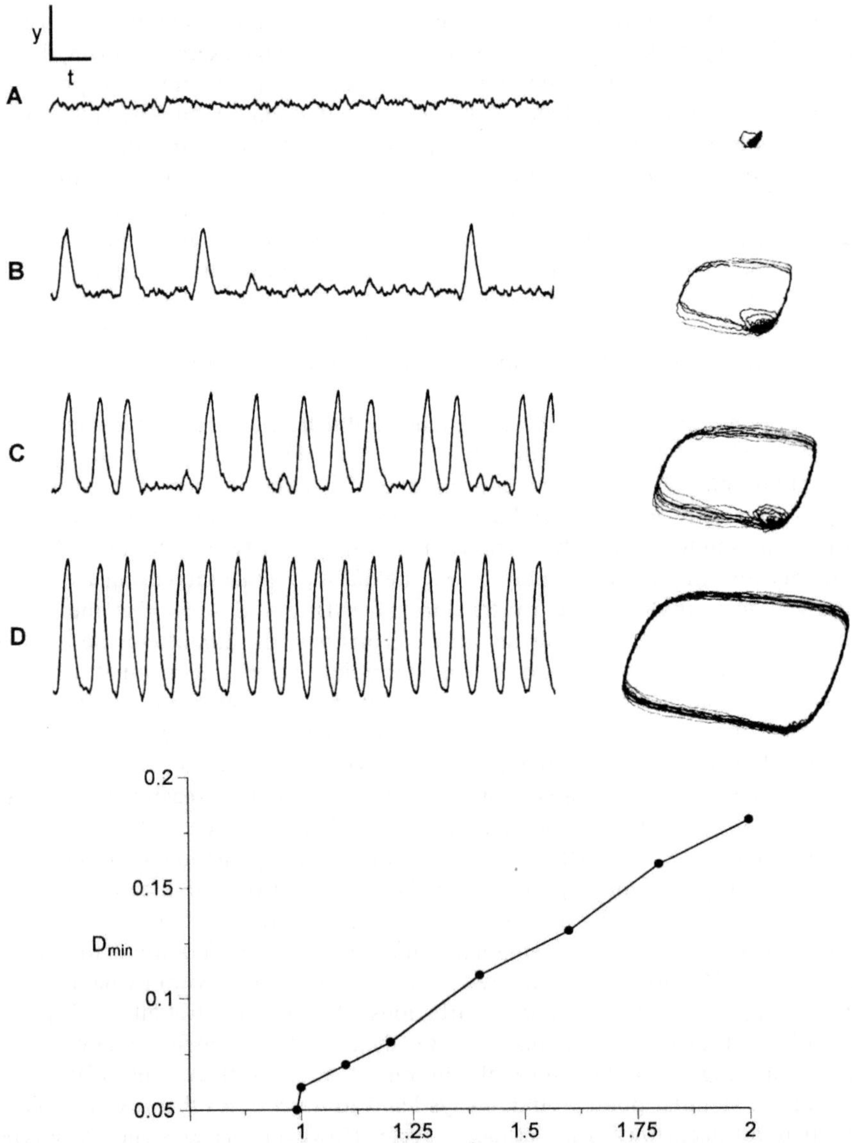

Figure 6. Spontaneous dysrhythmia: effect of changing *d*. *Top*: *y*(*t*) and *x,y* trajectories of the computed BvP equations (z=-0.34) with noise (η=0.02) for *d* value of 0.2 (*run A*), 0.5 (*run B*), 1.0 (*run C*) and 2.0 (*run D*). Increasing *d* results in conversion of the activity from continuous dysrhythmia (*A*) to episodic dysrhythmia (*B,C*) and regular rhythm (*D*). Unit bars: y=1, t=15. *Bottom:* Plot of the distance from the singular point to the most proximate portion of the attractor-cycle (without noise) for values of *d* from 1 to 2. Lower values of *d* result in no spontaneous rhythm in noiseless system.

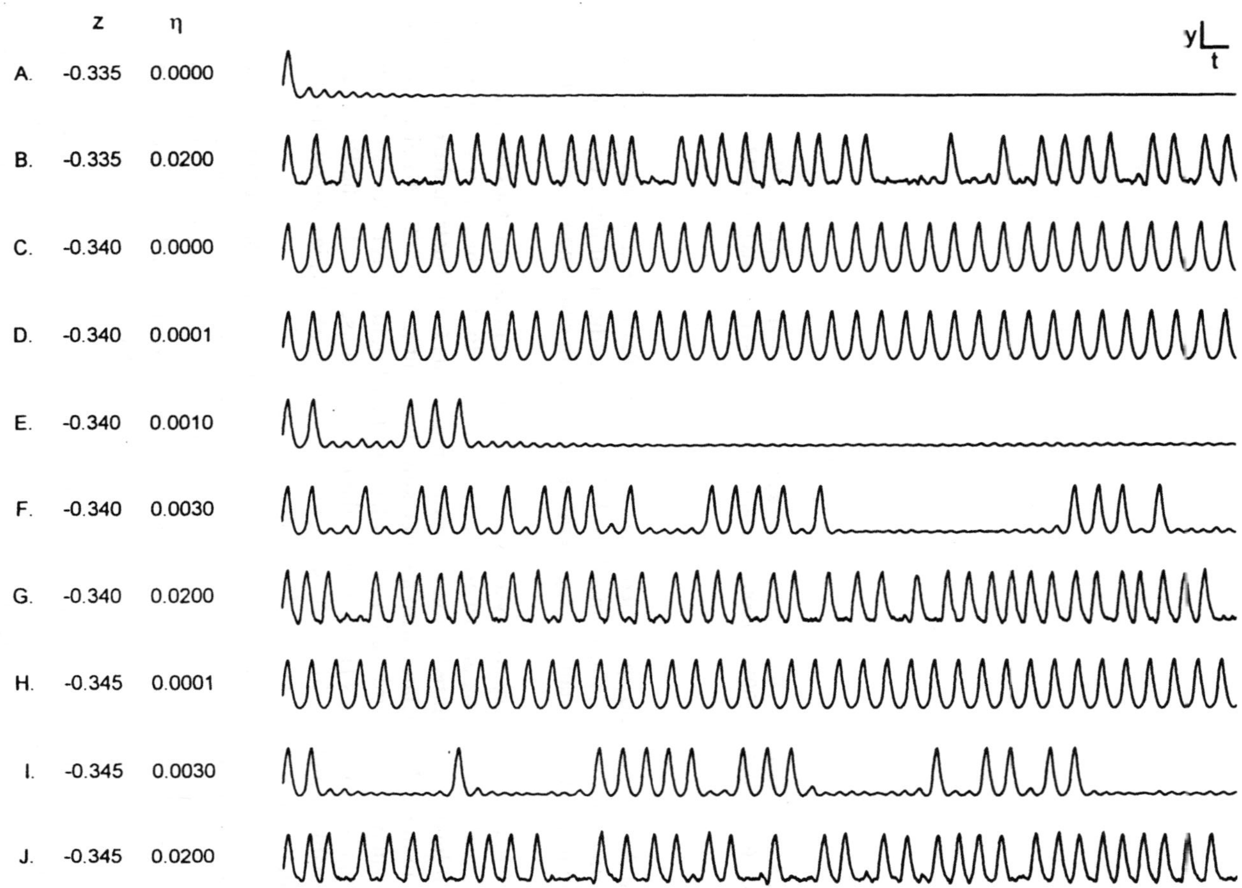

Figure 7. Critical effects of noise (η) on rhythmicity of the BvP model. Noise converts non-oscillatory sytem to one that exhibits spontaneous cyc.es (*A, B*). The Oscillatory system (*C-G, H-J*) exhibits increased dysrhythmia with certain levels of noise further increases above this critical level results in decreased dysrhythmia. Unit bars: $y=1$, $t=15$.

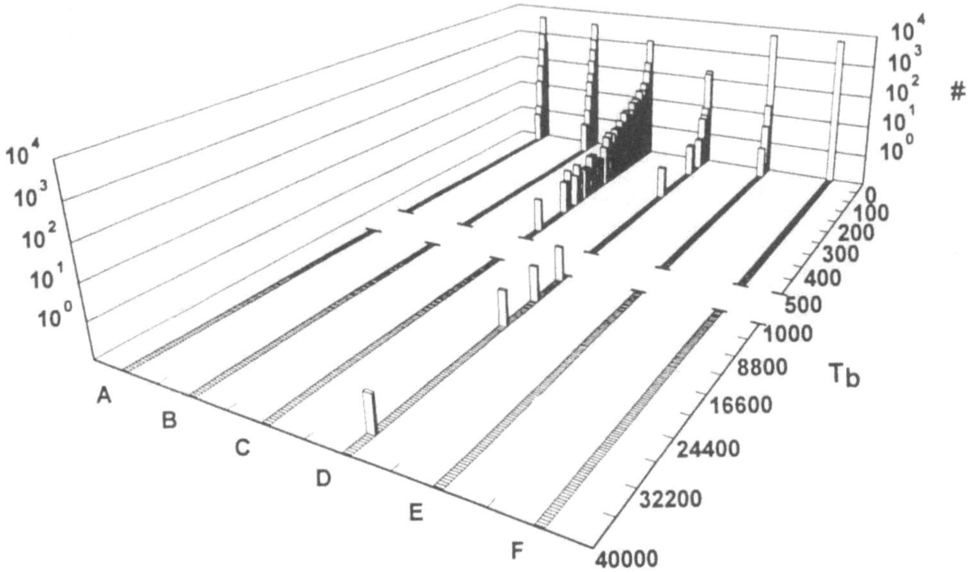

Figure 8. Histograms of T_b for different amounts of noise in the BvP model (z=-0.340). Noise levels (η): *A.* 0.05 *B.* 0.02 *C.* 0.003 *D.* 0.001 *E.* 0.0003 *F.* 0.00006. The noiseless cycle has period=13, T_a=4, T_b=9.

In run A, there are no spontaneous oscillations (z=-0.335) and a steady state is reached, reflecting a stable singular point. Run B shows that addition of noise (η=0.02) causes the system to exhibit spontaneous regenerative cycles with spontaneous episodes of dysrhythmia.

In the oscillatory BvP system increasing noise leads to increases in the incidence and duration of dysrhythmia up to a critical level of noise. Further increases in noise above this critical level results in reduction in dysrhythmia. Runs C-G illustrate the critical effects of noise in the BvP oscillator with a stable singular point (z=-0.3400). A noise level of 0.001 results in permanent attenuation of rhythm; the tiny fluctuations reflect small orbits around the singular point. Runs H-J shows similar effects of noise even for the BvP oscillator with an unstable singular point (z=-0.3450), although the dysrhythmias are much less prolonged than those associated with a stable singular point.

Figure 8 shows the effects of changing noise on the cycle periods in the BvP oscillator with a stable singular point (z=-0.340), using 6 different noise levels. The histograms represent the distribution of times (T_b) spent below an activity threshold (y_t) midway between the maximum and minimum y value (transecting the attracter in half). For each noise level, computation proceeded for 75,000 time units, roughly equivalent to 5,800 cycles of the noiseless rhythm (cycle period=13 time units) with the same parameters, and analogous to 6-8 hours of real breathing. The histograms show several features of the effects of noise on periodicity: 1) criticality, as illustrated in the shorter runs of Fig. 7; 2) skew distribution of the variability in T_b for a given noise level; 3) asymmetry of noise effects on T_b below and above the critical noise level, i.e., below the critical noise level, a small increase in noise causes transition of the pattern from very regular to highly dysrhythmic, whereas above the critical noise level, reduction in dysrhythmia by increases in noise are gradual rather than abrupt. Histograms of the times above the activity threshold (T_a) for the same set of runs show very little variability.

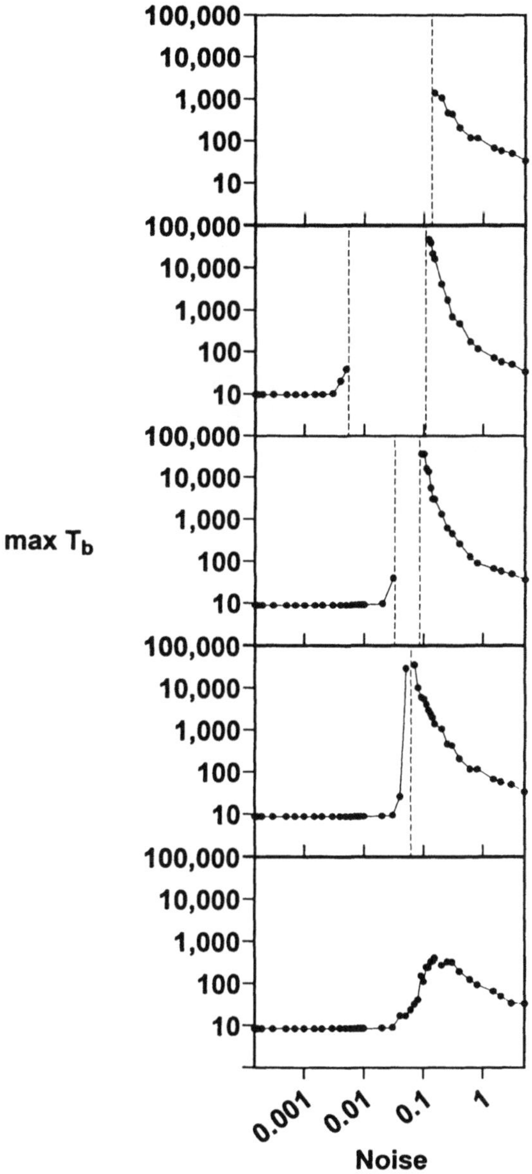

Figure 9. Relationship between max T_b and h for five z values. A. z = -0.3341 B. z = -0.3364 C. z = -0.3401 D. z = -0.3421 E. z = -0.3451. Vertical lines demarkate noise values for which there are no cycles of activity throughout the entire period of computation (t = 75,000).

Let max T_b be the longest T_b of the run, analogous to the longest apnea of a continuous 6-8 hour respiratory tracing. Figure 9 shows the relationship between max T_b and noise level for different levels of excitability (z) including non-oscillatory (9A), oscillatory with stable singular point (9B-D) and oscillatory with unstable singular point (9E). Criticality and asymmetry in the relationship between noise and max T_b are shown in these plots, similar to the full distribution of the T_b histograms. Another feature is noteworthy: the longest

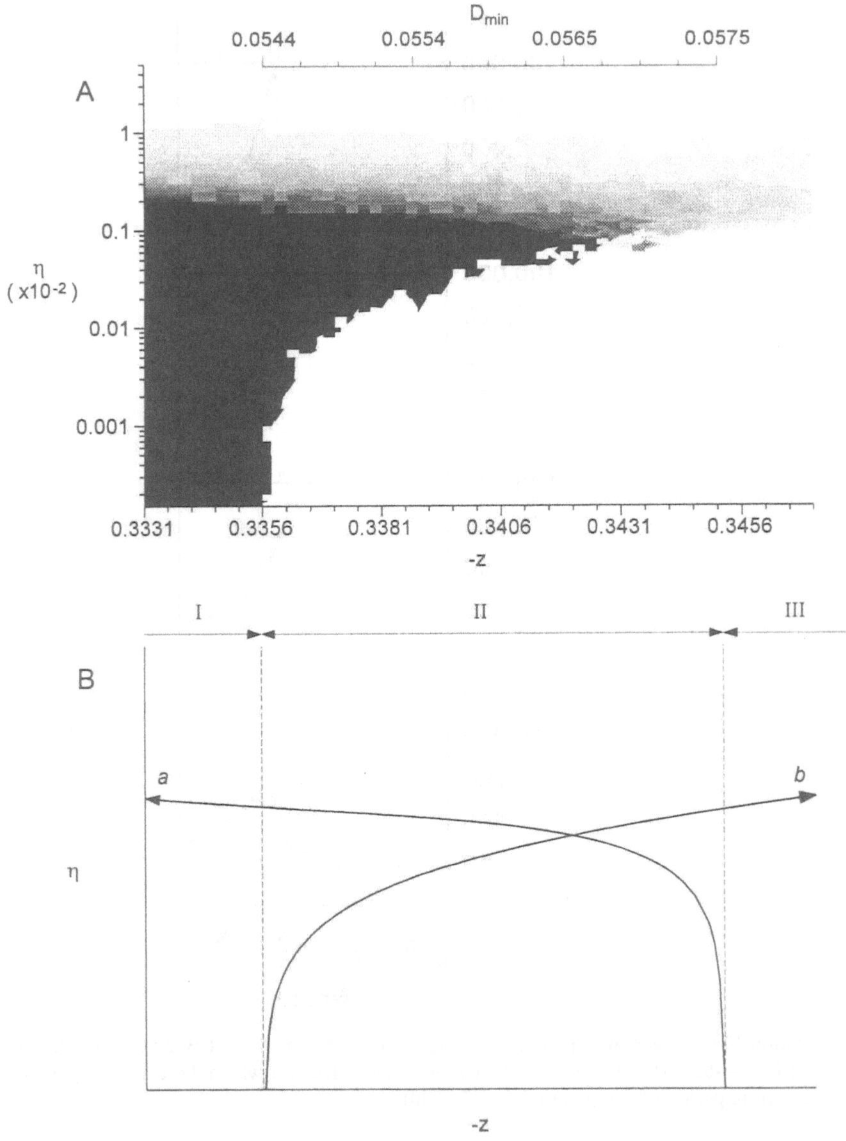

Figure 10. Relationships among max T_b, noise (η) and excitability (z). *A*. Gray-scale plot of max T_b (t=75,000) for 1,935 combinations of η and z. Gray-scale White represents max T_b < 50, Black is max T_b > 12,000 units of time. Top scale shows D_{min}, defined as the distance from the singular point to the most proximate portion of the attractor-cycle. *B*. Estimated threshold curves for noise levels η(z) that result in initiation and termination of BvP activity cycles. Curve *a* is threshold below which activity cycles never take place if the initial x,y values are at the singular point. Curve *b* is threshold below which oscillation is never interrupted by dysrhythmic activity. The scales for z and η are the same as in *A*.

"apnea" is nearly the same at the highest noise levels for all plots. This underscores the finding that addition of noise to the non-oscillatory BvP system (z>-0.3356 or z<-1.4118) in sufficient amounts (0.05 or more in these examples) can produce an activity pattern that is indistinguishable from the pattern of the spontaneous oscillator in the presence of the same amount of noise.

In Figure 10A, we present the relationships among max T_b, noise and excitability for 1,935 combinations of η (range 1.5×10^{-6} to 5×10^{-2}) and z (range -0.3331 to -0.3471). The maximum T_b for each η, z combination is depicted as a shade of gray, with white being shortest (< 50 time units, equivalent to approximately 5 cycles of oscillation) and black being longest (> 12,000 time units). For z>-0.3356 is no rhythmic activity up to a threshold of noise; increasing η above this threshold results in increased rhythmicity and progressive shortening of max T_b. For z< -0.3356 the plot shows an abrupt transition from eurhythmia to permanent dysrhythmia for intermediate noise levels and graded reductions in dysrhythmia with further increases in noise. The blending of gray tones across the top of the plot reflects the fact that in the presence of sufficient amounts of noise, the activity patterns of the BvP equations over a broad range of z values are indistinguishable and exhibit the same distribution of T_b values.

In Figure 10B, we have estimated the thresholds for initiation and termination of activitiy cycles of the BvP equations, computed for t=75,000. The z domains are labeled to indicate noiseless BvP solutions with a stable singular point & no oscillation (I, z>-0.3356), stable singular point & attractor-cycle (II, -0.3356>z>-0.3452) and unstable singular point & attractor-cycle (III, -0.3452>z>-1.4023). Curve a is the threshold $\eta_a(z)$ below which activity cycles never take place if the equations are initially set at the singular point. Above this threshold, noise is of sufficient magnitude to eventually cause a large enough displacement away from the stable singular point to form a regenerative cycle. Below this threshold, noise results in small fluctuations of x,y activity around the singular point, i.e., dysrhythmia. Curve b is the threshold $\eta_b(z)$ below which oscillation is never interrupted by dysrhythmic activity because noise does not displace the activity cycles far enough away from the attractor-cycle. Permanent dysrhythmia of the oscillatory BvP is identified in the η,z domain only if two conditions are satisfied: noise is large enough to initiate dysrhythmia ($\eta>\eta_b(z)$) and too small to terminate dysrhythmia ($\eta<\eta_a(z)$).

4. DISCUSSION

Irregularities in the pattern of breathing are most often pathological during sleep[19], a time when respiratory drive is reduced. In adults, the most common form is obstructive apnea, due in part to sleep-induced reduction in inspiratory phasic and tonic activities of the upper airway muscles, allowing for them to occlude the airway. These muscle activities are sleep state dependent, so that once occlusion is initiated, arousal causes upper airway muscle activation thereby relieving the obstruction. Recurrent cycles, not necessarily regular, of obstructive apnea terminated by arousal can take place throughout the night.

Another type of sleep apnea, most often seen in the infant, is sudden loss of phasic inspiratory effort[1], suggesting that there is loss of rhythmicity of the brainstem respiratory controller. Identifiable neurological causes of recurrent non-obstructive infant apnea include focal cortical seizure activity[20] and posterior fossa lesions[21]. However, in current clinical practice most cases remain unexplained after an extensive evaluation[1].

Apnea of central neural origin as a cause sudden infant death syndrome (SIDS) has been speculated for a long time, but definitive evidence remains elusive. A number of neuropathological studies have shown brainstem abnormalities in some SIDS victims[22]. Specific findings include increases in dendritic spine density, delayed myelination and adrenergic alterations that are suggestive of brainstem immaturity. These abnormalities are not easily recognized on routine neuropathological examinations and even in the investigational studies there is overlap between the normal age matched population and SIDS victims. Direct investigation of apnea in SIDS has so far proven extraordinarily difficult because of the rarity of observing the life-threatening event under conditions of intensive monitoring. A recent prospective epidemiological study has found increased risk of SIDS in infants sleeping in the prone position[23].

Our study on respiratory dysrhythmias should be viewed in this broader context of apnea and SIDS findings. The computational studies illustrate how an apparently normal respiratory pattern can be interrupted by apnea of the non-obstructive variety. These studies suggest that the apneic state can be induced by stimuli with specific combinations of strength and timing that impinge upon or arise within the respiratory oscillator. Dysrhythmias manifest as inspiratory apnea (apneusis) or expiratory apnea depending upon the dynamical structure of the respiratory oscillator: apneusis results when the singular point is situated near the inspiratory side of the attractor-cycle, whereas expiratory apnea occurs when the singular point happens to be below the inspiratory threshold. Small fluctuations in respiratory activity, clearly dysrhythmic but not arrhythmic, reflect small orbits expanding away from or contracting towards the singular point. Because we consider respiratory dysrhythmias as a disorder of state, a variety of parameter changes can result in the same pathological state, and there is no well-defined cutoff between "normal" and "abnormal". This may be relevant in the interpretation of clinical and pathological studies showing considerable overlap between "normal" infants and SIDS victims or those who are felt to be at high risk for SIDS.

4.1. Respiratory Rhythm Viewed as an Attractor-Cycle

Attractor-cycle oscillators have been the basis for many mathematical models of respiratory rhythm[11,24-26]. Experimental support for a respiratory attractor-cycle was published in the mid-1980s in which phase resetting of respiratory rhythm was studied by using respiratory perturbations with various combinations of stimulus strength and stimulus timing[9-11]. The data were analyzed using Winfree's definitions and topological classifications of phase resetting. The functional relationship of cophase was determined with respect to old phase and stimulus strength, defining the respiratory oscillator's resetting surface. The experimental data suggested that this surface is a helicoid that winds around a vertical axis and has contours of cophase that converge to this axis. Thus the axis represents a singularity in graphical terms. It is a phase singularity in dynamical terms because rhythmicity becomes more variable as stimuli are given closer to the point of convergence of the contours.

In the present investigation, we have developed the BvP metaphor to illustrate how attractor-cycle systems can exhibit the main features of dysrhythmia observed in the animal studies: 1) Critical stimuli, with only certain combinations of strength and timing, can induce dysrhythmias, and the stimuli that characterize the dysrhythmic response are intermediate in strength to stimuli that identify type 1 and type 0 resetting[10]; 2) As shown by Lumsden[27-29] spontaneous apneustic breaths are of varying duration and are often immediately preceded and followed by apparently normal breaths. This same pattern is seen in the newborn animal except the dysrhythmic state is expiratory rather than inspiratory apnea; 3) Rapid low amplitude oscillations of phrenic activity during expiratory apnea can be seen in newborn animals (see Fig. 4). As illustrated by Euler[30], apneustic neural inspiratory activity can also exhibit low amplitude oscillatory behavior having periodicity that is a fraction of the normal rhythm, approximately 1/3 in Euler's example.

The Bonhoeffer–van der Pol model of excitability was selected for study mainly because of its analytic simplicity and the broad range of excitable behaviors that it exhibits. The BvP singular point can be positioned off-center near one side of the attractor or the other (relative to an arbitrary threshold) by changing a single parameter (z). This proved convenient for illustrating dysrhythmic activity on either side of the cycle, analogous to inspiratory versus expiratory apnea. It is important to note that for a given z value, dysrhythmia occurs at or near a single state of activity. Indeed in respiratory recordings, spontaneous expiratory and inspiratory apneas rarely co-exist in the same experiment, supporting the idea of a unique singular point.

Other dynamical schemes have been proposed for the respiratory oscillator, notably integrate-and-fire and discrete models which have discontinuous dynamics. These models would be expected to exhibit discontinuities in phase resetting plots of cophase against old phase[32]. Indeed, some respiratory resetting plots appear discontinuous[9,31,33] and experimental limitations prevent finer resolution of the data in some experiments. Other experiments, particularly those with midbrain reticular stimulation[10], show the full series of resetting plots with much greater resolution and the appearance of continuity. To our knowledge, the characteristics of respiratory dysrhythmia reviewed in the present study have not been demonstrated in integrate-and-fire or discrete models.

4.2. Importance of Noise

The existence of stochastic processes over a broad time scale is a ubiquitous feature of neural systems[34-36], demonstrated in experimental work that has identified irregular fluctuations in the activity of neural systems due to membrane, synaptic and network properties. Noise has been proposed to have critical effects on neural sensory transduction[36] and phase locking of biological oscillators[37].

In the present study, we have shown that stochastic fluctuations of the BvP system can lead to spontaneous dysrhythmias of varying durations. Our demonstration of critical levels of noise that are most pathological, i.e., produce the longest spontaneous dysrhythmias, can be viewed as noise levels in which stochastic combinations of strength and timing are most likely to perturb the trajectory away from the idealized attractor to the singular point or its dysrhythmic vicinity. If the level of noise is too large, stochastic fluctuations in state promote dispersion away from the singular point. Very small increases in noise can, for a certain range of excitability (z), cause a very dramatic transition in the activity pattern from highly regular rhythm to a permanent dysrhythmic state. Increases in noise above the critical level(s) lead to progressive reductions in dysrhythmic periods. In our computations, pseudorandom (deterministic) perturbations were uniformly distributed over a specified interval of amplitude and given for each iteration. We speculate that use of truly random perturbations with other distributions of amplitude and timing could shift the η,z domains of maximum dysrhythmia but criticality would still be found. Whether noise within or impinging upon the oscillator is deterministic or truly random has little bearing on the actual pathological state of dysrhythmia in the BvP system. More important is the distribution of amplitudes and timings of the fluctuations that perturb the oscillator.

The critical effects of noise on the initiation and termination of dysrhythmias is shown to be dependent on the level of excitability (z). The dysrhythmias are most prominent, *i.e.*, are of greatest duration and initiated with smallest amplitudes of noise, for z values corresponding to a singular point situated closest to the attractor-cycle. In this regard, our findings are consistent with those of Kurrer and Schulten[38], who studied the influence of Gaussian white noise on stability of the BvP oscillator, using statistical approximation methods. They concluded that noise normal to the attractor can have large effects on the trajectories and produce large shifts in the phase of the attractor-cycle ("quasistationary behavior"), whereas noise tangential to the attractor-cycle induces diffusion-like dispersion of trajectories. These shifts were most prominent for "critical" z values (i.e., close to values corresponding to a stable stationary point) and for portions of the attractor most proximate to the singular point.

4.3. Swallowing: a Physiological Perturbation of Respiratory Rhythm

The human studies have shown that in healthy subjects, swallowing causes phase resetting of respiratory rhythm with a pattern that we have characterized topologically as type 0. The latency from the swallow to the next breath (cophase) was largest for swallows initiated near the inspiratory-expiratory (I-E) transition, and smallest for swallows initiated near the expiratory-inspiratory (E-I) transition. We have related these respiratory phase effects with the timing of bolus flow through the pharynx during deglutition. The latency between departure of the bolus from the larynx to onset of next inspiration is also largest for I-E swallows and smallest for E-I. This finding raises the possibility that aspiration due to inspiration of swallowed material into the lung is most likely to occur for swallows initiated near the E-I transition, especially in pathological conditions that weaken the impact of swallowing on respiratory rhythm or slow the transport of the bolus through the pharynx.

In adults, aspiration normally causes rapid expiratory activation and coughing, aiding the expulsion of aspirated material from the airway. The infant, however, often lacks a vigorous cough reflex and can exhibit a different response to aspiration: prolonged apnea[5]. This response is thought to be due to stimulation by the aspirant of laryngeal and tracheal receptors that in turn causes reflex inhibition of respiration. Obstructive apnea due to laryngospasm, prolonged tonic closure of the vocal folds and cords, can also result from irritation of the laryngeal mucosa by the aspirant.

The phase of transition between expiration and inspiration was also shown in the animal studies to be the time at which irregular resetting was demonstrated with superior laryngeal nerve stimulation having critical strength, suggesting that this response represented the respiratory oscillator's phase singularity.

We propose that swallowing causes perturbations of the respiratory oscillator that are similar to strong SLN stimuli. Two forms of respiratory dysrhythmia could then result from conditions that weaken the normal (type 0) resetting effect of deglutition: apnea triggered by aspiration, or apnea due to perturbation off the respiratory attractor toward the singular point.

4.4. Prospects

What are the biological parameters that influence the frequency and duration of dysrhythmias of the respiratory oscillator? One factor that has already been implicated is respiratory drive which is reduced during sleep, a well know period of vulnerability for the infant. The preliminary experiments in newborn animals suggest that maturation of the neural circuits that make up the respiratory oscillator might have some bearing on the dynamical structure of the respiratory attractor and its singular point. In the BvP oscillator, dysrhythmia was readily induced by small perturbations if the basin of attraction to the singular point was large and the basin was near the attractor-cycle. Such might be the case in some infants. Neural maturation might result in migration of the singular point away from the attractor cycle and conversion of the singular point from one with a basin of attraction to a strongly repelling singular point, rendering the mature oscillator much less vulnerable. The spectrum of neural noise impacting on or arising within the newborn respiratory oscillator might be much closer to the critical domain for dysrhythmia.

Therapeutic interventions for infant apnea can be proposed based on the postulated effects of noise. If infant apnea is initiated by perturbation of the respiratory trajectory away from the respiratory attractor-cycle towards a singular point, then increasing the ambient level of noise impinging upon the respiratory oscillator could increase the likelihood of terminating the apneic episode. Noise therapy would therefore aim at shortening the duration of apnea analogous to the effects of noise on the BvP system for $\eta>\eta_a$ and $\eta>\eta_b$. On the other hand, lower levels of noise might be sufficient to trigger but not terminate apnea. Methods of introducing external noise to the respiratory control system include tactile vibration[39] and auditory stimulation[2,3,40], both of which can cause perturbation of respiratory rhythm. The major challenge would be selecting the optimum stimulus frequencies and amplitudes that provide a therapeutic effect, *i.e.*, reduction of respiratory dysrhythmia, without disrupting the sleep cycle.

ACKNOWLEDGMENTS

This chapter has been reprinted with modification from the original article, which appeared in *Chaos* 5(1): 18-29, 1995, published by American Institute of Physics, Woodbury, NY.

REFERENCES

1. Southall, D. P., and D. G. Talbert. Mechanisms for abnormal apnea of possible relevance to the sudden infant death syndrome. *Ann. NY Acad. Sci.* 533: 329-349, 1988.
2. Gerhart, H. -J., H. Wagner, I. Thomschke, and B. Pasch. Zur Beeinflussbarkeit der Atmung durch rhythmische akustische Reize. *Z. Laryngol. Rhinol. Otol.* 46: 235-247, 1967.
3. Heron, T. G., and R. Jacobs. A physiological response of the neonate to auditory stimulation. *Int. Aud.* 7: 41-47, 1968.
4. Wolf, S. Sudden death and the oxygen conserving reflex. *Am. Heart J.* 71: 840-841, 1966.
5. French, J. W., B. C. Morgan, and W. G. Guntheroth. Infant monkeys - A model for crib death. *Amer. J. Dis. Child.* 123: 480-484, 1972.
6. Downing, S. E., and J. C. Lee. Laryngeal chemosensitivity: a possible mechanism for sudden infant death. *Pediatrics* 55: 640-649, 1975.
7. Wilson, S. L., B. T. Thach, R. T. Brouillette, and Y. K. Abu-Osba. Coordination of breathing and swallowing in human infants. *J. Appl. Physiol.* 50: 851-858, 1981, See p. 856.
8. Winfree, A. T. *The Geometry of Biological Time.* New York: Springer-Verlag, 1980.
9. Paydarfar, D., F. L. Eldridge and J. P. Kiley. Resetting of mammalian respiratory rhythm: existence of a phase singularity. *Am. J. Physiol.* 250: R721-R727, 1986. (See also Kitano, S. and A. Komatsu. Central respiratory oscillator: phase-response analysis. *Brain Res.* 439: 19-30, 1988; Lewis, J., Bachoo, M., Polosa, C., and L. Glass. The effects of superior laryngeal nerve stimulation on the respiratory rhythm: phase-resetting and aftereffects. *Brain Res.* 517: 44-50, 1989.)
10. Paydarfar, D., and F. L. Eldridge. Phase resetting and dysrhythmic responses of the respiratory oscillator. *Am. J. Physiol.* 252: R55-R62, 1987.
11. Eldridge, F. L., D. Paydarfar, P. G. Wagner, and R. T. Dowell. Phase resetting of respiratory rhythm: effect of changing respiratory drive. *Am. J. Physiol.* 257: R271-R277, 1989.
12. Paydarfar, D., R. J. Gilbert, C. Poppel, and P. Nassab. Respiratory phase resetting and airflow changes induced by swallowing in humans. *J. Physiol.* 483.1: 273-288, 1995.
13. Smith, D. M., R. R. Mercer, and F. L. Eldridge. Servo control of end-tidal CO_2 in paralyzed animals. *J. Appl. Physiol.* 45: 133-136, 1978.
14. Eldridge, F. L. Relationship between respiratory nerve and muscle activity and muscle force output. *J. Appl. Physiol.* 39: 567-574, 1975.
15. Fitzhugh, R. Impulses and physiological states in theoretical models of nerve membrane. *Biophys. J.* 1: 445-466, 1961.
16. Also called the Fitzhugh–Nagumo equations. See Nagumo, J. S., S. Arimoto, and S. Yoshizawa. An active pulse transmission line simulating nerve axon. *Proc. IRE.* 50: 2061-2070, 1962.
17. Bifurcation analysis has led to the same conclusion (see Hadeler, K. P., U. An Der Heiden, and K. Schumacher. Generation of nervous impulse and periodic oscillations. *Biol. Cybernet.* 23: 211-218, 1976;

Baer, S. M., T. Erneux, and J. Rinzel. The slow passage through a Hopf bifurcation: delay, memory effects, and resonance. *SIAM J. Appl. Math.* 49: 55-71, 1989.
18. Eldridge, F. L. (personal correspondence to D. Paydarfar, 1990). See also reference 11 for similar scaling factor for the van der Pol equation.
19. Remmers, J.E., DeGroot, W.J., Sauerland, E.K., and Anch, A. Pathogenesis of upper airway occlusion during sleep. *J. Appl. Physiol.* 44: 931-938, 1978.
20. Southall, D. P., V. Stebbens, and N. Abraham. Prolonged apnoea with severe arterial hypoxaemia resulting from complex partial seizures. *Dev. Med. Child Neurol.* 29: 784-804, 1987.
21. Brazy, J. E., H. C. Kinney, and W. J. Oakes. Central nervous system lesions causing apnea at birth. *J. Pediatr.* 111: 163-175, 1987.
22. Kinney, H. C., J. J. Filiano, and R. M. Harper. The Neuropathology of the Sudden Infant Death Syndrome. A Review. *J. Neuropathol Exp Neurol* 51: 115-126, 1992.
23. Dwyer, T., A. -L. B. Ponsonby, N.M. Newman, L. E. Gibbons. Prospective cohort study of prone sleeping position and sudden infant death syndrome. *Lancet* 337: 1244-1247, 1991.
24. Feldman, J. L., and J. D. Cowan. Large-scale activity in neural nets II: A model for the brainstem respiratory oscillator. *Biol. Cybern.* 17: 39-51, 1975.
25. Botros, S. M., and E. N. Bruce. Neural network implementation of a three-phase model of respiratory rhythm generation. *Biol. Cybern* 63: 143-153, 1990.
26. Lewis, J., L. Glass, M. Bachoo, and C. Polosa. Phase resetting and fixed-delay stimulation of a simple model of respiratory rhythm generation. *J. Theor. Biol.* 159: 491-506, 1992.
27. Lumsden, T. Observations on the respiratory centers. *J. Physiol. Lond.* 57: 354-367, 1923.
28. Lumsden, T. Observations on the respiratory centers in the cat. *J. Physiol. Lond.* 57: 153-160, 1923.
29. Lumsden, T. The regulation of respiration. Part I. *J. Physiol. Lond.* 58: 81-91, 1923.
30. von Euler, C. Brain stem mechanisms for generation and control of breathing pattern. In: *Handbook of Physiology. The Respiratory System II*, edited by N. S. Cherniack and J. G. Widdicombe. Bethesda, MD: American Physiological Society, 1986, p. 43, fig. 33.
31. Eldridge, F. L., and D. Paydarfar. Phase resetting of respiratory rhythm studied in a model of a limit-cycle oscillator: influence of stochastic processes. In: *Respiratory Control*, edited by G. D. Swanson, F. S. Grodins, and R. L. Hughson. New York: Plenum Publishing Corporation, 1989, p. 379-388.
32. Glass, L., and A. T. Winfree. Discontinuities in phase-resetting experiments. *Am. J. Physiol.* 246: R251-R258, 1984.
33. Lewis, J., M. Bachoo, C. Polosa, and L. Glass. The effects of superior laryngeal nerve stimulation on the respiratory rhythm: phase-resetting and aftereffects. *Brain Res.* 517: 44-50, 1989.
34. Moore, G. P., D. H. Perkel, and J. P. Segundo. Statistical analysis and functional interpretation of neuronal spike data. *Annu. Rev. Physiol.* 28: 493-522, 1966.
35. Croner, L. J., K. Purpura, and E. Kaplan. Response variability in retinal ganglion cells of primates. *Proc. Natl. Acad. Sci. (USA)* 90: 8128-8130, 1993.
36. Douglass, J. K., L. Wilkens, E. Pantazelou, and F. Moss. Noise enhancement of information transfer in crayfish mechanoreceptors by stochastic resonance. *Nature* 365: 337-340, 1993.
37. Glass, L., C. Graves, G. A. Petrillo, and M. C. Mackey. Unstable dysnamics of a periodically driven oscillator in the presence of noise. *J. Theor. Biol.* 86: 455-475, 1980.
38. Kurrer, C., and K. Schulten. Effect of noise and perturbations on limit cycle systems. *Physica D* 50: 311-320, 1991.
39. Shannon, R. Reflexes from respiratory muscles and costovertebral joints. In: *Handbook of Physiology The Respiratory System II*, edited by N.S. Cherniack and J.G. Widdicombe. Bethesda MD: American Physiological Society, 1986, p. 431-447.
40. Bradford, L. J. Respiratory Audiometry. In: *Physiological Measures of the Audio-Vestibular System*, edited by L. J. Bradford. New York: Academic Press, 1975, p. 249-317.

8

ASSESSING DETERMINISTIC STRUCTURES IN PHYSIOLOGICAL SYSTEMS USING RECURRENCE PLOT STRATEGIES

Charles L. Webber, Jr. and Joseph P. Zbilut

Departments of Physiology
Loyola University of Chicago, Stritch School of Medicine
Maywood, Illinois 60153
Rush Medical College
Rush-Presbyterian-Saint Luke's Medical Center
Chicago, Illinois 60612

1. INTRODUCTION

The purpose of this paper is to review the basic principles of recurrence plot analysis (RPA) as applied to complex systems. Recurrence plots were first introduced in the physics literature by Eckmann et al.[1] in 1987. Seven years later, Webber and Zbilut[2] enhanced the technique by defining five nonlinear variables that were found to be diagnostically useful in the quantitative assessment of physiological systems and states. Starting with the working assumption that breathing patterns are inherently complex, we carefully define what is meant by the presence of determinism (constraining rules) in physiological systems. Then, as an instructive example, we illustrate how RPA can reveal deterministic structuring at the orthographic level of a well-known children's poem. Next, we show how the multi-dimensional, nonlinear perspective of RPA can localize otherwise hidden rhythms in physiological systems and disambiguate between time series that are deceptively similar. Finally, we conclude with a discussion on new applications of RPA to nondeterministic systems and DNA orthography.

2. COMPLEX BREATHING PATTERNS

It is a sizable task for the central nervous system to regulate patterns of breathing expressed in a multiplicity of automatic and voluntary modes. The fact that mammalian systems can survive a wide variety of environmental challenges underscores the flexibility and adaptability of successful regulatory schemes. As evidenced by the voluminous information compiled in a recent book on the regulation of breathing,[3] it is obvious that degrees of complexity are high at many levels of design: neuroanatomical, neurophysiological, and

Bioengineering Approaches to Pulmonary Physiology and Medicine, edited by Khoo
Plenum Press, New York, 1996

neuropharmacological. The problem for investigators, however, is how to extract logical meaning from complicated patterns of breathing. The standard approach has been to deeply anesthetize the preparation, regularizing the breathing while decreasing the system's degrees of freedom. A newer approach has been to reduce the preparation, simplifying the participant nuclei and variables along the neuraxis.[4] But a third and more integrative approach is to allow the ventilatory system to express its full repertoire of state-dependant behaviors and to analyze those patterns using modern methods from nonlinear dynamics. Complex breathing patterns demand complex modes of analysis.

3. DETERMINISTIC SYSTEMS

Dynamical systems that consist of rule-obeying, interacting components are said to be deterministic. By this definition, simple equations of motion are indeed periodic (e.g. sine waves), but their repetitive, stereotypical and uninteresting trajectories fail to mimic the rich patterns observed for numerous physiological systems. Greater success has been enjoyed with more complicated equations such as the van der Pol limit cycle in modeling breathing,[5] but the details of frequency scaling are still lacking. It is no wonder that mathematically-inclined physiologists found new ideas stemming from deterministic chaos very attractive. Contrasted with simple periodic systems, chaotic systems have the ability to generate very complex, yet bounded, patterns, using very few components. Such deterministic chaos, driven entirely by mathematical rules, has enabled even more complex modeling of the respiratory system with heteroclinic orbits,[6] for example.

Progressively, it is becoming more apparent that physiological systems are not necessarily best thought of as being fully deterministic in design. Independent of the number of degrees of freedom present, completely deterministic systems are simply fixed and rigid whether they be periodic or chaotic. Such dynamical structures are quite unprepared to exist, let alone succeed, in noisy environments extant in the real world. Quoting Albert Einstein[7] in this context, "So far as the laws of mathematics refer to reality, they are not certain. And so far as they are certain, they do not refer to reality." The distinction is further underscored when it is understood that mathematical systems are closed and noise-free, whereas biological systems are open and constantly challenged by noise. These differences are of such importance that one might conclude global mathematical descriptions of biological "wiggles" are categorically precluded. One satisfying solution to this dilemma between constrained mathematics and open-ended physiology may reside in the realm of nondeterminism.[8] Such terminal dynamics are characterized by piecewise determinism (trajectories) interposed with periods of stochasticity (pauses). We expand on the functionality and workability of this intriguing perspective toward the end of this paper.

4. RECURRENCE PLOT ANALYSIS

One evidence for logical design and deterministic structuring of dynamical systems is the presence of recurrent (recurring) patterns. Recurrence plots have the ability to locate unexpected rhythms in otherwise benign time series despite nonstationarity or quasi-periodicity in the dynamic (two noted characteristics of physiological systems). The concept of recurrence (in the first dimension) has a long mathematical history,[9] and our description of RPA will start by demonstrating how recurrence plots can define rhythmic repeating patterns encountered in one famous poem from children's literature. Even a cursory reading of only the first several lines (207 characters, excluding spaces) of the poem *Green Eggs and Ham* by the late Dr. Seuss (Theodor Seuss Geisel) gives the sense of rhythm and rhyme present in the verse.[10]

- I am Sam, I am Sam, Sam I am.
- That Sam-I-am! That Sam-I-am!
- I do not like that Sam-I-am!
- Do you like green eggs and ham?
- I do not like them, Sam-I-am.
- I do not like green eggs and ham.
- Would you like them here or there?
- I would not like them here or there.
- I would not like them anywhere.

The vocabulary of this 812-word poem is limited to only 50 words, which means that many words must recur throughout the poem. At the orthographic level (spelling), parts of words also recur: "... am ... Sam ... ham..." Even entire phrases recur: "I do not like" (31 times), "Sam I am" (14 times), and "green eggs and ham" (11 times). Analysis of this poem by RPA is now described.

For pedagogical purposes, Fig. 1 schematizes the recurrence of letters in the word "green" while ignoring intervening characters (blanked out for simplicity). The same snippet of text is lined up on both horizontal and vertical axes of the graph. Whenever letters on the

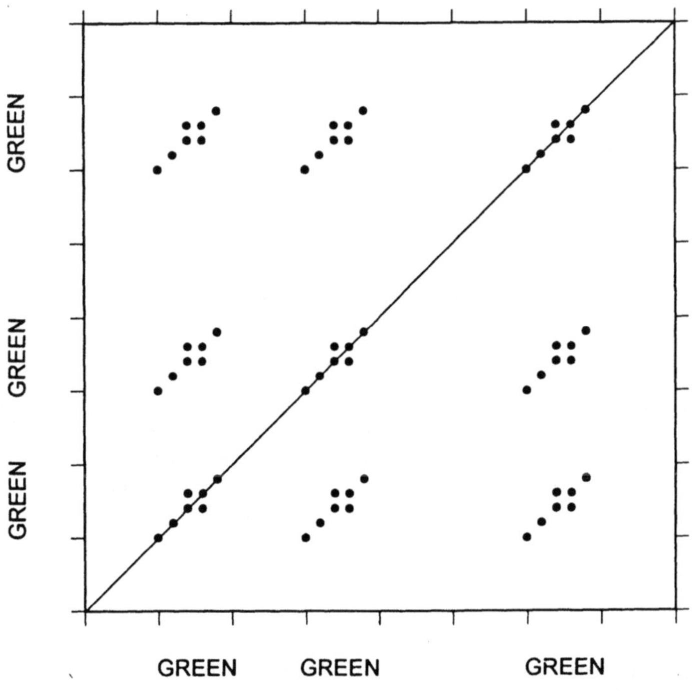

Figure 1. Schematized recurrence plot of the five characters in the word "green." The identical text is positioned along both horizontal and vertical axes, and 63 recurrent points (•) are found at the intersections of all letter matches. Note that 45 recurrent points form 9 diagonal line structures each 5 units long. Also, 18 additional recurrent points are isolated because of the double-"e" construction in the text. Since the same text is represented on both axes, recurrent points always form a long central diagonal, bisecting the recurrence plot into two triangles with recurrent symmetry.

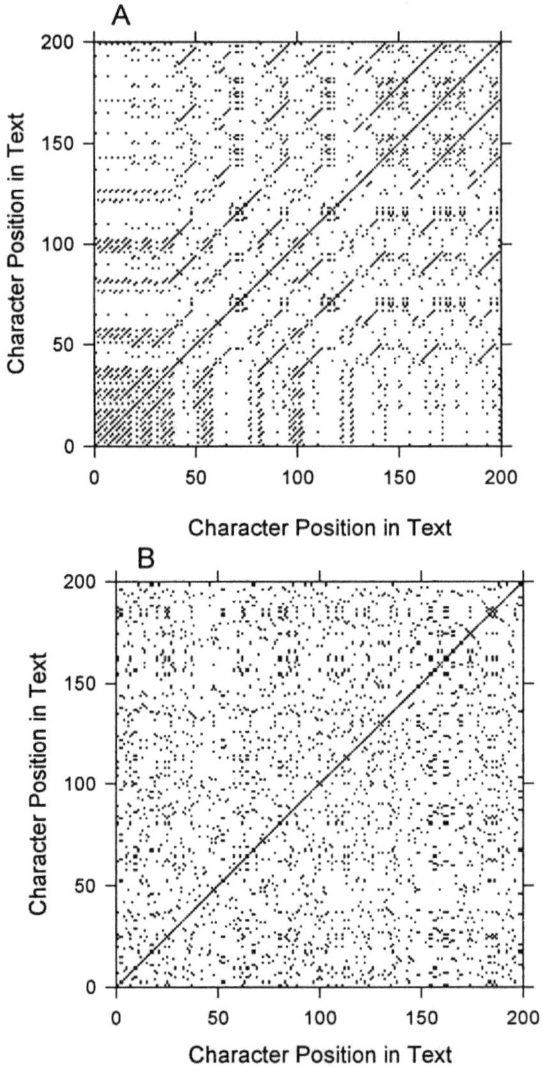

Figure 2. Actual recurrence plots of the first 200 characters of Dr. Seuss' children's poem, *Green Eggs and Ham,*[10] before (A) and after (B) pseudo-random shuffling of the text orthography. Diagonal line segments of varying length are indicative of deterministic information in the native text which can be disrupted by shuffling procedures. For both recurrence plots, RPA parameter settings include: embedding = 1, lag = 1, cutoff = 0, line = 2.

orthogonal axes strike an exact match ("g" for "g" or "r" for "r" etc.), a point is positioned at the corresponding horizontal and vertical intersection. In the example shown, each "green" on the horizontal axis scores three sequence matches vertically (with itself and the other two occurrences of the same string), resulting in the formation of nine diagonal line structures. Eighteen additional matches between the "e" characters at different positions within each sequence also cause every diagonal line to bulge. Text characters matching at identical horizontal and vertical positions form a central line of identity (ubiquitous for all recurrence plots). All other recurrent points form symmetrical patterns on either side of this central diagonal.

The recurrence plot for the first 200 characters of the poem *Green Eggs and Ham* is shown in Fig. 2A. As expected, recurrent points are distributed along diagonal line segments of different lengths, depending upon the span of consecutive matching letters. The actual analysis was performed on numerical values after encoding the entire poem: "a" = 1, "b" = 2, ..., "z" = 26 (case insensitive). All spaces, dashes and punctuation marks were consistently ignored, leaving the legal character codes to be strung together as an orthographic string. Determinism in this plot is carried in the line segments: their number, length, and distribution. Complete shuffling of text characters both destroys the intelligibility of the verse and disrupts the characteristic line patterning in the resultant recurrence plot as shown in Fig. 2B.

5. QUANTIFICATION OF RECURRENCES

Webber and Zbilut[2] introduced the concept of "quantification of recurrences" by deriving five RPA variables that quantify the patterns of recurrent point distributions. Again, the logic for this approach is based on the diagonal line structures found in recurrence plots. These five variables include %recurrence, %determinism, ratio, entropy and trend (numerical values specified are from Fig. 2A). Since recurrence plots are symmetrical, computations need only be performed on the upper triangle above the central line of identity. First, %recurrence is defined as the number of recurrent points per total triangular area (7.322% = 1,457/19,900), excluding the long central diagonal. Second, %determinism is defined as the number of recurrent points forming diagonal line structures (minimum of 2 points per line) per total number of recurrent points (47.014% = 685/1,457). Third, ratio is defined as the quotient of %determinism divided by %recurrence (6.421 = 47.014%/7.322%) and is useful for detecting transitions between states. Fourth, entropy is computed as the Shannon entropy of the deterministic line segment lengths distributed in a histogram (2.104 bits) and represents the information content of the dynamic. Fifth, trend is the linear slope of %recurrent points per unit distance away from the central diagonal (-26.665%/1,000 points) and addresses the stationarity of the system (trends near zero) or system drift (non-zero trends beyond 10 units from zero). After shuffling these five RPA variables were as follows: %recurrence = 7.322% (unchanged); %determinism = 12.148% (decreased); ratio = 1.659 (decreased); entropy = 0.320 bits (decreased); trend = 13.114%/1,000 points (decreased absolute amplitude).

Repeated recurrence computations within an epoch window sliding down a longer data set enable the expression of RPA variables as a function of position (or time) within the data structure. This is illustrated for Dr. Seuss' poem in Fig. 3. The 2,438 encoded characters plotted in panel A constitute the RPA input. In this case, the epoch was selected as 200 characters, offset by a single character on each new RPA computation, resulting in the generation of 2,239 values for each of the six variables shown. (Note that each variable is plotted at the end of the data set contributing to that point so as to prohibit false anticipatory responses.) In panel B, the mean and standard deviation of the encoded characters are plotted, neither of which possess any remarkable rhythmic structuring. By contrast, however, panels C and D show oscillations in four RPA variables including %recurrence, %determinism, ratio and entropy. Repeating the analysis on the completely shuffled text decreases the %determinism and reduced the entropy (not shown), inferring that the patterns in the original native text were governed by non-random linguist rules at the orthographic level.

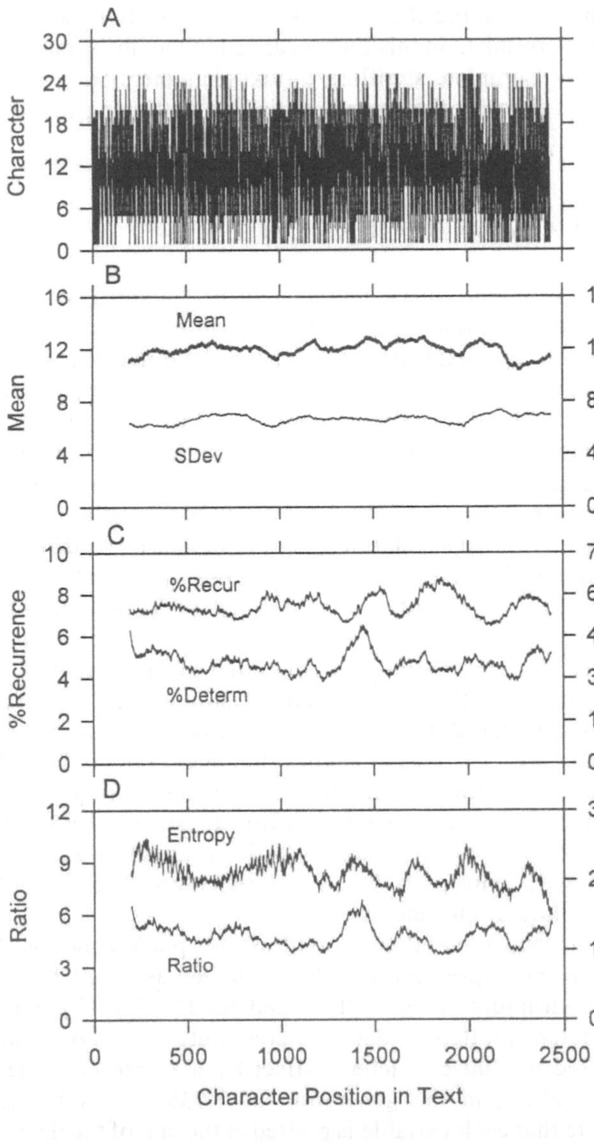

Figure 3. Recurrence plot analysis (RPA) of the entire 2,438 characters of the poem, *Green Eggs and Ham*, epoch by epoch. Individual alphabetic characters ("a" to "z") were encoded into integers (1 to 26) and plotted as a function of character position in the text (A). Using a sliding epoch window of 200 characters, the running mean and standard deviation values of the encoded characters were relatively stable throughout the text (B). Nonlinear RPAvariables of %recurrence, %determinism, ratio, and entropy, however, displayed rhythmic structuring within the textual dynamic at the orthographic level (C,D). RPA parameter settings include: embedding = 1, lag = 1, cutoff = 0, line = 2, epoch = 200, offset = 1.

6. RECURRENCE PLOTS IN HIGHER DIMENSIONS

In the linguistic example above, points were recurrent only when they formed exact matches at the character level. Thus near matches (e.g. "a" is close to "b" alphabetically) were categorically excluded. For chaotic systems of equations (mathematics) and semi-periodic patterns of breathing (physiology), exact matches rarely occur in the first dimension, except by chance alone, causing recurrent points to fall at random and disallowing significant diagonal line structuring. This effect on non-linguistic systems can be attributed to two facts: the dynamic may live in higher dimensional space, and/or noise may be perturbing the system. Eckmann et al.[1] handle these two conditions, respectively, by embedding the dynamic in higher dimensional n-space and by relaxing the constraint of exact matches. Although it is known that systems governed by mechanistic and biophysical rules do possess fractal structures in space (e.g. alveoli in the lung) and periodicities in time (e.g. cyclic breathing from birth to death),[11] full appreciation of the more subtle recurrent structures is oftentimes reserved for the multidimensional perspective.

Recurrence plots in higher dimensions emerge from the wedding of embedding theory with single-dimension recurrence theory. It has been proven mathematically, for example, that if a system has five coupled, interacting variables (n=5), that knowledge of the time series of but one variable (any one) is adequate to describe the entire system.[12] Theoretically, the single variable can be embedded in 5-space (m=5) by the method of time delays, and recurrence analysis can be performed on the new embedded variable. Practically speaking, however, the embedding needs to be carried out at a higher dimension (e.g. m=11) to satisfy the topological relationship m=2n+1, accounting for real or induced dynamical noise in the system.[13] What this means for physiological systems is that surrogate variables can be reconstructed from a single observable (e.g. intrapleural pressure) and probed for recurrent structures graphically, revealing higher dimensional rhythmicities not observed in the original time series.

Finally, visualization of recurrence plots in higher dimension depends upon the setting of a cutoff parameter. If cutoff equals zero, then only exactly matching vectors in n-space will score recurrent points (too conservative). If cutoff equals the maximum vector distance, then every point in the recurrence plot will be recurrent (too liberal). Since we are practically interested in discovering deterministic rules in an embedded, but noisy, dynamic, it is necessary to allow the cutoff parameter to be set somewhat above zero, but well below the maximum distance. When the cutoff parameter is selected correctly, RPA diagonal lines form delicate lace-like patterns, indicative of deterministic structuring of the embedded dynamic within local neighborhoods. Further details regarding proper recurrence plot implementation are described by us elsewhere,[2,14] and explicit instructions and example time-series data are also included with our freely distributed RPA software (see the end of this paper).

Representative recurrence plots of rodent breathing patterns in higher dimensional space (m=10) are shown in Fig. 4 for 12 overlapping epochs during hypercarbic challenge.[15] In this study we evaluated the hysteretic responses of respiratory chemoreflexes to ramp increases and decreases in environmental carbon dioxide (P_iCO_2). While recording intrapleural pressure, awake rats were placed in a sealed, oxygen-filled chamber into which CO_2 was first added (20 mm Hg/min) and then removed (-17 mm Hg/min). As expected, breath amplitude and frequency variables both increased during chemoreceptor stimulation and returned to normal as the CO_2 challenge was diminished. The figure illustrates qualitative changes in recurrence plot patterns (256-point epochs) for breathing periods (1/frequency) during a single CO_2 up-ramp. The recurrences are

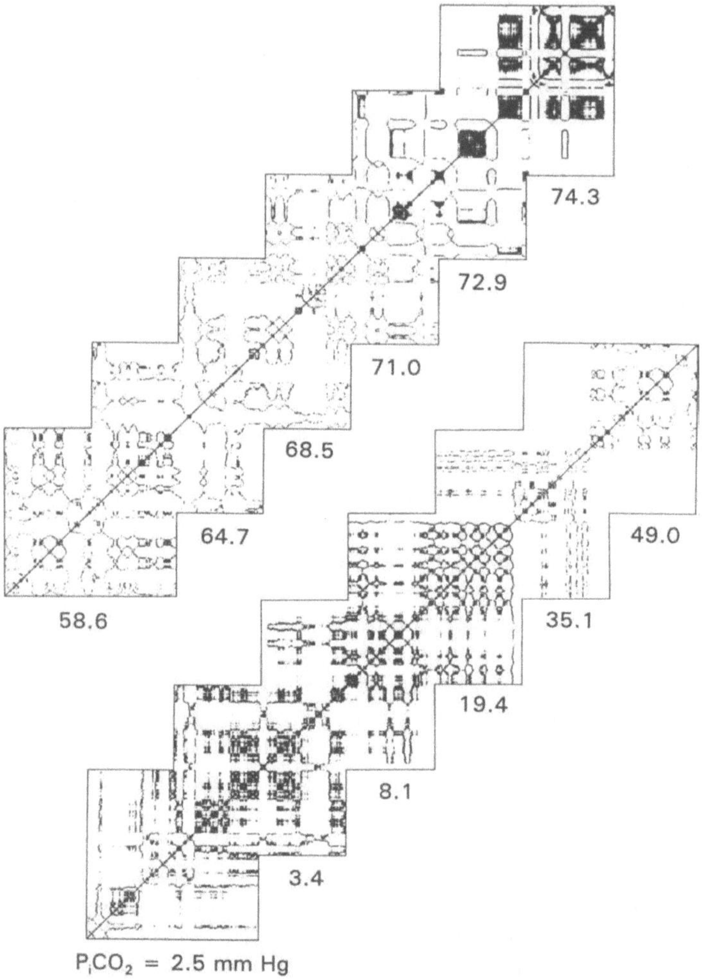

Figure 4. Multi-dimensional recurrence plots of twelve overlapping epochs of cycle-by-cycle breathing periods during chemoreceptor activation in a conscious, unanesthetized rat. Starting at a low level of inspired carbon dioxide partial pressure of 2.5 mm Hg, the chamber P_iCO_2 was increased to 74.3 mm Hg (20 mm Hg/min) while recording intrapleural pressure. During the transition from low frequency breathing (long-breath periods) to high frequency breathing (short-breath periods), the lace-like texture of the recurrence plot was changed. RPA parameter settings include: embedding = 10, lag = 1, cutoff = 1.0, line = 3, epoch = 256, offset = 128.

higher (darker patterns) during quasi-steady state CO_2 challenges (2.5 and 74.3 mm Hg), but rarer (lighter patterns) during the transient ventilatory response to the rapid CO_2 increase. Summary results are presented in Fig. 5 for this representative rat. Panels A and B show the one-dimensional and quasi-linear responses of mean breath periods and amplitudes which exhibit very little hysteresis. By way of contrast, panels C and D show that increasing the dimensional perspective on the system by RPA unveils unexpected nonlinearities and hysteresis in %determinism. Such results infer that many variables participate in the chemoreflexes governing respiratory control and that the interaction of variables is highly nonlinear and complex. Thus, the dynamical

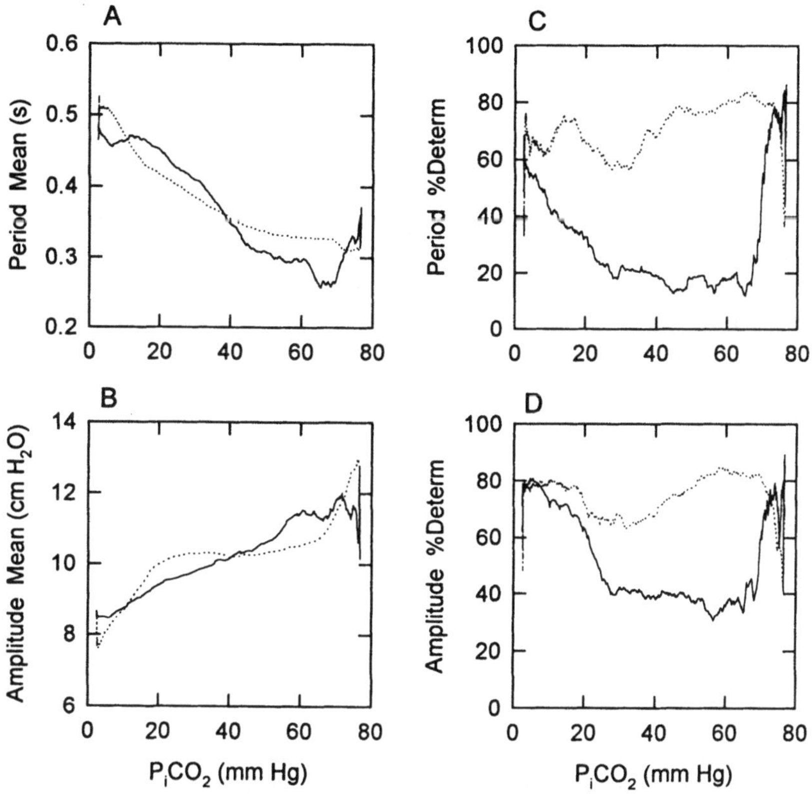

Figure 5. Hysteresis plots of respiratory variables (ordinates) recorded from a conscious, unanesthetized rat subjected to increases (solid lines) and decreases (doted lines) in P_iCO_2 (abscissa). Epoch means for periods and amplitudes are quasi-linear and exhibit little hysteresis (13.6% and 13.7% respectively) (A,B). Epoch %determinism for periods and amplitudes are curvi-linear and exhibit much hysteresis (68.4% and 54.0% respectfully) (C,D). RPA parameter settings include: embedding = 10, lag = 1, cutoff = 1.0, line = 3, epoch = 256, offset = 1.

details of chemoreceptor gain are more complicated than implied by the simple linear relationship between mean respiratory periods or amplitudes and CO_2 tension.

7. RECURRENCE PLOTS IN PHYSIOLOGY

Since recurrence plots are particularly useful for studying the behavior of non-stationary time series,[16] it is not surprising that there is a growing literature on the successful application of this non-linear tool to dynamical systems in physiology. RPA has been used to identify quasi-deterministic firing patterns in neuronal spike trains of the cerebellum and red nucleus,[17] to reveal very slow brain waves in the abnormal electroencephalogram,[18] and to investigate higher-dimensional heart beat dynamics.[19] A couple of studies have compared the performances of RPA and the fast Fourier transform (FFT) on physiological systems. First, Keegan et al.[20] found that the pupillary dynamics of narcoleptic patients and over-tired individuals could be disambiguated by RPA, but not FFT. Second, Webber et al.[21] reported an enhanced sensitivity of RPA over FFT in detecting the time-course of fatigue in electromyographic recordings from the human biceps muscle during isometric loading. Results such as these suggest that a broader perspective is to be gained by analyzing complex, nonlinear physiological systems, not only with standard spectral techniques, but also with sophisticated, nonlinear tools.

Finally, Webber and Zbilut[2] illustrate the utility of recurrence plots in extracting information from dynamical breathing patterns not available from linear metrics. While recording breath-by-breath periods of breathing, active rats were intra-peritoneally injected with pentobarbital sodium in order to study the time course of anesthetization. Within a few minutes the breathing pattern slowed and the 200-breath mean (epoch window) stabilized, as did the standard deviation of breaths. RPA analysis of the same data in 10 dimensions, however, revealed sustained oscillations in %recurrence, %determinism and entropy even 40 minutes after the pentobarbital injection (the data closely resemble Fig. 3). One interpretation of these results was that the pharmacodynamics of the administered drug had not achieved a steady state, despite the early stabilization of the mean periods. Evidently, the embedded dynamic is far richer in structure than suggested by the one-dimensional time-series data.

8. NONDETERMINISTIC SYSTEMS

Our conceptualization of physiological systems has matured from one of deterministic dynamics to nondeterministic dynamics. Nondeterministic systems are described as being piecewise deterministic in the sense that the trajectories evolve deterministically, but when the system is near a singularity in the equations of motion, noise can cause unique non-deterministic jumps to unique trajectories.[22] Differential equations with fractional exponents fulfill such criteria, creating systems which have multiple deterministic trajectories serially coupled with stochastic pauses of indeterminable duration.[23] Piecewise deterministic systems not only permit the modeling of breathing frequency, for example, but they allow for rule-obeying structures to operate properly within noisy environments, surviving even unanticipated challenges (surprises). This type of design gives added flexibility and adaptability to the basic physiological "rules" in operation. And with experience, of course, comes learning and anticipatory responsiveness as evidenced by the large repertoire of behaviors possible for various systems. Nondeterminism can also be studied from the RPA perspective in which recurrence is punctuated by variable pauses as seen in the normal human electrocardiogram.[24]

9. DNA LINGUISTICS

There have been many clever schemes devised for analyzing DNA codes, including 1/f analysis showing long-range DNA correlations[25] and a "chaos-game" representation of gene structure which generates unique patterns for different DNA sequences.[26] DNA strings can likewise be processed by RPA methodology by treating the coded information as a 4-letter grammatical text. Therefore, a hypothesis can be posited that encoding regions of DNA possess deterministic, rule-driven base sequences that can be disrupted by shuffling procedures. The RPA test of this hypothesis is much like the analysis of the Dr. Seuss text described above. For example, the 9,770 base pairs of the human virus HIV were downloaded from *GenBank* and encoded into integers: A = 1, C = 2, G = 3, T = 4 (purines: A=adenine and G=guanine; pyrimidines: C=cytosine and T=thymine) . As shown in Fig. 6., RPA was run on both the native HIV code and the shuffled HIV code using an epoch window size of 1,000 base pairs. The results indicate that complete shuffling reduces both the level of deterministic structuring in the code as well as the information entropy, implying that the

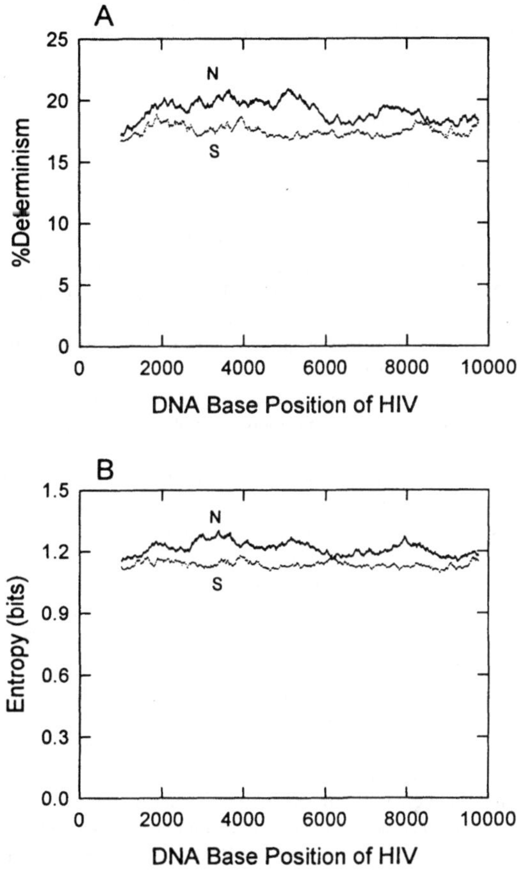

Figure 6. Recurrence plot analysis of the 9,770 base sequences in the DNA code (deoxyribonucleic acid) for the human immunodeficiency virus (HIV) using repeated RPA computations, epoch by epoch. RPA variables %determinism (A) and entropy (B) were both higher for the native base sequences (N), whereas complete shuffling of the code (S) led to decreases in both variables. These data indicate that DNA codes can be treated like a 4-character linguist text, the deterministic information content of which can be discovered by complete shuffling procedures. RPA parameter settings include: embedding = 1, lag = 1, cutoff = 0, line = 3, epoch = 1,000, offset = 1.

original DNA sequence is carrying deterministic information at the orthographic level. Interestingly, if it can be demonstrated that the non-coding regions (junk DNA) which interrupt the deterministic exons (coding DNA) are stochastic, then DNA strands could appropriately be described as being nondeterministic. Further details on how information theory is being applied to the field of molecular biology are available elsewhere.[27] RPA analysis of DNA codes is simply another example of the utility and portability of recurrence plots across disciplines.

10. CONCLUSIONS

Recurrence plot analysis (RPA) seems ideally suited for diagnosing complex patterns commonly found in physiological systems. RPA possesses several advantages over alternative approaches: 1) RPA can operate on either linear or nonlinear data sets; 2) RPA is not stymied by nonstationary data drifts and can actually detect dynamical transitions between states; 3) the size of RPA input data sets can be relatively small (50 to 200 cycles); and 4) the statistical distribution of RPA input data can assume any form (Gaussian, non-Gaussian). As with any new technology, however, care must be taken when applying this mathematical tool to physiological data. Each new user must experience a personal learning curve while mastering the proper implementation of RPA in parameter space.[28] With time, the lesson that must be learned is that what one obviously sees (in the first dimension) is not what one necessarily has (in multiple dimensions).

REFERENCES

1. Eckmann, J.-P., S.O. Kamphorst, and D. Ruelle. Recurrence plots of dynamical systems. *Europhys. Lett.* 4: 973-977, 1987.
2. Webber, C.L., Jr., and J.P Zbilut. Dynamical assessment of physiological systems and states using recurrence plot strategies. *J. Appl. Physiol.* 76: 965-973, 1994.
3. Dempsey, J.A., and A.I. Pack. *Regulation of Breathing.* Lung Biology in Health and Disease, Volume 79, Edition 2. New York: Marcel Dekker, Inc., 1995.
4. Berger, A.J. Recent advances in respiratory neurobiology using *in vitro* methods. *Am. J. Physiol.* 259: L24-L29, 1990.
5. Eldridge, F.L. Phase Resetting of Respiratory Rhythm: experiments in animals and models. In: *Rhythms in Physiological Systems,* edited by H. Haken and H.-P. Koepchen. Berlin: Springer-Verlag, 1991, p.165-175.
6. Sammon, M. Geometry of respiratory phase switching. *J. Appl. Physiol.* 77: 2468-2480, 1994.
7. Einstein, A. Geometry and experience. In: *Sidelights on Relativity.* New York: Dover, 1983, p. 27-56.
8. Zak, M. Introduction to terminal dynamics. *Complex Systems* 7: 59-87, 1993.
9. Feller, W. *An Introduction to Probability Theory and its Applications.* Chapter 12, Recurrent Events: Theory, New York: John Wiley & Sons, Inc., 1950, Volume 1, Chapter 12, p. 238-306.
10. Seuss, Dr. (pseud.) *Green Eggs and Ham.* New York: Beginner Books, Random House, Inc., 1960.
11. Bassingthwaighte, J.B., L.S. Liebovitch, and B.J. West. *Fractal Physiology.* New York: Oxford University Press, 1994.
12. Grassberger, P., T. Schreiber, and C. Schaffrath. Nonlinear time sequence analysis. *Int. J. Bifurcation Chaos* 1: 521-547, 1991.
13. Parker, T.S., and L.O. Chua. *Practical Numerical Algorithms for Chaotic Systems.* New York: Springer-Verlag, 1989, p. 191-197.
14. Webber, C.L., Jr. Rhythmogenesis of deterministic breathing patterns. In: *Rhythms in Physiological Systems,* edited by H. Haken and H.-P. Koepchen. Berlin: Springer-Verlag, 1991, p. 177-191.
15. Webber, C.L., Jr., and J.P. Zbilut. Complex breathing pattern responses to carbon dioxide challenges in unanesthetized rats. *Amer. J. Respir. Crit. Care Med.* 151: A637, 1995.
16. Koebbe, M., and G. Mayer-Kress. Use of recurrence plots in the analysis of time-series data. In: *Nonlinear Modeling and Forecasting, SFI Studies in the Sciences of Complexity,* Proc. Vol. XII, edited by M. Casdagli and S. Eubank. Redwood City, CA: Addison Wesley, 1992, p. 361-378.
17. Kalowzsy, P., and R. Tarrecki. Recurrence plots of neuronal spike trains. *Biol. Cybern.* 68: 527-534, 1993.
18. Babloyantz, A. Evidence for slow brain waves: a dynamical approach. *Electroenceph. Clin. Neurophysiol.* 78: 402-405, 1991.
19. Zbilut, J.P., M. Koebbe, H. Loeb, and G. Mayer-Kress. Use of recurrence plots in the analysis of heart beat intervals. In: *Proceedings Computers in Cardiology.* Los Alamitos, CA: IEEE Computer Society Press, 1991, p. 263-266.
20. Keegan, A.P., J.P. Zbilut, S.L. Merritt, and P.J. Mercer. Use of recurrence plots in the analysis of pupil diameter dynamics in narcoleptics. *Soc. Photo-Optical Inst. Engineers Proc.* 2036: 206-214, 1993.
21. Webber, C.L., Jr., M.A. Schmidt, and J.M. Walsh. Influence of isometric loading on biceps EMG dynamics as assessed by linear and nonlinear tools. *J. Appl. Physiol.* 78: 814-822, 1995.
22. Dixon, D.D. Peicewise deterministic dynamics from the application of noise to singular equations of motion. *J. Physics A.* 28:5539-5551, 1995.
23. Zbilut, J.P., M. Zak, and C.L. Webber, Jr. Physiological singularities in respiratory and cardiac dynamics. *Chaos, Solitons, and Fractals* 5: 1509-1516, 1995.
24. Webber, C.L., Jr., and J.P. Zbilut. Quantification of cardiac nondeterminism using recurrence plot strategies. *Soc. Neurosci. Abstr.* 21: 1402, 1995.
25. Voss, R.F. Evolution of long-range fractal correlations and 1/f noise in DNA base sequences. *Physical Rev. Lett.* 68: 3805-3808, 1992.
26. Jeffrey, H.J. Chaos game representation of gene structure. *Nucleic Acids Res.* 18: 2163-2170, 1990.
27. Yockey, H.P. *Information Theory and Molecular Biology.* New York: Cambridge University Press, 1992.
28. Zbilut, J.P., and C.L. Webber, Jr. Embeddings and delays as derived from quantification of recurrence plots. *Physics Lett. A* 171: 199-203, 1992.

9

MEASURES OF RESPIRATORY PATTERN VARIABILITY

Eugene N. Bruce

Center for Biomedical Engineering
University of Kentucky
Lexington, Kentucky 40506-0070

INTRODUCTION

The possible causes of breath to breath variability in the pattern of breathing have been discussed in a recent review[1]. This variability may be due to stochastic factors (e.g., random disturbances), to the dynamical behaviors of chemoreflex and mechanoreflex feedback loops, or to interactions between these stochastic and dynamical non-stochastic mechanisms. The recognition that the type and structure of breath to breath variability in respiratory pattern may reflect the actions of various underlying respiratory control mechanisms has motivated the desire to quantify this variability. Although some prior work has attempted to separate respiratory variability into its stochastic and various non-stochastic components[2], such a separation is difficult in general and is not considered here. A more fundamental question is whether a change in state or condition of a subject alters overall respiratory pattern variability. This report will discuss this more fundamental question and compare several methods for quantifying overall breath to breath variability in breathing pattern using a specific example.

BACKGROUND

In recent studies[3,4] on anesthetized rats we demonstrated that spontaneous respiration was visually more variable when the rat was exposed to a negative transrespiratory pressure (PTR), which should cause a reduction in end-expiratory volume (EEV). On the other hand, respiratory pattern variability was greatly reduced by exposing the rat to a positive PTR and was virtually absent after bilateral cervical vagotomy. Based on these observations we proposed that EEV in the rat is regulated by pulmonary vagal afferents which alter their activities in response to a decrease in lung volume and evoke compensatory responses in the activation of chest wall and upper airway muscles, as well as in respiratory cycle timing. We further hypothesized that nonlinear behavior and time delays inherent in both central neural respiratory mechanisms and in the coupling between muscle activities, mechanical motion

Bioengineering Approaches to Pulmonary Physiology and Medicine, edited by Khoo
Plenum Press, New York, 1996

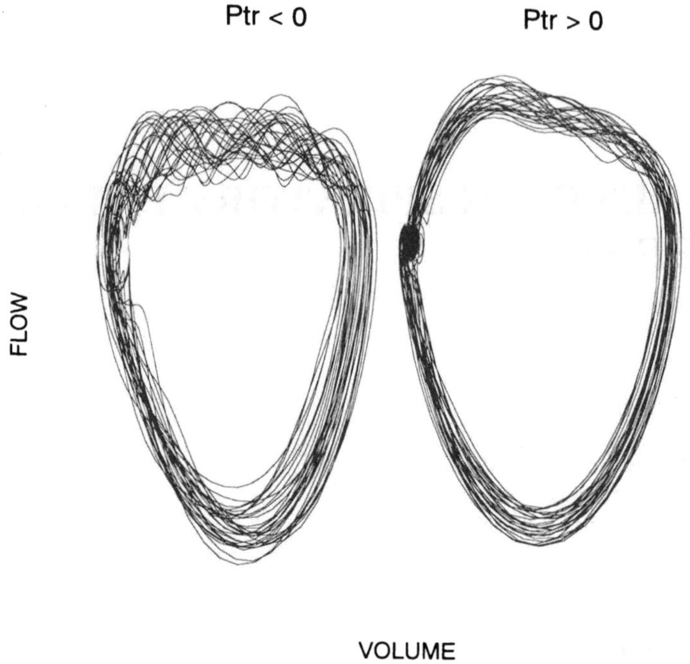

Figure 1. Flow-volume plots from an anesthetized, tracheotomized rat exposed to positive (right) and negative (left) transrespiratory pressure. Breathing proceeds in a clockwise manner with inspiration occupying the top half of the loop. Each variable is normalized by the magnitude of its range in the data set in order to emphasize relative variability. Each graph shows 20 seconds of data. Note both breath to breath and within-breath variability with deflationary pressure (left) and more regular pattern with inflationary pressure.

of the lungs, and lung receptor activation, could introduce nonstochastic variability in the respiratory pattern. Consequently the respiratory pattern would be highly variable with a negative PTR but quite regular when EEV was raised by a positive PTR. Contributing vagal afferent modalities could include pulmonary stretch receptors, rapidly adapting receptors, and slowly-adapting deflation-sensitive receptors[4,5]. An example of the dependence of variability on PTR is given in Figure 1. A potential limiting factor to the above hypothesis is that the rats that were studied were tracheotomized. It is well known that many species utilize upper airway muscles to regulate expiratory airflow[4,6] and it is conceivable that respiration was variable in these animals because this mechanism was bypassed. To address this possibility we analyzed the respiratory pattern at different EEV levels in rats with intact upper airways.

EXPERIMENTAL METHODS

Adult rats were anesthetized with 1200 mg/kg of urethane I.P. and placed in a head-out, temperature-compensated plethysmograph, as previously described[7]. Pressure in the plethysmograph (PBOX) was measured with a Validyne transducer. Pressure fluctuations were calibrated in terms of volume by injecting a known volume into the plethysmograph. A pump which could produce adjustable positive or negative pressures was connected to a port of the plethysmograph. By adjusting this pump we produced control (PBOX = PTR =

0), inflationary (PBOX = -3 cm H_2O, PTR = +3 cm H_2O), or deflationary (PBOX = +3 and +6 cm H_2O, PTR = -3 and -6 cm H_2O) mean plethysmograph pressures. The first minute of breathing at each pressure was discarded to allow for any transient changes in respiratory pattern due to the abrupt change in EEV. Then a 2-minute record of the respiratory fluctuations of the PBOX signal was obtained by sampling at 150 Hz with a laboratory microcomputer. After each data acquisition PBOX was returned to control for at least 3 minutes before changing to another positive or negative pressure.

DATA ANALYSIS

A continuous respiratory volume signal, V, was derived by applying the calibration factor to the pressure fluctuations of the PBOX signal. Numerical differentiation was used to calculate the first, second, and third derivatives of V[3] (referred to as DV, DDV, and DDDV respectively) and phase plots were constructed at the different mean transrespiratory pressures using various combinations of these four signals.

Breath to breath variability in breathing pattern was quantified in several ways: by coefficients of variation of cycle parameters - e.g., tidal volume (V_T) and breath duration (T_{TOT}); by calculating quantitative measures of recurrence plots of cycle parameters; by calculating approximate entropy[8,9] of the continuous volume signal and of its derivatives; by determining a spectral degrees of freedom measure[10] for V and its derivatives. Each of these analyses is described in detail below.

Coefficients of Variation

Breath by breath values of V_T, T_{TOT}, inspiratory and expiratory durations (T_I and T_E), mean inspiratory flow (i. e., V_T/T_I), and ventilation ($\dot{V}_I$) for all breaths in a 2-minute data record were calculated from the volume signal using an interactive computer program. Occasional breaths that were strikingly different from most breaths (e. g., sigh-like) were discarded. The mean, standard deviation, and coefficient of variation for each cycle parameter were determined for each data record.

Recurrence Analysis

Recurrence analysis examines the data record for sequences of data points from one part of the record that are correlated with sequences of data points later in the record. This approach was developed by Eckmann and Ruelle[11] and has been summarized and applied to respiratory data by Webber and Zbilut[12]. The method proceeds by first embedding the time series in m-dimensions[13] to approximate the phase space of the system. Then one calculates the distances between all possible pairs of data points using an appropriate distance measure such as the largest absolute value of the components of the difference vector between two points, or the second-order norm of the difference vector. Finally, a graph is constructed whose X and Y coordinates *both* represent the indices of the time series of embedded data points. One puts a dot on the graph at location (n,m) for every such n and m for which the distance between data points x(n) and x(m) is less than a specified distance, r. Typically 'r' is on the order of 5-20% of the maximum of the distances between data points. A sequence of k consecutive points parallel to the main diagonal of the graph indicates that k data points from one part of the record are point-by-point close to k data points which occur later. Consequently for uncorrelated random data a recurrence plot exhibits such diagonally-directed "lines" only randomly, and they comprise at most a few points.

Both correlated random and deterministic data exhibit longer diagonal lines and chaotic signals produce line segments whose lengths depend on the degree of correlation in the signal.

Although one has a visual impression from comparing two recurrence graphs, it is difficult to draw inferences without quantitative measures. Webber and Zbilut[12] have proposed three such measures: percent recurrence, percent determinism, and entropy. Define N as the number of points in the embedded time series, N_R as the number of "dots" on the recurrence plot, N_L as the number of "dots" comprising diagonal lines of 3 or more consecutive points, and P_i as the fraction of such line segments having length "i". Then the above measures are defined as:

1. percent recurrence = N_R/N^2
2. percent determinism = N_L/N_R
3. entropy = $-\Sigma\, P_i \log P_i$

Recurrence plots were determined from breath by breath values of respiratory cycle variables using an embedding in 5 dimensions. Each of the three quantitative measures was determined for 'r' ranging from 5% to 25% of the mean of the distances in a given data record.

Approximate Entropy (ApEn)

This measure of complexity of a discrete signal is described in qualitative terms in the paper by Pincus and Goldberger[9] and quantitatively in an original paper from Pincus[8]. Approximate entropy measures the logarithmic conditional probability that two data points from different parts of a data record will be close together given that the 'm' data points preceding each of these points are themselves close together. The calculation of ApEn is based on embedding the time series in m-dimensions, counting the number of (embedded) data points that are within distance 'r' of each individual data point, then calculating the average log of these counts over all individual data points, $\Phi^m(r)$. If N is the total number of points in the data record, then

$$\mathrm{ApEn}(m,r,N) = \Phi^{m}(r) - \Phi^{m+1}(r)$$

The actual value of ApEn depends on the values of m, r, and N, but for a sinusoid this value will be very close to zero and for mixtures of sinusoids and uncorrelated random noise it is larger than zero. ApEn has been shown to increase when a chaotic system bifurcates to a more complex behavior[8] but nonzero approximate entropy does not imply chaos.

Approximate entropy was calculated for the DDDV signal, sampled at 30 Hz until 1024 points were obtained. We chose m = 2 and r = 15% of the standard deviation of the sampled data points. DDDV was chosen because it is sensitive to variability in the volume signal near end-inspiration and end-expiration, and approaches zero elsewhere. Previously we had observed[3,4] that the breath to breath variability was particularly evident near these respiratory phase transitions.

Rdof

The final measure of complexity that was applied to these data is the rate of increase of spectral degrees of freedom, Rdof, which was proposed by Farmer, *et al.*[10]. If one defines the degrees of freedom of a power spectrum, P(f), as

$$\mathrm{dof}(P,N) = [\sum_{n=1}^{N} P(f_n)]^2 / \sum_{n=1}^{N} [P(f_n)]^2$$

then a pure sine wave has one degree of freedom, independent of N. Uncorrelated random noise has N degrees of freedom and deterministic chaos has $1<\mathrm{dof}<N$. In the latter case dof increases with the observation time (i. e., N), but less so than for random noise. Farmer, *et al.*, originally proposed this measure to characterize slowly evolving chaotic systems. The rate of increase of degrees of freedom per data sample, Rdof, is found by calculating dof for the first half of a data set, then for the entire data set, and finally dividing their difference by N/2. In order to compare data sets sampled at different rates, one would have to include the sampling interval in the divisor.

Rdof was calculated for DDDV, as was approximate entropy. Based on preliminary tests of reproducibility, we analyzed 4096 data points of DDDV acquired by sampling at 37.5 Hz. Consequently each record represented about 109 seconds of data, or typically about 120-150 breaths.

RESULTS

As we had seen with tracheotomized rats, lowering of EEV by exposure to a negative transrespiratory pressure elicited a high degree of variability in the respiratory pattern (Figure 2). In contrast to the earlier studies, raising EEV by positive transrespiratory pressure did not lead to observably increased regularity in the breathing pattern. In most cases, however,

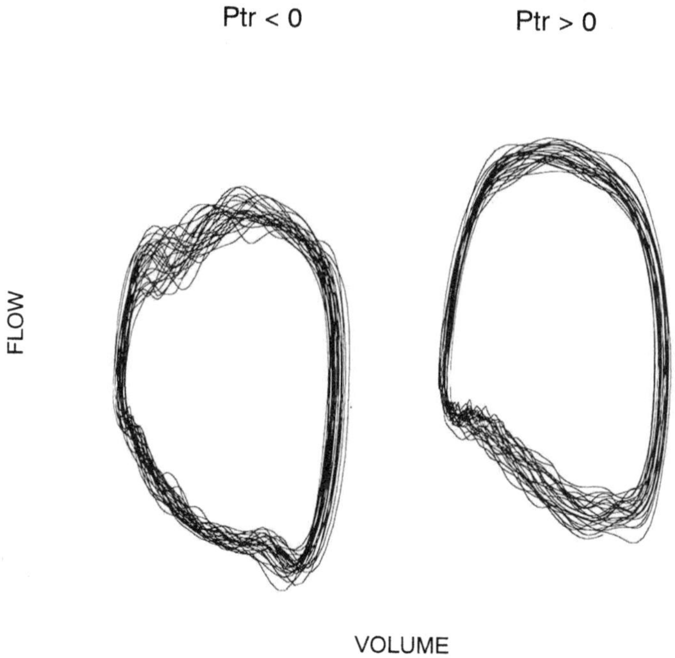

Figure 2. Flow-volume plots, prepared as described in figure 1, from an anesthetized rat breathing through an intact upper airway when exposed to positive (right) and negative (left) transrespiratory pressure. Note more similar breath to breath variability with deflationary and inflationary pressures than was the case in the tracheotomized rat (figure 1).

there was a visual impression that breath to breath variability was greater with negative PTR and that the variability occurred at different times in the respiratory cycle with the different transrespiratory pressures.

Coefficients of Variation

The coefficients of variation (CV) of respiratory cycle parameters related to inspiratory volume and flow were larger during negative PTR, whereas those related to expiratory flow and to timing of the respiratory cycle were not consistently different. The CV was larger during negative transrespiratory pressure for V_T (10/11 rats), peak inspiratory flow (11/11), and rate of change of flow at end-inspiration (9/11). On this basis we concluded that in rats breathing through an intact upper airway the respiratory pattern is more variable from breath to breath with a sustained deflating transrespiratory pressure than with a sustained inflating pressure, just as we had observed in tracheotomized rats. The difference in variability was not as obvious visually compared to the tracheotomized rats.

Recurrence Analysis

Recurrence plots often exhibited evidence of correlation in the breath by breath records of respiratory cycle variables when transrespiratory pressure was negative (Figure 3). In order to compare these analyses between positive and negative PTR using the three proposed quantitative measures, however, it is necessary to standardize the embedding and the specified distance, r. Specifying r is problematic because the mean values and standard deviations are different at the two pressures. Using the same absolute value for r often resulted in too few points on the recurrence plot at

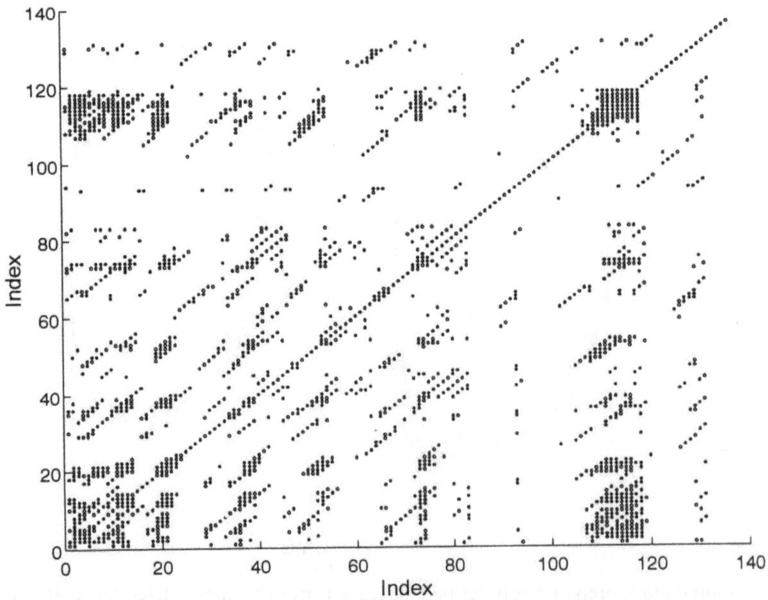

Figure 3. Recurrence plot for tidal volume (135 breaths) from a rat with intact upper airway exposed to deflationary mean transrespiratory pressure. Embedding dimension equals 5.

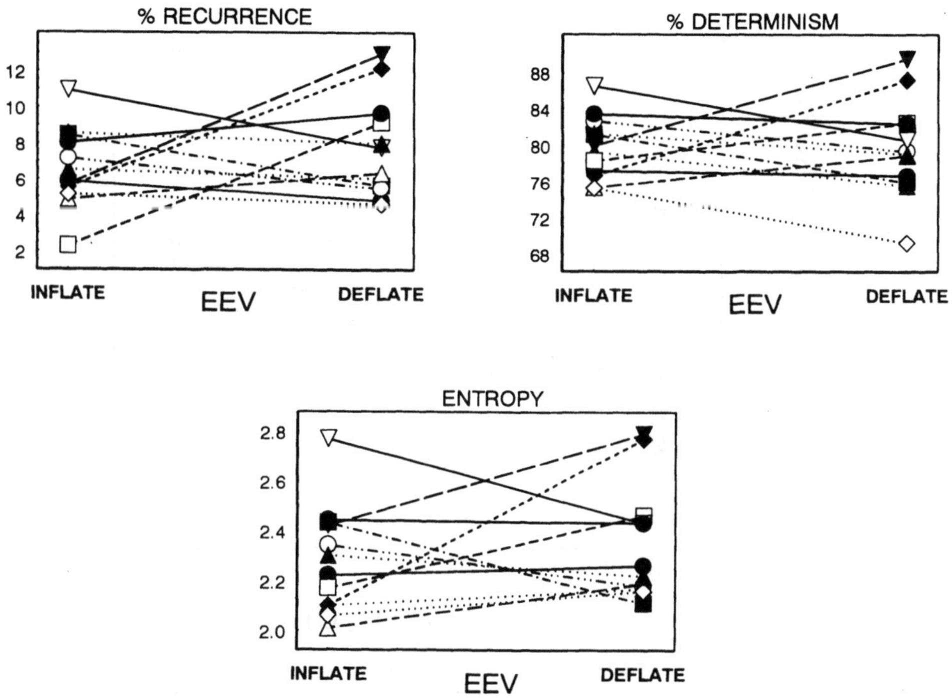

Figure 4. Quantification of recurrence plots of breath by breath tidal volume from eleven anesthetized rats breathing through intact upper airways. Neither percent recurrence, nor percent determinism, nor entropy exhibit consistent differences between inflationary and deflationary transrespiratory pressures.

one of the pressures. Consequently we calculated the three measures for r equal to one-half of the mean value of the cycle parameter. The results are shown in figure 4. None of the three quantification methods yielded a consistent difference between positive and negative transrespiratory pressures.

Approximate Entropy

ApEn based on DDDV provided a clear discrimination between the respiratory patterns during the inflationary and deflationary transrespiratory pressures (Figure 5) for 6 of 7 animals tested. For individual animals the larger value of ApEn during deflation implies that the breathing pattern is more complex than during inflation.

Rdof

The rate of increase of spectral degrees of freedom of DDDV also indicated a greater complexity of the respiratory pattern during deflation than during inflation (Figure 6). In this instance Rdof was greater during deflation for all 7 animals. As with approximate entropy, comparisons between animals were not possible because of the wide range of Rdof at a given transrespiratory pressure. Rdof was calculated also using shorter data records but the reproducibility within a data record was poor when only 1024 points were used. Because we acquired only two minutes of data at each pressure, we could not address reproducibility when using 4096 points.

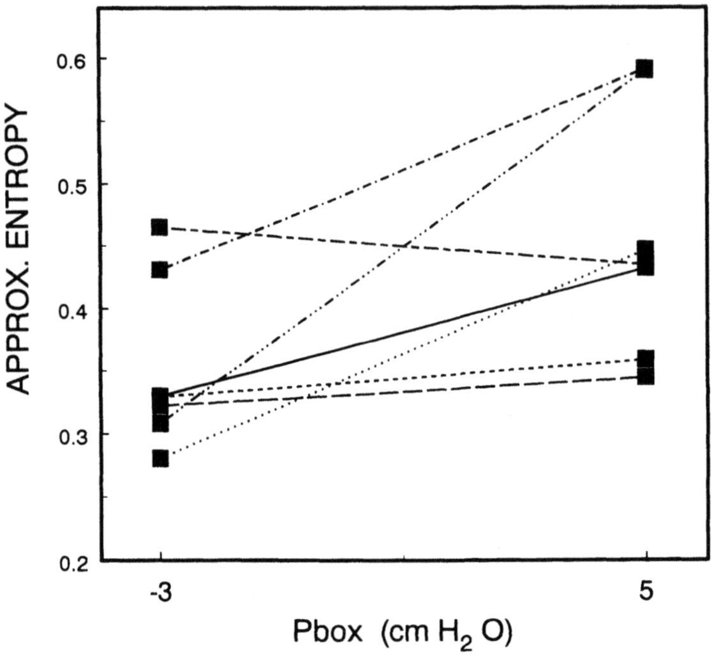

Figure 5. Approximate entropy of DDDV during inflationary (Pbox = -3) and deflationary (Pbox = 5 cm H_2O) transrespiratory pressure. Each line represents one rat. Analysis based on sampling DDDV at 30 Hz for 34.13 sec.

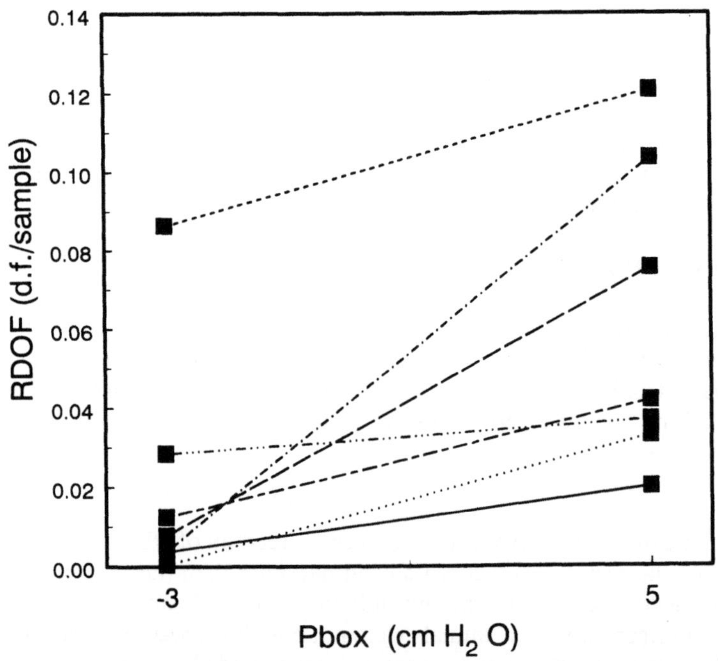

Figure 6. Rate of increase of spectral degrees of freedom of DDDV during inflationary (Pbox = -3) and deflationary (Pbox = 5 cm H_2O) transrespiratory pressure. Each line represents one rat. Analysis based on sampling DDDV at 37.5 Hz for 109.23 sec.

DISCUSSION

The initial objective of this study was to determine whether the breathing pattern of anesthetized rats with intact upper airways would be more variable when the animals were exposed to a negative transrespiratory pressure than when exposed to a positive transrespiratory pressure. Because the breathing pattern was still variable in the latter situation, another objective was to establish a method for quantitative analysis of the degree of breath to breath variability of respiratory pattern. By visual observation, and on the basis of the coefficients of variation of certain respiratory cycle variables, we concluded that the pattern was indeed more variable during the deflationary pressure. Several potential quantitative measures of variability were assessed and two - approximate entropy and rate of increase of spectral degrees of freedom - gave results consistent with expectations derived from visual observations and CV analysis.

The qualitative finding that respiratory pattern is more variable during deflation maneuvers is consistent with our previous observations in tracheotomized rats and interpretations of those results have been presented[1,3,4,7]. Here the focus will be on the quantitative analysis of respiratory pattern variability. The first consideration is the question of what aspect of respiratory pattern should be used as an index of breath to breath variability. Single-point indices such as tidal volume may not always sensitively reflect variability in the time course of a breath. For example, the observation that the CV of peak expiratory flow was not different between the two pressures disguises an actual change in variability of the expiratory flow pattern. With positive PTR peak expiratory flow occurred at nearly the same time in expiration on every breath, but with negative PTR the time at which expiratory flow peaked was highly variable (although the values of peak expiratory flow were similar under both pressures and had similar CVs). Furthermore we had observed previously that the variable pattern during deflation was associated with variability in the breath trajectory near the respiratory phase transitions. This variability was more observable in the flow signal, DV, than in volume itself and even more obvious in the second derivative of flow, DDDV. Therefore we chose to use this latter signal to represent the variability in the respiratory pattern. Although similar results were obtained from analyses of DV, differences in ApEn and Rdof were smaller. We did not, however, determine whether reproducibility was higher using DV, in which case smaller differences still might have been significant.

Recurrence analysis has been used to discriminate patterns of heart rate variability[14] and individual recurrence plots of respiratory cycle variables often exhibited evidence of repeated local correlations in the data. However we found it very difficult to compare one recurrence plot to another when the means and standard deviations of the two data sets differed. Initially we calculated percent recurrence, percent determinism, and entropy over a wide range of specified distance, r, and tried to identify an absolute value for r which could apply to both data sets. When this approach failed to produce any consistent results, we reasoned that setting r to be the same fraction of the mean might be more appropriate. However this approach also failed to produce consistent findings. It appears that more analyses of the properties of these measures are needed. For example, there is an interaction between r and the amplitudes of deterministic and stochastic signal components and it is unclear how to select an optimal r. Also these methods may be sensitive to changes in the frequencies of any periodic variations in the data. At any rate we were not successful in using them as quantitative measures of respiratory pattern variability.

Approximate entropy, which has been used to discriminate patterns of heart rate variability in neonates[15], identified a difference in the complexity of respiratory pattern in 6 of 7 rats. On the other hand ApEn values could not be compared between animals even though the data analysis parameters m, r, and N were standardized. This measure of variability may be useful for within-animal comparisons, but several issues need clarification before its use can be recommended. A major issue is the sensitivity of ApEn to changes in variability of different types. Pincus[8,9] analyzed his MIX processes to demonstrate that approximate entropy increased when the stochastic proportion of the signal increased relative to the deterministic component. We have observed a similar result for the addition of increasing levels of uncorrelated noise to a sine wave. Pincus[8] also tested chaotic signals of increasing dimension. It is still necessary to test the behavior of ApEn when it is applied to various combinations of deterministic signals with added correlated noise. It should be noted that ApEn is *not* an index of whether a signal is chaotic or fractal, *per se*.

We did not systematically assess reproducibility of ApEn on either synthetic or real data and the question is open regarding how to select the appropriate data length, N. The results from Pincus[8] suggest that 500 points may be too few and 1000 may be sufficient, but one needs to address this issue using synthetic data having properties similar to those expected in real data. Furthermore it is clear that the actual value of approximate entropy changes with sampling frequency; thus, it probably is necessary to sample data to be compared at the same frequency. Several methods of testing results of analyses of unknown data using surrogates have been proposed[16,17] and it is suggested that these tests should be applied whenever any of the methods considered here are utilized.

Assessment of the rate of increase of spectral degrees of freedom was proposed by Farmer, *et al.*[10] to distinguish slowly-evolving, low-dimensional chaotic systems from nonchaotic deterministic systems on the basis that the apparent spectral content of the former would increase if one observed the system for a longer time. If one implements this approach using an FFT-based algorithm, it is necessary to consider the potential complication from the effect of the data window on the FFT. Thus, for quasi-periodic signals such as respiratory signals it might be prudent to analyze an additional surrogate - a sine wave having the same mean frequency as the data - in order to evaluate the windowing effect.

For our data Rdof identified the respiratory pattern during deflation as being more complex in all 7 rats. Nonetheless, the issues raised above regarding ApEn also limit the enthusiasm for recommending general applicability of this measure. It is clear that sampling frequency and data length considerably influence the actual value of Rdof and its current use requires standardizing these parameters between data sets. Through trial and error it was found that reproducibility on real data was better using 4096-point data records sampled at 37.5 Hz than using data records of this same length sampled at the original rate of 150 Hz. Also, reproducibility suffered when records of 1024 points were used.

CONCLUSIONS

1. Both ApEn and Rdof, when applied to a waveform such as DDDV, may provide useful quantification of the degree of breath to breath variability of respiratory pattern.
2. The concept of recurrence analysis is intuitively appealing but in practice it is difficult to compare data sets quantitatively using proposed measures.
3. All of these indices require further testing before they can be recommended for general use as quantitative measures of respiratory pattern variability.

REFERENCES

1. Bruce, E. N., and J. A. Daubenspeck. Mechanisms and analysis of respiratory variability. In: *Control of Breathing, 2nd ed.*, Dempsey, J., and A. I. Pack, ed.; Marcel Dekker, New York, 1995; p. 285-314.

2. Modarreszadeh, M., E. N. Bruce, and B. Goethe. Nonrandom variability in respiratory cycle parameters of humans during stage 2 sleep. *J. Appl. Physiol.* 69: 630-639, 1990.
3. Sammon, M. P., J. R. Romaniuk, and E. N. Bruce. Bifurcations of the respiratory pattern associated with reduced lung volume in the rat. *J. Appl. Physiol.* 75: 887-901, 1993.
4. Sammon, M. P., J. R. Romaniuk, and E. N. Bruce. Role of deflation-sensitive feedback in control of end-expiratory volume in rats. *J. Appl. Physiol.* 75: 902-911, 1993.
5. Tsubone, H. Characteristics of vagal afferent activity in rats: three types of pulmonary receptors responding to collapse, inflation, and deflation of the lungs. *Exp. Neurol.* 92: 541-552, 1986.
6. Bartlett, D., Jr. Respiratory functions of the larynx. *Physiol. Rev.* 69: 33-57, 1989.
7. Sammon, M. P., and E. N. Bruce. Vagal afferent activity increases dynamical dimension of respiration in rats. *J. Appl. Physiol.* 70: 1748-1762, 1991.
8. Pincus, S. M. Approximate entropy as a measure of system complexity. *Proc. Natl Acad. Sci. USA* 88: 2297-2301, 1991.
9. Pincus, S. M., and A. L. Goldberger. Physiological time-series analysis: what does regularity quantify? *Am. J. Physiol.* 266 (*Heart Circ. Physiol.*) H1643-H1656, 1994.
10. Farmer, D., J. Crutchfield, H Froehling, N. Packard, and R. Shaw. Power spectra and mixing properties of strange attractors. *Annals NY Acad. Sci.* 357: 453-472, 1980.
11. Eckmann, J.-P., and D. Ruelle. Recurrence plots of dynamical systems. *Europhys. Lett.* 4: 973-977, 1987.
12. Webber, C. L., Jr., and J. P. Zbilut. Dynamical assessment of physiological systems and states using recurrence plot strategies. *J. Appl. Physiol.* 76: 965-973, 1994.
13. Parker, T. S., and L. O. Chua. Chaos: A tutorial for engineers. *Proc. IEEE* 75: 982-1008, 1987.
14. Zbilut, J. P., M. Koebe, H. Loeb, and G. Mayer–Kress. Use of recurrence plots in the analysis of heart beat intervals. In: *Proc. IEEE Computers in Cardiology*, edited by A. Muray and K. Ripley. Los Alamitos, CA: IEEE Comput. Soc., 1991, p.263-266.
15. Pincus, S. M., T. R. Cummins, and G. G. Haddad. Heart rate control in normal and aborted-SIDS infants. *Am. J. Physiol.* 264 (*Regul. Integr. Comp. Physiol.*): R638-R646, 1993.
16. Theiler, J., B. Galdrikian, A. Longtin, S. Eubank, and J. D. Farmer. Using surrogate data to detect nonlinearity in time series. In: *Nonlinear Prediction and Modeling*, edited by M. Casdagli and S. Eubank, Addison Wesley, Redwood City, CA, 1991.
17. Schiff, S. J., and T. Chang. Differentiation of linearly correlated noise from chaos in a biologic system using surrogate data. *Biol. Cybern.* 67: 387-393, 1992.

10

FRACTAL NOISE IN BREATHING

Bernard Hoop, Melvin D. Burton, and Homayoun Kazemi

Pulmonary and Critical Care Unit, Medical Services
Massachusetts General Hospital, Harvard Medical School
Boston, Massechusetts 02114

1. WHY WE BREATHE WITHOUT HAVING TO THINK ABOUT IT

Our understanding of respiration derives from applications of a variety of physical and life science disciplines, methods, and models to a critical physiological process: exchange and balance of oxygen and carbon dioxide. We know that breathing at rest arises from a diversity of interrelated and interactive physical and chemical mechanisms involving molecular and cellular processes in the brainstem which include — among other phenomena common to the central nervous system — metabolism, synaptic transmission of neurochemicals, neurochemical-mediated alteration of neural cell membrane potential, transmembrane ion conductance, neural electrical signal propagation, and neuromodulation by afferent chemoreceptive and mechanoreceptive inputs.

2. BREATHING FLUCTUATES

As with many recurring natural events, the resulting depth, shape, and duration of breaths *fluctuate* in time. Even at rest, during sleep, or during altered states of consciousness, the shape of each breath we take, as well as each time interval of recurrence, is never an exact replicum of any previous breath. Fluctuations observed (measured) in breathing may often reflect *uncertainties* in measured quantities which are usually assumed to be normally distributed and therefore characterized by a mean value and a variance defined for a specific interval of time over which a number of measurements are made. Changes in mean values of these measured quantities during stimulation are often interpreted in the context of physical models of central respiratory control as *regulator* and/or as *central pattern generator*. Until recently, a key feature of such control models is *linearity*, characterized by two properties: proportionality and independence[1]. Underlying the interpretation of breathing and other living processes as *linear* phenomena is a powerful and persuasive axiom of 20th-century medical and life sciences — the principle of *homeostasis* — articulated by Walter B. Cannon of Harvard Medical School[2]. Homeostasis means that the normal operation of a physiological *system* is to reduce variability and to maintain a constant internal *function*. That is, fluctuations occur about a presumed normal steady state which may be represented (modeled) as a continuous function. The concept that a mathematical function and the geometric representation of a line are related—a seminal idea of the 17th century underlying the method of calculus — has thus promoted development of theories of *linear systems* as mathematical representations (models) of physiological control processes. In models of the neural motor act of breathing, homeostasis and linearity are served and sustained by a structurally well-organized linear network or control system with well-defined and identifiable afferent and efferent neural pathways, attendant control processes, and a controller or control center[3,4,5,6,7].

3. WHY BREATHING FLUCTUATES

Alternatively, fluctuations observed in breathing may be interpreted as transient responses to a fluctuating environment over a wide range of *scales* [1]. Upon initial inspection, a time sequence of normal breath-to-breath fluctuations in magnitude and rate of resting respiration, except for sighs and other isolated events, appears irregular and completely random. By *random* is meant that one fluctuation is independent of another, i.e., there is no apparent dependence or correlation among fluctuations. If, however, several minutes of breathing are considered, the fluctuations may look remarkably like those observed over an hour or more. That is, breath-to-breath fluctuations on different time scales may appear to be *self-similar* * like the branches of a geometric *fractal* object [9]. In other words, magnitude and rate of breathing may fluctuate over many scales, even in the absence of fluctuating external stimuli, rather than relaxing to a homeostatic steady state. This concept — termed *homeodynamic* by West [1] — arises out of recent applications of nonlinear dynamics to living systems and is based on the premise that there exists a complex of multiple states which determines the behavior of healthy living systems. This may be viewed as a spectrum of small but superimposed fluctuations in internal control mechanisms in living systems over a wide range of temporal and spatial scales. A characteristic of this new paradigm is flexibility of response and *tolerance of error*. In this paradigm, the single scale steady state of homeostasis is replaced by a multiplicity of nonequilibrium states which are *correlated* over many scales of time and distance.

4. FLUCTUATIONS IN BREATHING ARE CORRELATED

Long-term correlations at the microscopic level are observed in a variety of structures and processes ranging from DNA sequences[10] to action potentials[11]. Consider, for example, ion channel proteins. These cell membrane proteins can have different shapes called conformational states. There is sufficient thermal energy in their environment to cause these proteins to switch spontaneously between different conformational states[12]. The switching times between these states have fractal properties. That is, kinetic *rate constants* and related parameters which describe the probability per unit time that neurotransmitter-activated membrane ion channels will change from a closed to an open state are *power law* functions of the time resolution used to determine that probability[13]. A rate constant which describes ion channel switching serves as an example of one of a number of classical linear representations of a cellular process which underlies the concept of a central respiratory pacemaker, i.e., a well-defined

* Self-similarity in fractal phenomena is strictly defined under a scaling in which the variables scale by the same ratio, in distinction to *self-affinity*, in which the variables scale by different ratios [8].

internal timing mechanism which can influence the rate or set the pace of breathing[14]. However, respiration is not only generated by a single pacemaker at the cellular level, but rather by a conditional process consisting of a sequence of motor neuronal events extending over a time scale which is long compared with cellular physiological events such as ion channel switching, but which is short compared to neuromodulatory processes[15]. A fractal relationship in resting respiration is therefore the consequence of physio-chemical processes acting over short times and distances at the cellular level, and which are correlated with neural processes acting simultaneously over long times and distances.

It is clear that different experimental methods which alter respiration and respiratory neural output have quite different physiological effects on resipration, some of which are quite specific and some which are quite general. For example, exercise hyperpnea and hypoxia have a general effect on many, if not all underlying metabolic processes in the central nervous system, whereas application of a specific neurochemical or a specific agonist or antagonist of a neurochemical receptor site affects, by definition and experimental design, a specific neurochemical system and its underlying metabolic pathways. For example, the classical neurotransmitter acetylcholine is a powerful ventilatory stimulant when applied centrally and is important in generating respiratory output in the isolated brainstem preparation[16]. That is, cholinergic transmission plays a critically important role in central chemoreception, which has a major influence on respiratory rhythm generation. It is, however, *correlation* within and between this and other respiratory-related phenomena over a wide range of time and distance scales which underlies the internal regulation of breathing and which implies that breathing is fractal.

5. FLUCTUATIONS IN BREATHING ARE FRACTAL NOISES

Fractal series or sets are those whose characteristic form or degree of irregularity is the same through a succession of *scale* changes[17]. The geometrical structure of sets of many naturally occurring phenomena, like tree ring widths, water table levels, lightning bolt paths, etc., can be characterized deterministically by a fractal dimension rather than by a topological dimension. By *deterministic* is meant that the dimension of the set defines the set completely. The *measure* of a set may be the values of any physical observable with units of length, area, volume, etc. For a fractal structure the measure of the set increases with an increase in the *resolution* of the measuring instrument. In this instance, the double log plot of measure against resolution is always a straight line, the slope of which is related to the fractal dimension. A temporal fractal is a process that does not have a characteristic scale of time, analogous to a fractal structure that lacks a characteristic scale of length. In the case of respiration, for example, the time intervals of breathing in fetal lambs[18] have self-similar bursts of activity, power law distributions of the duration of intervals, and a power law form of the power spectral density. * Depth of breathing (tidal volume) at uniform breathing rate in the fully innervated mammal and in stimulated brainstem respiratory neural output are positively correlated[19,20]. In human subjects breathing hyperoxic and hypoxic gas mixtures at rest, a large percentage of spectral power is not harmonic but exhibits a power law dependence[21]. These properties of temporal correlation, self-similarity, power law distributions of intervals, and power law form of the power spectrum are examples of processes with properties of fractal *noise* [22].

* In this context, it should be clear that the word *power* is being used with two distinct meanings.

6. FRACTAL PROCESSES ARE ERROR TOLERANT

Why are fractal processes ubiquitous in living phenomena, including breathing? As suggested above, one reason is that fractal processes are more tolerant to error than are classical processes. This may be demonstrated with a simple model of error response which introduces a random fluctuation into the parameters of a classical model and of a fractal model and then by averaging (measuring) over an ensemble of these fluctuations. We follow the argument proposed by West and Deering[23]. Consider the measurement of two time-dependent phenomena: a classical process represented by an exponential law, $\varphi_C(t) = \exp(-\gamma t)$ and a fractal process represented by a *power law*, $\varphi_F(t) = t^{\alpha}$. The functions φ_C and φ_F are taken to represent any number of physical observables, either directly measured quantities (e.g., ion channel switching times), or derived from directly measured quantities, where the variable t represents generations of time scales over which measurements are made. We now assume that the scaling factor or *rate constant* γ for the classical process consists of two parts: a constant γ_0 and a *random* part ξ, and that values of measured quantities are distributed according to a Gaussian probability density function with zero mean and variance σ^2. The measured classical quantity is therefore given by $\langle\varphi_C(t)\rangle_\xi = \exp(-\gamma_0 t)\,\langle\exp(-\xi t)\rangle_\xi$ where the brackets denote the average over the distribution of time scales. In the case of Gaussian distributed quantities, carrying out the average yields $\langle\varphi_C(t)\rangle_\xi = \langle\varphi_C(t)\rangle_0 \exp(\sigma^2 t^2/2)$ which demonstrates that the error grows as $\exp(\sigma^2 t^2/2)$. In the same way, we assume that the index α for the fractal process consists of two parts: a constant α_0 and a *random* part ξ. Again assuming the same probability density function, we find that

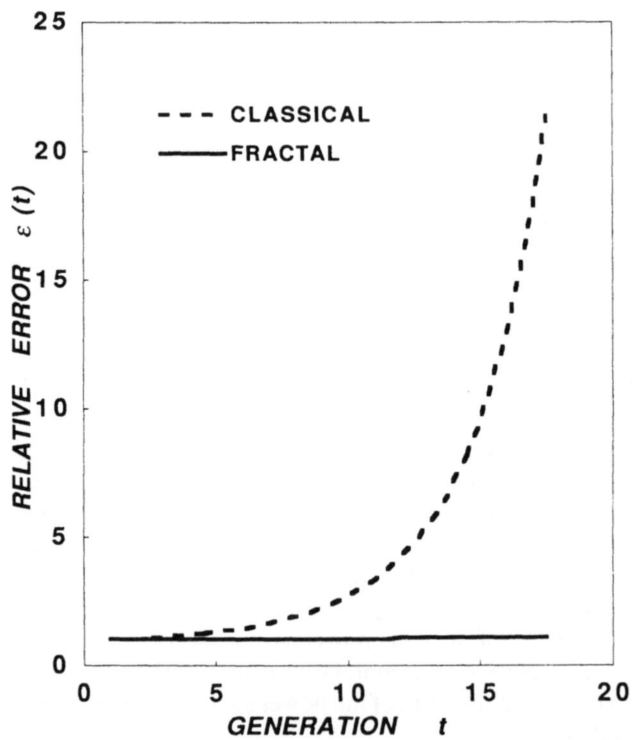

Figure 1. Comparison of relative error $\varepsilon(t)$ over generation t with random noise for classical and for fractal scaling (From West and Deering, 1994 with permission).

the measured fractal quantity is $\langle\varphi_F(t)\rangle_\xi = t^{\alpha_0}\langle\exp(-\xi \text{in} t)\rangle_\xi$ which yields $\langle\varphi_F(t)\rangle_\xi = \langle\varphi_F(t)\rangle_0 \exp(\sigma^2(\text{in}t)^2/2)$, in which the error grows as $\exp(\sigma^2(\text{in}t)^2/2)$. The relative error $\varepsilon(t) = \langle\varphi(t)\rangle_\xi / \langle\varphi(t)\rangle_0$ in both the classical and fractal models for $\sigma^2 = 0.02$ is plotted in Figure 1.

In Figure 1, it is clear that as t increases, relative error in the classical model rises rapidly, compared to the fractal model. For $t > 20$, relative error in the classical model is $\varepsilon(t) > 100$, whereas relative error in the fractal model is only 1.10. This illustration suggests that a fractal process is essentially unresponsive to error and very tolerant of variability in the physiological environment. Therefore, a process such as breathing with a diversity of inputs over a wide range of temporal and spatial scales and with concomitant sensitivity to error incurred by this spectrum of inputs must have the error-tolerant properties of fractals.

7. FRACTAL NOISES ARE POWER LAWS

Whereas events which recur with constant frequencies appear as peaks (harmonics) in a power spectrum, *power laws* may dominate the power spectra of *stochastic* processes such as *noise* [24]. A stochastic process is a process which involves a variate at each moment of time, where a *variate* is a variable that may take on any of the values of a specified set with a specified probablility. Sequences of breath heights (tidal volumes), time intervals between openings and closings of ion channels, and step lengths of a Brownian particle are all examples of stochastic processes. Among the many examples of fluctuating natural processes in space and time which the word *noise* may invoke, two familiar concepts are those of a random walk and Brownian motion[25,26]. The distribution of step lengths in such processes is often modeled as a Gaussian or normal probability distribution. If in these processes, each event (step) is independent of another, with unpredictable direction, length and/or duration, we refer to the processes as *ordinary.*

Consider the position of a Brownian particle which is moving only along the x-axis from position $x(t_0)$ at time t_0 to $x(t)$ at a later time t. Following the argument reviewed by Feder[8], if we assume that the step length of the Brownian particle is represented by a normalized independent Gaussian random process, then the increments in the position of the Brownian particle are defined by $x(t) - x(t_0) \sim \xi\,|\,t - t_0\,|^{\,H}$ for the two times t and t_0, where $t \geq t_0$. What this means is that, given the position $x(t_0)$, one determines the position $x(t)$ by choosing a random number ξ from a Gaussian distribution, multiplying it by an appropriate normalization constant and by the time increment $|t - t_0|^H$, and adding the result to the position $x(t_0)$. Here, the exponent $H = 1/2$ for ordinary Brownian motion. A generalization of this to any H in the range $0 < H < 1$ is called fractional or *fractal Brownian motion* (fBm) (cf. APPENDIX). If the position of the Brownian particle at time t is $B_H(t)$, the *derivative* of fractal Brownian motion is called fractal Brownian noise (fBn) or, alternatively, *fractal Gaussian noise* (fGn). The derivative of $B_H(t)$ with respect to t is formally defined as $\lim_{\delta\to 0}\{[B_H(t+\delta) - B_H(t)]/\delta\}$, familiar from the differential calculus for continuous functions. For a discrete set of uniformly spaced points, this definition reduces to taking the differences between adjacent positions. Fractal Gaussian noise has a variance that does not diverge with time and is therefore *stationary*. (In common usage, *stationary* means that something is not changing in time. In this context, stationary means only that the moments (e.g., variance) of the process are defined. Often, if the variance in fluctuations of a time series is determined for increasing intervals of time, the value of the variance will increase. That is, the sample variance may not reach a limiting finite value but may increase indefinitely with the length of the time interval used to evaluate it. In the statistical literature, this is known as *heteroscedasticity*[27]. A time series with such moments is said to be *nonstationary* .)

For fGn, the power spectral density or power vs frequency f is of the form $1/f^{(2H-1)}$. That is, the average power spectrum of fGn varies inversely with frequency to the power $2H - 1$, where $0 < H < 1$. The power spectrum is thus a power law with exponent $2H - 1$. For $H = 1/2$, the exponent is zero, which is to say that all frequency components are equally represented, from which the term "white noise" comes, and which is known as ordinary Gaussian noise. For other values of H, we have what is known as fractal Gaussian noise. For $0 < H < 0.5$, this noise is negatively correlated and for $0.5 < H < 1$, it is positively correlated. As we shall demonstrate below, fractal noise in breathing is found to be positively correlated.

8. SIMULATING FRACTAL GAUSSIAN NOISE

In order to determine the fractal dimension of a measured sequence of respiratory events, we must first be able to simulate noise with *known* fractal dimension and to recover that fractal dimension with our analytic methods. Fractal Gaussian noise may be simulated to an arbitrary resolution using one particular method known as successive random additions, proposed by Voss [28]. As reviewed by Feder[8] we consider a horizontal axis t, where $0 \leq t \leq 1$, on which we wish to construct a set of *vertical* positions $x(t)$ equally spaced on this interval, such that the variance of increments in the vertical positions is given by $|t|^{2H}\sigma_0^2$, where the chosen value of H in the range $0 < H < 1$ is the exponent related to the fractal dimension of the noise, and where σ_0^2 is the initial variance of the random additions described below. We begin by setting the vertical positions $x(t_1)$, $x(t_2)$, $x(t_3)$ of the end-points and mid-point of the interval $0 \leq t \leq 1$ equal to zero, i.e., at $t_1 = 0$, $t_2 = 1/2$, $t_3 = 1$. Next, we add to each of these three positions random vertical amounts chosen from a normal distribution with zero mean and an initial variance of unity, i.e., $\sigma_0^2 = 1$. We then interpolate between these three vertical positions to obtain the vertical positions at the midpoints of the two sub-intervals, which now give us five equally spaced vertical positions on the entire interval. Each of these five vertical positions are again given random additions from a normal distribution with zero mean and a variance *reduced* by the factor r^{2H}, where $0 < r < 1$ is a chosen scaling factor and $0 < H < 1$ is the chosen exponent of the simulated fractal noise. These five positions are again interpolated to the midpoints of their subintervals to give nine positions, etc. After n applications of this algorithm we will have defined a set of $(2^n + 1)$ vertical positions. The variance of the addition in the nth generation of this process is $\sigma_n^2 = r^{2H}\sigma_{n-1}^2 = r^{2nH}\sigma_0^2$. To obtain normally distributed random numbers, we employ the direct method[29] which selects two random numbers uniformly distributed on the interval [0,1], from which a single normally distributed random number is determined.* The differences between adjacent values (derivative) of the resulting set of $(2^n + 1)$ vertical positions are then taken to obtain fractal Gaussian noise.

9. FRACTAL METHODS OF ANALYSIS

Time series of respiratory events lend themselves to analysis with fractal methods characterized by a single parameter: the exponent H, which is related to the fractal dimension of the time series. The magnitude of the exponent H is therefore expected to be a sensitive indicator of fractal noise in respiration. Fractal methods of analysis include autocorrelation

* Other choices for generating and for testing the adequacy of random deviates from a normal distribution have also been described[29].

and extended range correlation[17], Fano factor analysis[30], detrended fluctuation analysis[31], power spectral, rescaled range, and relative dispersional analysis[17]. In this tutorial, we use rescaled range and relative dispersion as examples of fractal methods, and we compare these methods with power spectral analysis.

Rescaled range analysis was first described in 1950 by Hurst[32]. Its importance was recognized at once by Feller[33] and summarized by Mandelbrot in his classic treatise on the fractal geometry of nature[9]. Let $\langle V \rangle_T$ be the mean value of a sequence of respiratory events — for example resting tidal volume — over T breaths, i.e., the sum, divided by T, of all breath heights (tidal volumes) $V(t)$ indexed by integer t, where $1 \leq t \leq T$. Let us now define $X(t, T)$, the accumulated departure of $V(t)$ from $\langle V \rangle_T$ as the sum of the differences $[V(u) - \langle V \rangle_T]$ taken over all breaths indexed by integer u where $1 \leq u \leq t$ for each t on the interval $1 \leq t \leq T$. On this interval (where T is often called the lag), the difference between the maximum X_{max} and the minimum X_{min} is called the range $R(T)$. If $R(T)$ is divided by the estimate of the standard deviation $S(T)$ for the same number of breaths on the interval $1 \leq t \leq T$, then the quantity R/S is called the rescaled range and is described by the power law, $R/S \sim T^H$, where the exponent H is called the Hurst exponent. As we have seen above, in the absence of long-term statistical dependence, R/S becomes asymptotically proportional to $\sqrt{T}$ (i.e., $H = 1/2$) for sequences generated by statistically independent processes with finite variances. Correction for linear trend[34] is made by determining values of the range $R(T) = X_{max} - X_{min}$ from the difference between actual cumulative departure $X(t, T)$ and linear regression of $X(t, T)$ with time t for each lag T.

Dispersional analysis or relative dispersion, *RD*, first described in 1988 by Bassingthwaighthe[35], makes use of the variance or standard deviation of a sequence of events determined at a succession of different levels of resolution. What is meant by *resolution* is the number or size t of a group of consecutive data points, over which a mean and a standard deviation S are determined. As reviewed by Bassingthwaighte and Raymond[36], for fractal phenomena there is a power law relating S to t, i.e., $S(t) = S(t_0)(t/t_0)^{(H-1)}$, where t_0 is a reference size, and where H is the Hurst exponent. This expression is an example of a power law which describes a fractal relationship, because for positively correlated noise with positive values, the possible slopes of the power law relationship on a log-log plot are bounded by the extreme values $H = 0.5$, for which the data series is indistinguishable from random uncorrelated noise, and a value of H near unity, which represents high near-neighbor correlation and uniformity of the signal over all size scales. Glenny et al.[37] were the first to publish a simple spreadsheet for practical calculation of relative dispersion. In brief, for a given size t, one calculates the mean μ and S and plots $RD = S / \mu$ for a range of sizes t. The slope of log (*RD*) vs log t is (H - 1).

There are several practical methods for determining whether a measured value of H for a given time series of respiratory events differs significantly from a value of 0.5. One method is to randomly reorder, i.e., shuffle the series n times and determine the standard deviation in the mean of H derived from the n shuffled series. One may then employ a test for the significance of the difference between mean H determined from the n shuffled series with the mean H determined from the original series. Another method is to reduce the sampling rate by one-half by taking every other data point in the series and redetermining H for comparison with H determined from the original series. Such tests are essential, particularly when one is working with short time series, as has been repeatedly emphasized and demonstrated elsewhere[20,34,36] and as is shown below. Of equal importance is correction for biases in measured values of H introduced by different methods of fractal analysis. As first demonstrated by Schepers et al.[38] and stressed by Bassingthwaighte and Raymond[34,36], such corrections depend critically not only on the choice of analysis, but on the length of the time series to be analyzed.

10. COMPARING FRACTAL AND SPECTRAL METHODS

To illustrate rescaled range and relative dispersion methods of fractal analysis and to compare them with power spectral analysis, we use the method of fractal Gaussian noise simulation described above to generate ten fGn series, each of 64 points, for each of $H = 0.2$, 0.4, 0.6, 0.8, and 1.0, with a scaling ratio of $r = 0.5$. From power spectral densities of each of these series, slopes $2H - 1$ were determined from unweighted linear regression over the entire frequency range of log power vs log frequency. Rescaled range analysis, uncorrected and corrected for trend, as well as relative dispersion analysis were also applied to each simulated fGn series and compared with power spectral analysis, as shown in Figure 2.

Figure 2 shows values of H determined, respectively, from trend-uncorrected rescaled range (*R/S*), trend-corrected *R/S*, and relative dispersion (*RD*), plotted vs. corresponding values of H determined from power spectra in the upper, middle, and lower panels. For *RD* analysis, each simulated fGn series was set to positive values by adding the absolute value of the largest negative value in the series to all members of the series. It is clear from Figure 2 that for the short time series (64 points) analyzed, there is a considerable scatter among estimates of H from the two fractal methods. However, there is no significant effect on estimates of H with linear trend correction in rescaled range analysis. Furthermore, linear regressions of H from the two fractal methods vs H from power spectra, as shown by the heavy lines in Figure 2, agree with that reported elsewhere[20] and by Schepers et al.[38]. Namely, bias correction in measured H from rescaled range is positive for $H > 0.7$ and negative for $H < 0.7$ and from relative dispersion, positive for $H > 0.5$. Lowen and Teich[39] have made

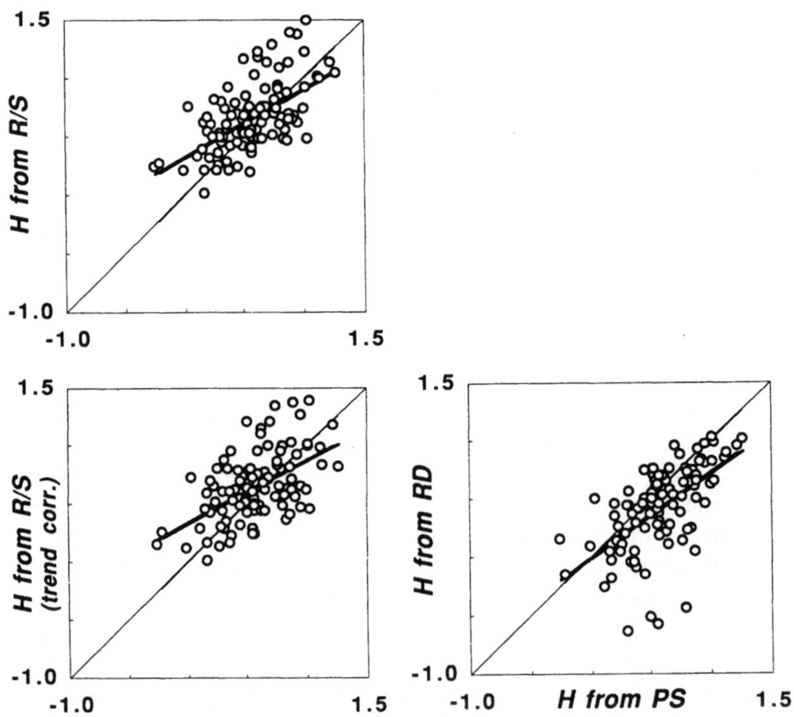

Figure 2. Hurst exponent H from rescaled range (*R/S*) uncorrected and corrected for linear trend and relative dispersion (*RD*) vs. H from power spectra *PS* for simulated fractal Gaussian noise. Heavy lines are linear regressions; light lines have unit slope.

similar comparisons of Fano factor and power spectral density analyses for several different stochastic processes. In this regard, it must be emphasized that the type of fractal noise observed in a time series depends on the specific experimental observable. For example, time series of voltages recorded across the cell membrane of a human T-lymphocyte appear to be a random walk of the form of fractal Brownian motion[40].

11. FRACTAL ANALYSIS OF TIDAL VOLUME

Examples of fractal analyses of simulated noise and of tidal volume noise are shown in Figure 3.

In Figure 3, the three time series each consist of 512 points. The upper part of Figure 3 shows an example of tidal volume simulated as fractal Gaussian noise (fGn), in which the first 128 and final 256 points are generated with a value of $H = 0.8$, and the remaining series of 128 points are generated with a value of $H = 0.5$ (ordinary Gaussian noise), as indicated at the top of Figure 3. The ordinate axis is in μLiters, for which the mean and variance of simulated fGn were chosen to be approximately that of actual tidal volume measured in a 4-day old, 4-g mouse during hyperoxia (30% O_2 breathing) and hypoxia (8.7% O_2 breathing), as shown in the middle and lower parts of Figure 3. In the lower part of Figure 3, the animal breathed 30% O_2 for the first 64 breaths and 8.7% O_2 for the remainder of the series. Below each of the time series shown in Figure 3, bar charts show values of H from rescaled range R/S and relative dispersion RD analyses, respectively. Each bar represents mean H (±SE) derived from linear regression of respective log R/S vs log lag and log RD vs log size plots for 64-point series. Successive bars represent H values determined from 64-point series overlapped by 8 points. That is, the solid bars show H values of contiguous non-overlapping 64-point series, positioned at the center of each series plotted above the bar charts.

In the case of simulated tidal volume (Fig. 3, top), both R/S and RD analyses clearly exhibit the simulated transition from $H = 0.8$ to $H = 0.5$, as shown by the transient decrease in H. The magnitude of standard error and the variation in mean H confirms the findings of Bassingthwaighte and Raymond[34,36] that for short time series (< 200 points), H can be determined to an accuracy of no greater than 10%. Compared to hyperoxia in a 4-day old mouse (Fig. 3, middle) there is a transient decrease in H from both R/S and RD during hypoxia (Fig. 3, bottom), although the standard error in H from RD becomes relatively large during the transient phase. A reason for this may be that hypoxia adds ordinary Gaussian noise to all the central and peripheral cardiorespiratory neural mechanisms and their underlying metabolic pathways. As pointed out by Bassingthwaighte and Raymond[34,36], RD is much more sensitive than R/S to the addition of ordinary Gaussian noise to a process for which $H < 0.5$ than for which $H > 0.5$. This superposition of ordinary Gaussian noise by hypoxia may very well be sustained, although the sequence in the lower part of Figure 3 (mean breathing rate: 1.84 Hz) represents only ca. 15 minutes of breathing. We again caution that these results are derived from very short (64 points) series. However, the results for even short series are encouraging because log-log plots do indeed exhibit power law distributions.

In summary, the Hurst exponent H determines the degree of correlation in time series of respiratory noise, specifically resting tidal volume. For $0 < H < 0.5$, such correlations are negative. That is, increases in the values of the time series are more likely to be followed by decreases. When $H = 0.5$, there are no correlations. That is, increases in the values of the time series are just as likely to be followed by increases as by decreases. When $0.5 < H < 1$, correlations are positive. That is, increases in the values of the time series are more likely to be followed by increases. Based on our analyses, noise in breathing is positively correlated.

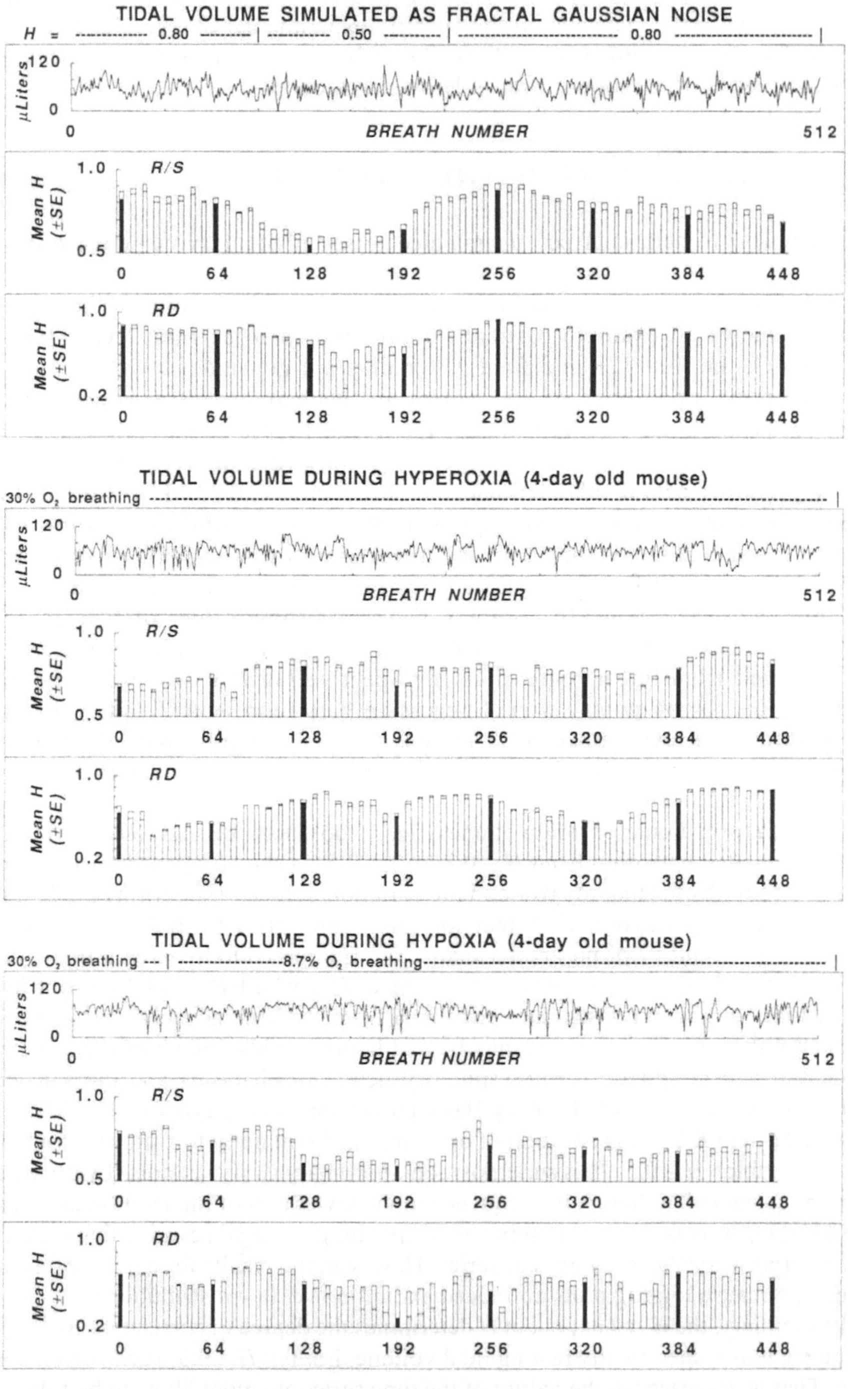

Figure 3. Sequences of tidal volumes simulated as fractal Gaussian noise (top), tidal volume measured during 30% O_2 breathing (middle), and during 8.7% O_2 breathing in a 4-day old mouse (bottom). Bar charts show H values from rescaled range (*R/S*) and relative dispersion (*RD*) analyses (cf. text). [N.B.: Mean H from *R/S* is plotted on an ordinate scale $0.5 < H < 1.0$, whereas H from *RD* is plotted on $0.2 < H < 1.0$.]

From comparisons of rescaled range and relative dispersion analyses with power spectral analysis of simulated fractal Gaussian noise generated with known values of H, the power law dependence of power spectra with exponent $2H - 1$ applies as well to tidal volume fluctuations and is satisfied to within the accuracy of determination of the slope of the power spectrum. It must be emphasized that the slope $2H - 1$ of log power vs log frequency is valid only for fractal Gaussian noise[41]. However, we can not rule out the possibility that deviations from this relationship may also be due to the fact that tidal volume fluctuations may not necessarily be of the form of fractal Gaussian noise.

APPENDIX: FRACTIONAL (FRACTAL) BROWNIAN MOTION

Consider a Brownian particle moving along a line in *increments* under the action of random impulses. Fractal Brownian motion is a generalization of ordinary Brownian motion and is defined by a zero mean random function $B_H(t)$, which is the position of the particle at time t, given by the expression

$$B_H(t) = \frac{1}{\Gamma(H + \frac{1}{2})} \int_{-\infty}^{t} (t - t')^{(H - \frac{1}{2})} dB(t')$$

where Γ is the gamma function and where the exponent H may take the value of any real number in the range $0 < H < 1$. This expression states that the value of the random function $B_H(t)$ at time t depends on all the previous increments $dB(t')$ *at* $t' < t$ of ordinary Brownian motion represented by the function $B(t)$ with zero mean and unit variance, and for which $H = 1/2$. By definition, $B_H(t)$ is a Gaussian random process in time which represents fractal Brownian motion on the interval $[t_1, t_2]$ if and only if

$$y = \int_{t_1}^{t_2} B_H(t) g(t) dt$$

is a Gaussian random variable for all $g(t)$ such that $\langle y^2 \rangle < \infty$, where $\langle y^2 \rangle$ denotes the average value of y^2. For y to have zero mean and unit variance we require the probability density function of y to have the form

$$f_Y(y) = \frac{1}{\sqrt{2\pi\sigma^2}} \exp(-y^2/2\sigma^2)$$

where $\sigma = 1$, which is the probability density function for individual increments $y = B_H(t) - B_H(t_0)$ with zero mean and unit variance. Variances diverge with time for both ordinary and fractal Brownian motion[8] (originally generalized as "fractional" by Mandelbrot [9]).

ACKNOWLEDGMENT

The authors thank L.S. Liebovitch for useful discussion and for communicating results prior to publication, and J. Beagle for technical assistance. One of us (BH) wishes to gratefully acknowledge the support of a Fulbright scholarship during which some of this work was carried out, as well as support, in part, by the U.S. Department of Health and Human Services of the Public Health Service.

REFERENCES

1. West, B.J. *Fractal Physiology and Chaos in Medicine.* Singapore: World Scientific, 1990.
2. Cannon, W.B. *The Wisdom of the Body.* New York: Norton, 1963.
3. Feldman, J.L. Neurophysiology of breathing in mammals. In: *Handbook of Physiology. The Nervous System. Intrinsic Regulatory Systems of the Brain.* Bethesda, MD: Am. Physiol. Soc., 1986, sect. 3, vol. II, chapt. 7, p. 463-524.
4. Richter, D.W., D. Ballantyne, and J.E. Remmers. How is the respiratory rhythm generated? A model. *News Physiol. Sci.* 1:109-112,1986.
5. Botros, S.M., and E.N. Bruce. Neural network implementation of a three phase model of respiratory rhythm generation. *Biol. Cybern.* 63:143-153,1990.
6. Poon, C.S. Respiratory models and control. In: *Physiologic Modeling, Simulation, and Control.* New York: CRC Press, 1995, pp. 2404-2421.
7. Khoo, M.C.K. (ed.). *Bioengineering Approaches to Pulmonary Physiology and Medicine.* New York: Plenum, 1996 (other contributions to this volume).
8. Feder, J. *Fractals.* New York: Plenum, 1988.
9. Mandelbrot, B.B. *The Fractal Geometry of Nature.* San Francisco: Freeman, 1983.
10. Ossadnik, S.M., S.V. Buldyrev, A.L. Goldberger, S. Havlin, R.N. Mantegna, C.K. Peng, M. Simons, and H.E. Stanley. Correlation approach to identify coding regions in DNA sequences. *Biophys. J.* 67:64-70,1994.
11. Turcott, R.G., S.B. Lowen, E. Li, D. Johnson, C. Tsuchitani, and M.C. Teich. A non-stationary Poisson point process describes the sequence of action potentials over long time scales in lateral-superior-olive auditory neurons. *Biol. Cybern.* 70:209-217,1994.
12. Goetze, T., and J. Brickmann. Self similarity of protein surfaces. *Biophys. J.* 61:109-118,1992.
13. Nogueira, R.A., W.A. Varanda, and L.S. Liebovitch. Hurst analysis in the study of ion channel kinetics. *Brazil. J. Med. Biol. Res.* 28:491-496, 1995.
14. Onimaru, H., A. Arata, and I. Homma. Intrinsic burst generation of preinspiratory neurons in the medulla of brainstem-spinal cord preparations isolated from newborn rats. *Exp. Brain Res.* 106:57-68,1995.
15. Bianchi, A.L., M. Denavit-Saubie, and J. Champagnat. Central control of breathing in mammals: neuronal circuitry, membrane properties, and neurotransmitters. *Physiol. Rev.* 75, 1-45,1995.
16. Burton, M.D., M. Nouri, and H. Kazemi. Acetylcholine and central respiratory control: perturbations of acetylcholine synthesis in the isolated brainstem of the neonatal rat. *Brain Res.* 670:39-47,1995.
17. Bassingthwaighte, J.B. , L.S. Liebovitch, and B.J. West, *Fractal Physiology.* New York: Oxford, 1994.
18. Szeto, H.H, P.Y. Cheng, J.A. Decena, Y. Cheng, D. Wu, and G. Dwyer. Fractal properties in fetal breathing dynamics. *Am. J. Physiol.* 263:R141-R147, 1992.
19. Hoop, B., H. Kazemi, and L. Liebovitch. Rescaled range analysis of resting respiration. *CHAOS* 3:27-29,1993.
20. Hoop, B., M.D. Burton, H. Kazemi, and L.S. Liebovitch. Correlation in stimulated respiratory neural noise. *CHAOS* 5:609-612,1995.
21. Tuck, S.A., Y. Yamamoto, and R.L. Hughson. The effects of hypoxia and hyperoxia on the 1/f nature of breath-by-breath ventilatory variability. In: *Modelling and Control of Ventilation,* edited by S.J.G. Semple and L. Adams. New York: Plenum, 1996 L.Adams and B. J. Whipp. New York: Plenum, 1996, p. 297-302.
22. Lowen, S.B., and M.C. Teich. Fractal renewal processes generate 1/f noise. *Phys. Rev. E.* 47:992-1001, 1993.
23. West, B.J., and W. Deering. Fractal physiology for physicists: Levy Statistics. *Physics Reports* 246:2-100,1994.
24. Schroeder, M. *Fractals, Chaos, Power Laws.* New York: Freeman, 1991.
25. Hausdorff, J.M., C.K. Peng, Z. Ladin, J.R. Wei, and A.L. Goldberger. Is walking a random walk: evidence for long-range correlations in stride interval of human gait. *J. Appl. Physiol.* 78:349-358,1995.
26. Berg, H.C. *Random Walks in Biology.* Princeton: Princeton University Press, 1983.
27. DeGroot, M.H. *Probability and Statistics* (2nd ed). Reading: Addison-Wesley, 1989.
28. Voss, R.F. Random fractal forgeries. In: *Fundamental Algorithms in Computer Graphics,* edited by R.A. Earnshaw, Berlin: Springer, pp. 805-835, 1985.
29. Abramowitz, M., and I.A. Stegun (eds). *Handbook of Mathematical Functions, AMS 55.* Washington DC: Natl. Bureau Stand., 1970 (9th printing), sect. 26.8.
30. Teich, M.C., and S.B. Lowen. Fractal patterns in auditory nerve-spike trains. *IEEE Engr. Med. Biol.* 13:197-202,1994.
31. Peng, C.K., S. Havlin, H.E. Stanley, and A.L. Goldberger. Quantification of scaling exponents and crossover phenomena in nonstationary hearbeat time series. *CHAOS* 5:82-87,1995.

32. Hurst, H.E. Long-term storage capacity of reservoirs. *Trans. Amer. Soc. Civ. Engrs.* 116:770-808, 1951.
33. Feller, W. The asymptotic distribution of the range of sums of independent random variables. *Ann. Math. Stat.* 22:427-432, 1951.
34. Bassingthwaighte, J.B., and G.M. Raymond. Evaluating rescaled range analysis for time series. *Ann. Biomed Engr.* 22:432-444,1994.
35. Bassingthwaighte, J.B. Physiological heterogeneity: fractals link determinism and randomness in structures and functions. *News Physiol. Sci.* 3:5-10,1988.
36. Bassingthwaighte, J.B., and G.M. Raymond. Evaluation of the dispersional analysis method for fractal time series. *Ann. Biomed. Engr.* 23:491-505,1995.
37. Glenny, R.W., H.T. Robertson, S. Yamashiro, and J.B. Bassingthwaighte. Applications of fractal analysis to physiology. *J. Appl. Physiol.* 70:2351-2367,1991.
38. Schepers, H.E., J.H.G.M van Beek, and J.B. Bassingthwaighte. Four methods to estimate the fractal dimension from self-affine signals. *IEEE Eng. Med Biol Mag.* 11(2): 57-64&71,1992.
39. Lowen, S.B., and M.C. Teich. Estimation and simulation of fractal stochastic point processes. *Fractals* 3:183-210,1995.
40. Churilla, M., W.A. Gottschalke, L.S. Liebovitch, L.Y. Selector, A.T. Todorov, and S. Yeandle. Membrane potential fluctuations of human T-lymphocytes have fractal characteristics of fractional Brownian motion. *Ann. Biomed Engr.* 24:1996 99 - 108, 1996.
41. Flandrin, P. On the spectrum of fractional Brownian motions. *IEEE Trans. Infor. Theor.* 35:197-199,1989.

11

NONLINEAR CONTROL OF BREATHING ACTIVITY IN EARLY DEVELOPMENT

Hazel H. Szeto

Department of Pharmacology
Cornell University Medical College
1300 York Avenue
New York, New York 10021

1. INTRODUCTION

Spontaneous breathing movements have been described in the fetal sheep, fetal baboon and the human fetus prior to birth. In the fetal lamb and baboon, breathing movements can be observed as electromyographic activity in the diaphragm coupled with negative changes in intratracheal pressure[1,2]. Movements of the fetal chest wall and diaphragm can readily be seen with the use of real-time ultrasound in the human[3]. Although they do not regulate fetal oxygenation, fetal breathing movements are similar to postnatal respiratory activity in that they involve phrenic nerve firing and are modulated by a variety of chemical signals including hypoxia, hypercarbia and pharmacological agents such as opiates. However, the dynamics of fetal breathing movements are quite unique and very different from the postnatal breathing pattern.

In all species studied, spontaneous breathing movements in the fetus tend to occur intermittently and do not become continuous until after birth. There is also a high degree of variability in instantaneous breathing rates, giving the impression that the dynamic pattern is random and unpredictable. In addition, there appears to be differences in fetal breathing dynamics across different species. The dramatic fluctuations in fetal breathing rate makes it unlikely that fetal breathing is regulated by a simple linear control system. Recent findings from our laboratory suggest that spontaneous breathing movements in the ovine fetus is a stochastic process with fractal properties[4], suggesting that fetal breathing may be better understood as a nonlinear control system. In this chapter, new evidence will be provided demonstrating that such fractal processes are ubiquitous in all species, and that these fractal properties undergo similar developmental changes in the different species. This similarity in the control of ventilatory dynamics across different species has not been apparent from previous investigations using conventional techniques. Finally, results from a variety of dynamical analyses will be presented which suggest that ventilatory cycling in early development has some of the characteristics of a chaotic system.

Bioengineering Approaches to Pulmonary Physiology and Medicine, edited by Khoo
Plenum Press, New York, 1996

2. TECHNIQUES FOR MONITORING FETAL BREATHING MOVEMENTS

Fetal breathing data presented in this chapter were collected from fetal sheep, fetal baboon, human fetus and premature human infants by a number of investigators using a variety of different techniques. Fetal sheep data (0.7-1.0 gestation) were obtained in our own laboratory at Cornell University Medical College using a pair of chronic indwelling electromyographic electrodes in the fetal diaphragm[4,5]. Breathing data from the fetal baboon (0.7-1.0 gestation) were provided by Dr. Raymond Stark, Department of Pediatrics, Columbia University College of Physicians & Surgeons. These data were collected using a chronic indwelling catheter implanted in the trachea of the fetal baboon[2]. Both fetal sheep and fetal baboon data were collected from unanesthetized, unrestrained animals at least 5 days after surgery. Human fetal breathing data (37-40 weeks gestation) were obtained using ultrasound by Dr. Eduard Mulder, Department of Obstetrics & Gynecology, University of Utrecht, The Netherlands[3]. Breathing data from premature human infants (31-39 weeks conceptual age) were collected by Dr. Carl Schultz, Department of Pediatrics, Columbia University College of Physicians & Surgeons.

3. DYNAMICS OF FETAL BREATHING MOVEMENTS

In both the primate and the sheep, fetal breathing movements occur in aggregates interspersed with periods of quiescence that can often be longer than 30 minutes (see Figure 1). Both the duration of the breathing aggregates, as well as the duration of the apneic intervals, can be highly variable even within the same fetus.

The problems encountered in studying the dynamic nature of fetal breathing movements are similar to those associated with quantitative analysis of any irregular time series. In the past, analyses of these signals have generally been limited to statistical description of the time intervals,

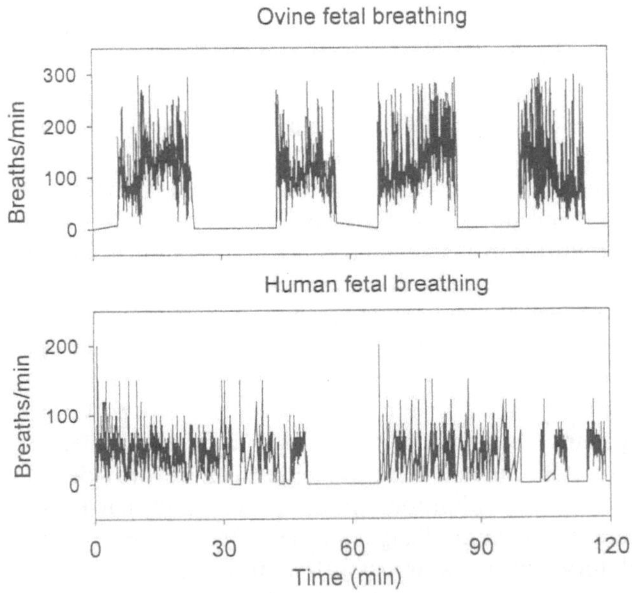

Figure 1. Time series of instantaneous breathing rates from an ovine fetus (*top panel*) and human fetus (*bottom panel*).

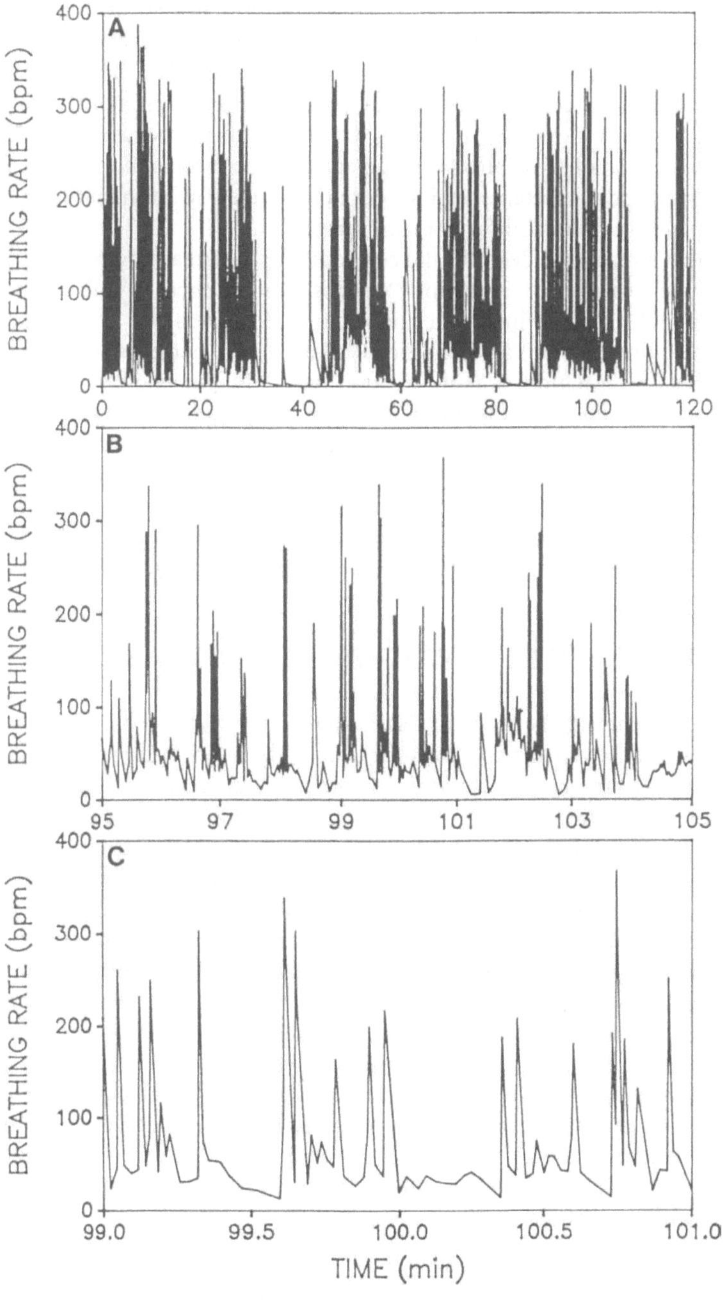

Figure 2. Time series of instantaneous breathing rate from an ovine fetus shown on three different time scales. A: a 2-h interval with clusters of breathing activity interspersed with periods of relative quiescence. B: an amplification of interval between 95 and 105 min. Note that clusters of faster breathing rates can be seen within original cluster of breathing activity. C: a further amplification of the time series, covering interval from 99 to 101 min. Appearance of clusters within clusters at all different time scales is suggestive of self-similarity. Reprinted with permission from Szeto et al.[4].

or a measure of short-term and long-term variability. The quantitation of fetal breathing activity has been particularly challenging because of its discontinuous and highly irregular pattern. Early attempts focused on describing the "incidence" of fetal breathing movements, which was defined as the percent of time in which breathing movements were present[1]. This was generally based on visual inspection of polygraphic records obtained from unanesthetized fetal lambs, and although the method was relatively crude, most investigators concurred that the "incidence" of fetal breathing movements under normal physiological conditions ranged from 40% - 60%. However, it was difficult to demonstrate transient changes in fetal breathing because of the high degree of variability in the duration of these breathing aggregates under normal conditions. Furthermore, this measure of "incidence" of fetal breathing does not provide any information on the continuity of the breathing pattern or the stability of breathing rates.

Subsequent attempts to describe the continuity of the breathing pattern have included the measure of the duration of breathing "epochs" and "apneas"[5,6]. In a study from our laboratory, an "epoch" was defined as a minimum of 3 breaths per 9s, while "apnea" was defined as an inter-breath interval (IBI) greater than 10s[5]. Using these classifications, we found that the breathing pattern was more continuous (longer epochs) after 0.8 term, but there was no further change after that until birth[5]. We also found, among the highly irregular pattern, the presence of small clusters of fairly regular breaths. These stable clusters were defined as group of successive breaths whose IBIs differ by <10%, although the mean instantaneous rate varied widely from cluster to cluster. The percent of breaths in stable clusters increased from 32% to 55% throughout the third trimester. Thus the major developmental change in breathing pattern appears to be an increase in stability of breathing rates.

4. SELF-SIMILARITY IN FETAL BREATHING PATTERN

A major problem with the above methods is their reliance on the initial arbitrary definitions of "epoch" and "apnea". As small variation in these definitions can result in vastly different results, we have decided to examine the entire breathing pattern without these definitions. The more closely the breathing patterns were inspected, the more details were revealed. Figure 2 shows the time series of instantaneous breathing rate in a fetal lamb using different time scales. Under normal physiological conditions, aggregates of breathing activity were separated by quiescent periods (Figure 2A). Within these aggregates of breathing, instantaneous breathing rates were highly variable, ranging from 0-300 breaths/min. Figure 2B shows the amplification of a short segment taken from within one of these breathing aggregates, and similar clusters of faster breathing rates were also observed on this time scale. Figure 2C shows a further amplification of the time scale, and further clustering became apparent. Thus the time series looked similar on all time scales, and this clustering of faster rates was observed at all levels, consistent with the idea of self-similarity.

Self-similarity or scale invariance is a property of fractal objects. The concept of fractals can also be applied to dynamic processes. Fractal processes tend to have many component frequencies and cannot be characterized by any one time scale. Our observation of self-similarity in the ovine fetal breathing pattern suggested that fetal breathing may be a fractal process without a characteristic time scale.

5. STATISTICAL EVIDENCE OF SCALE-INVARIANCE

Although such descriptive data may suggest the lack of a characteristic time scale, a more quantitative measure of scale invariance is necessary. One method is to examine the statistical distribution of IBIs in the fetal lamb. Figure 3A shows the distribution of IBI from a fetal lamb at 0.7 term. The distribution was highly skewed to the right, and although not visualized on this linear

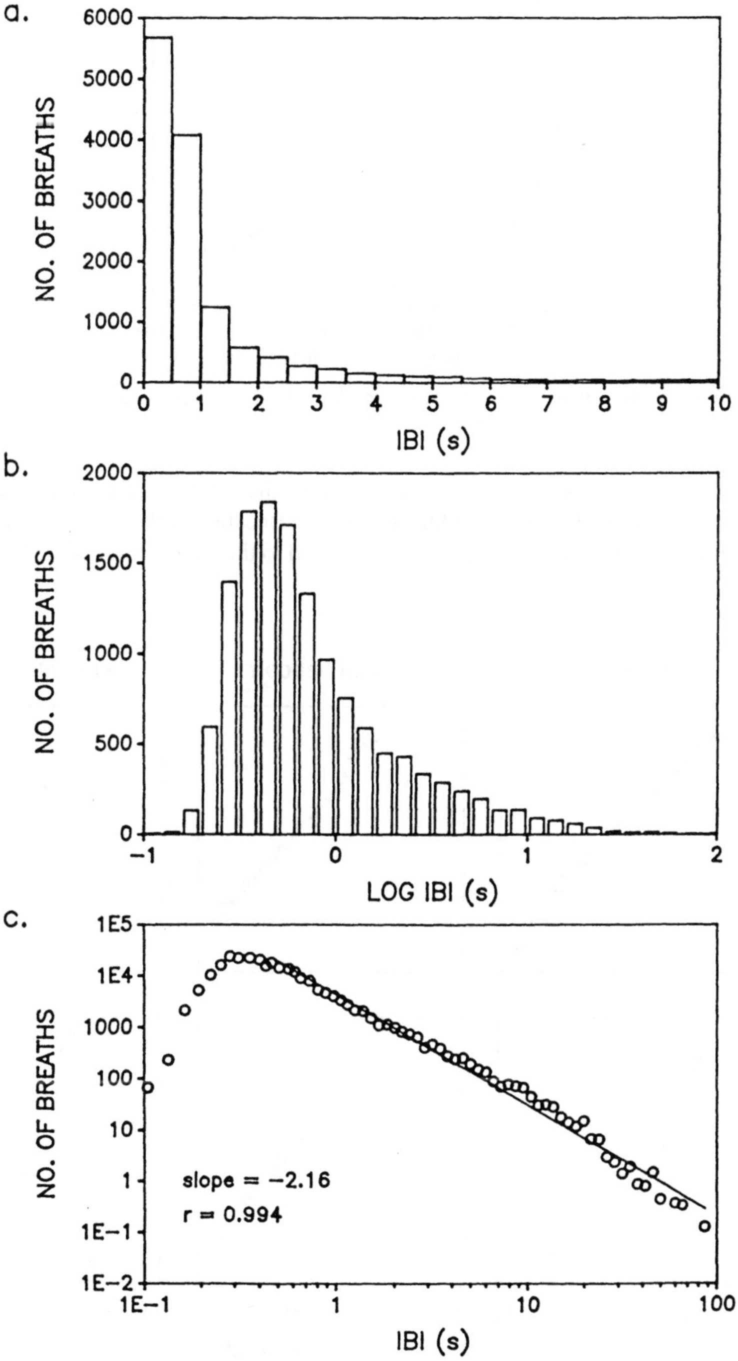

Figure 3. Distribution histograms of interbreath intervals (IBI) from a fetal lamb plotted on linear (A), logarithmic (B), and log-log (C) coordinates. Total number of breaths was 13,620. Reprinted with permission from Szeto et al.[4].

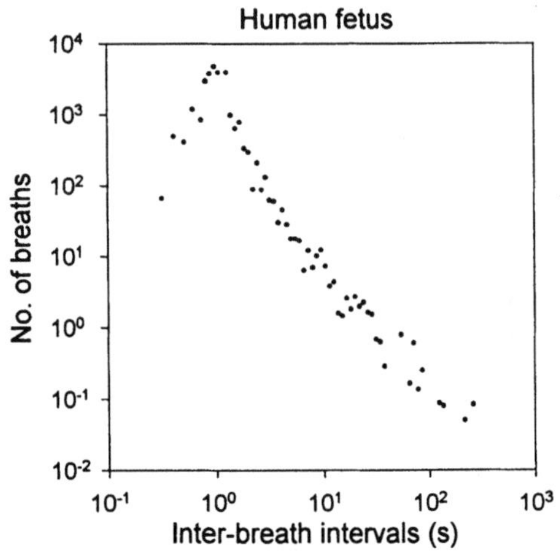

Figure 4. Distribution histograms of interbreath intervals (IBI) on log-log coordinates from a human fetus.

scale, 2-3% of the IBIs were >10s. The contribution of these longer IBIs to the entire dynamic pattern is greatly underestimated using linear analysis. Because of the extreme skewness in the distribution of IBI, a logarithmic transform was applied to the data (Figure 3B). The distribution

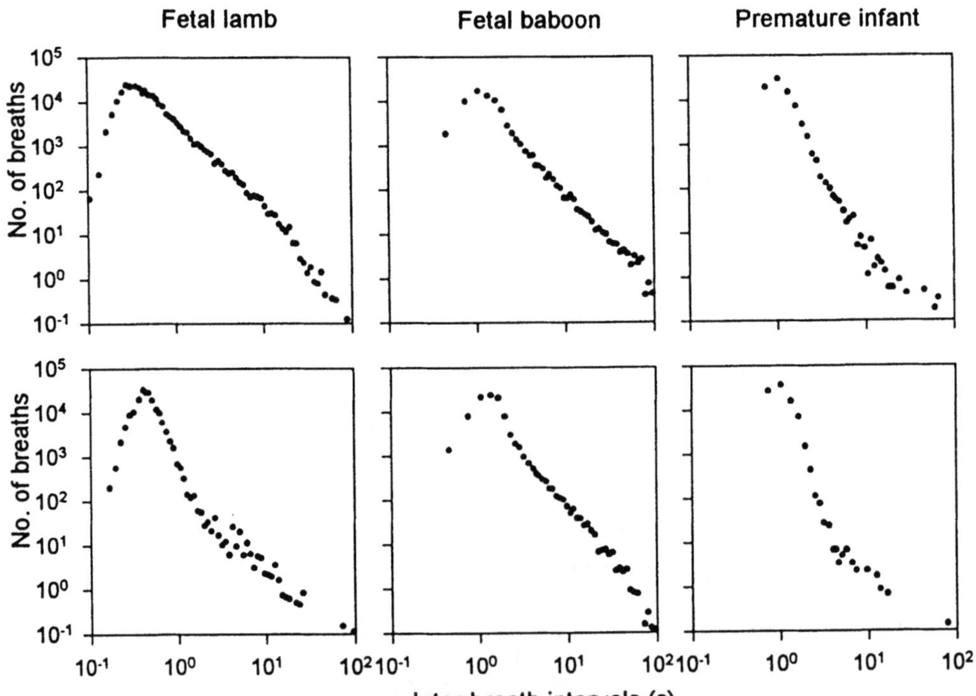

Figure 5. Distribution histograms of interbreath intervals (IBI) on log-log coordinates from fetal sheep (*left panel*), fetal baboon (*middle panel*) and premature human infants (*right panel*) at two different gestational ages: *top*, 0.7 - 0.8 term; *bottom*, 0.9 - 1.0 term.

histogram resembled the sum of two distribution profiles, a log-normal distribution for the shorter IBIs and an exponential distribution for the longer IBIs. The inverse power-law distribution for the long IBIs is more easily visualized on a log-log plot (Figure 3C). This power-law distribution indicates the lack of a characteristic time scale and supports the idea of a fractal process. Figure 4 shows that this power-law distribution of IBI was also observed in the human fetus.

Interestingly, the log-normal component becomes more prominent as a function of increasing gestational age in the fetal lamb, fetal baboon and premature human infant (Figure 5). We have previously shown that the ratio of log-normal to power-law distribution increases exponentially during the last two weeks before birth in the fetal lamb[4].

Montroll & Schlesinger[7] have shown that complex tasks that involve many subtasks are better described by log-normal rather than normal distribution. They further showed that as the tasks become even more complex, their distributions become broader, and they take on a 1/x tail. This progressive crossover from power-law distribution to log-normal distribution would therefore suggests a reduction in the complexity of the regulatory system for fetal breathing as a function of gestational age.

6. DYNAMIC EVIDENCE OF SCALE-INVARIANCE

Although histograms provide information regarding the various components that make up the dynamic pattern, they do not provide any information on the sequential order of IBIs. This can be achieved with the use of power spectral analysis. Fast Fourier transform

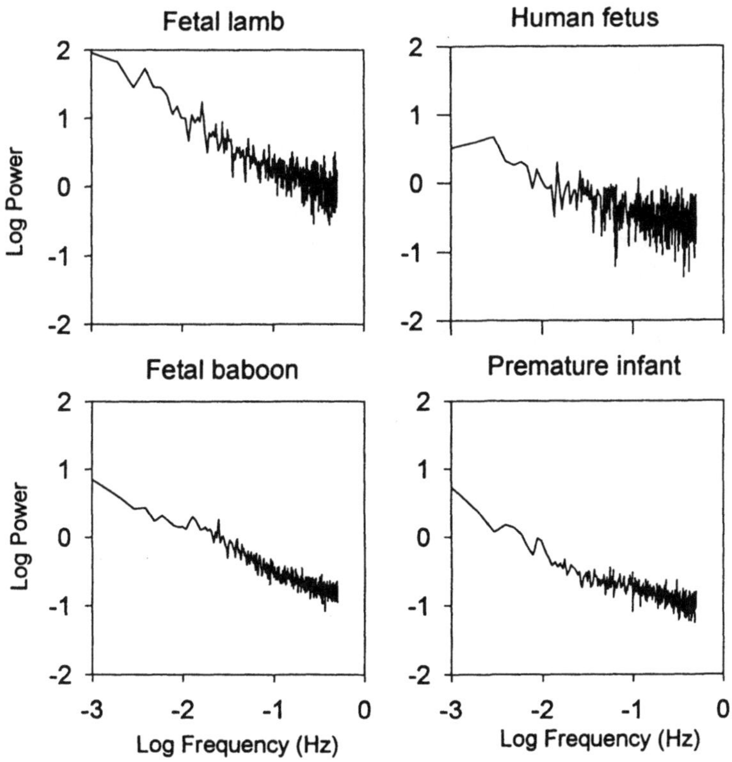

Figure 6. Power spectra of instantaneous breathing rate from a fetal lamb, fetal baboon, human fetus, and premature human infant.

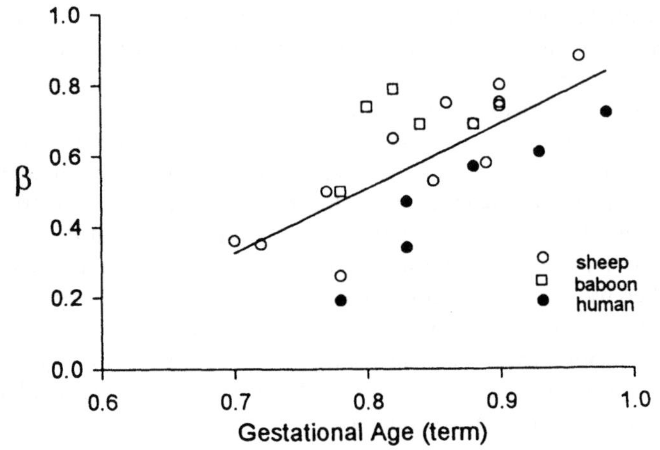

Figure 7. Plot of β (calculated from slope of power spectrum) as a function of gestational age.

was performed on consecutive series of 1024 detrended, windowed (Hamming) data points, and all power spectra were averaged to produce a mean power spectrum for each data set. Figure 6 shows the power spectra obtained from a fetal lamb, fetal baboon, human fetus and premature human infant.

Plotted on logarithmic coordinates, power was found to be inversely proportional to frequency for frequencies $<10^{-1}$ Hz in all four cases. Such power spectra have been referred to as $1/f^{\beta}$ power spectra. The slope of the power spectrum, the exponent β, increased as a function of gestational age in the sheep, baboon and human (see Figure 7), and did not reach a value of 1.0 until after birth in the lamb[4]. It was particularly interesting to find similar fractal properties in the ventilatory cycle of the premature human infant.

Although $1/f^{\beta}$ power spectrum has been reported for many physiological systems, this is the first system in which changes in fractal characteristics has been reported with normal growth and development. The age-dependent increase in the exponent of the $1/f^{\beta}$ power spectrum suggests a more correlated signal, and may indicate a less flexible, or more rigid, control mechanism. In terms of developmental processes, a loss of frequency response may be viewed as decreased plasticity with maturation.

7. DELAY PLOTS FOR ANALYSIS OF DYNAMIC SEQUENCE

Closer examination of the above power spectra revealed that the spectrum tends to bend and plateau at frequencies $> 10^{-1}$ Hz, suggesting that random breaths have a tendency to be clustered or bunched in groups, and the group sizes are distributed and arranged sequentially so that the global density modulation gives rise to a $1/f$-like power spectrum. To further examine the dynamic sequence of breath intervals, we used delay plots of sequential IBIs.

Figure 8 (*left panel*) shows a delay plot of IBIs from a fetal lamb. The same data can be examined by plotting the sequential change in direction and magnitude between successive breath intervals (Figure 8, *right panel*). The broad dispersion of points along the identity line suggests the presence of an infinite number of periodicities. Examination of the sequential pattern revealed that the breathing pattern tends to

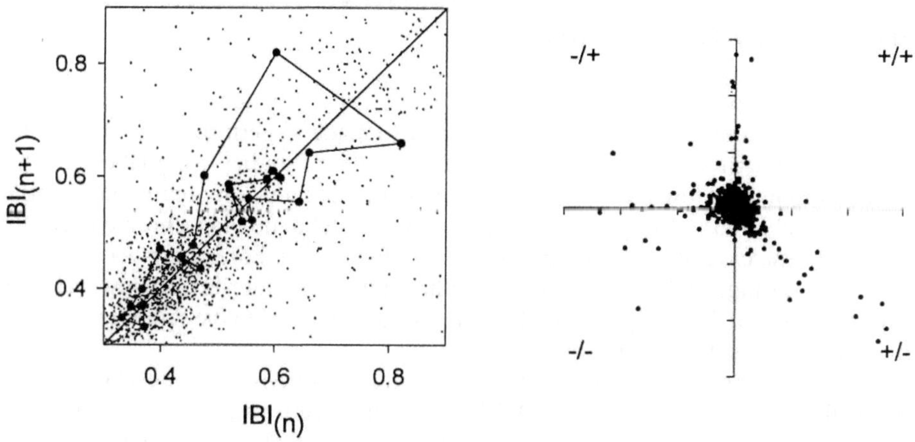

Figure 8. Left panel: Delay plot of IBI from a fetal lamb. *Right panel*: quadrant map showing direction and change of successive interbreath intervals.

spend brief intervals at a certain periodicity, and then undergo a transition to another periodicity. These are consistent with our earlier observation of "stable clusters" of breaths. Although it may take several breaths before the cycling falls into a relatively stable periodicity, the quadrant maps suggest that it is often possible for the system to correct a prolonged breath interval immediately. The data points along the diagonal line in the +/- quadrant suggests that a prolonged IBI is often immediately followed by a much shorter IBI so that the ongoing breathing rate can be sustained. A similar type of dynamical pattern was found in the human fetus (Figure 9). These data suggest that it is the excursions between these unstable fixed points that results in the $1/f$ power spectra, and is consistent with the loss of $1/f$ at frequencies > 10^{-1} Hz.

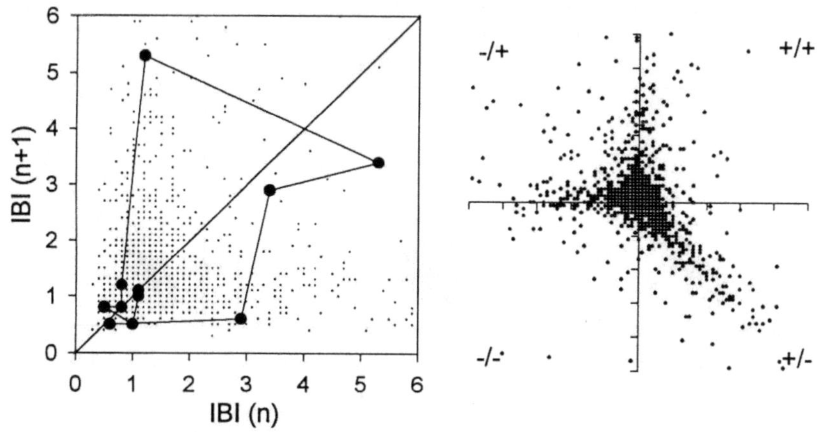

Figure 9. Left panel: Delay plot of IBI from a human fetus. *Right panel*: quadrant map showing direction and change of successive interbreath intervals.

8. PERTURBATION OF FETAL BREATHING DYNAMICS BY PHARMACOLOGICAL MEANS

One method to enhance our understanding of a control system is by perturbation of the system itself. In the fetal lamb, attempts were also made to perturb the normal spontaneous breathing pattern with pharmacological means. Previous studies in our laboratory have shown that morphine, at low doses, induced a continuous and much more regular breathing pattern in the fetal lamb[8,9]. Morphine was infused to the fetal lamb via a chronic indwelling catheter in the vena cava at a constant rate of 5 mg/h for 1h. Fetal breathing was monitored for 2h before, during, and for 3h after morphine infusion. Inter-breath intervals were determined for each of these experimental segments and subjected to the same statistical and dynamical analyses as described above.

The infusion of morphine to the fetal lamb resulted initially in the onset of rapid and regular breathing movements (Figure 10). This was followed by a progressive decrease in instantaneous rate until a relatively stable breathing rate was attained at steady state drug level. However, despite the appearance of a rather stable rate of ~100 breaths/min, there was some evidence of a slower rate (~60 breaths/min).

With the termination of morphine infusion and gradual decline in morphine levels, the slower rate became much more pronounced, and the delay plot clearly shows transitions between the two periodicities (Figure 11). The stabilizing effect of morphine on fetal breathing was completely reversible.

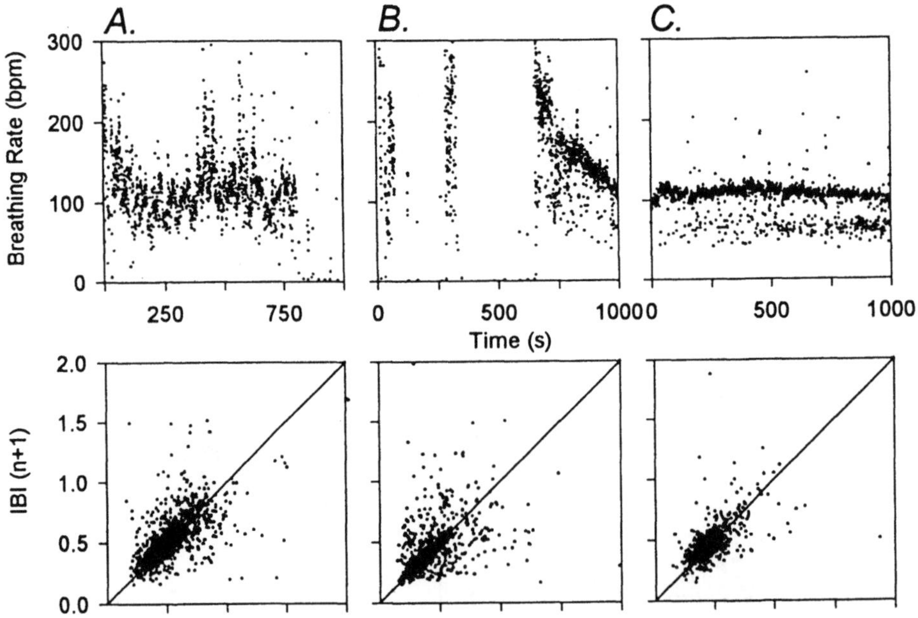

Figure 10. Top panel: Time series of instantaneous breathing rate from a fetal lamb before (*left*) and during (*middle and right*) morphine infusion (5 mg/h, iv). *Bottom panel*: delay plot of IBIs before (*left*) and during (*middle and right*) morphine infusion. Each panel represents total time of 1000s.

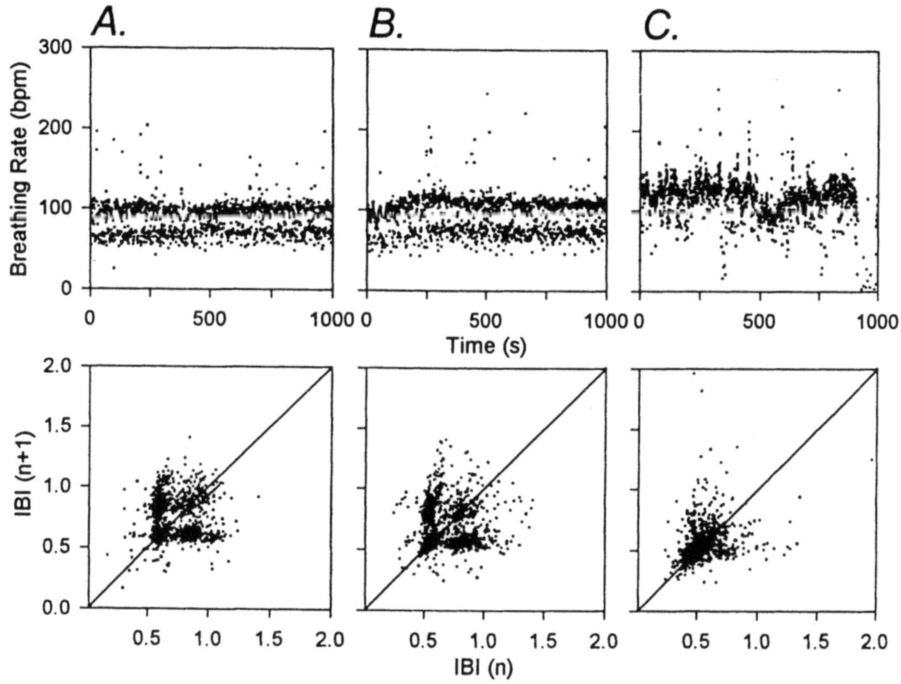

Figure 11. Top panel: Time series of instantaneous breathing rate from the same fetal lamb after termination of morphine infusion (5 mg/h, iv). *Bottom panel*: delay plot of IBIs before (*left*) and during (*middle and right*) morphine infusion. Each panel represents total time of 1000 s.

Morphine was capable of reducing the complexity of the system to a single periodicity of approximately 1.8 Hz. With gradual decline in morphine levels, a second periodicity can be observed at ~1.0 Hz. The breathing rate then cycles between these two predominant periodicities. As morphine levels continue to decline, the system reverts back to a multi-periodicity pattern.

9. CONCLUSIONS

The present studies with breathing data from both ovine and primate species demonstrate that fractal characteristics in ventilatory cycling are ubiquitous in early development. Fractal processes have no characteristic time scale, and tend to show clustering of events with prolonged periods of inactivity. The presence of long apneic periods is typical of fetal breathing patterns in both the ovine and primate species. These fractal processes undergo similar maturational changes as a function of normal growth and development. Maturation is associated with a loss of frequency response in the system, resulting in a more correlated signal, suggesting a more rigid control system with less plasticity. This highly complex breathing dynamics in early development can also be reduced to a stable breathing pattern with the introduction of a pharmacologic agent such as morphine.

These data also suggest that ventilatory cycling in early development has some of the characteristics of a chaotic system. Chaotic motion includes an infinite number of unstable periodic motions. A chaotic system never remains long in any of these unstable motions but continually switches from one periodic motion to another, thereby giving the appearance of randomness[10]. Our delay plots demonstrate that ventilatory cycling in the fetus continuously switches from one periodicity to another. The observation that very long inter-breath intervals tend to be followed by very short inter-breath intervals is also consistent with a flip saddle. Furthermore, it has been postulated that it is possible to stabilize a chaotic system by perturbations of initial conditions[10]. Recently, Garfinkel et al.[11] demonstrated that it was possible to convert cardiac arrhythmia to periodic beating with the use of electrical stimuli. In fetal breathing, perturbation with morphine stabilized the system and resulted in a stable periodicity. Furthermore, upon termination of morphine infusion, a series of bifurcation was observed en route to the multi-periodicity pattern. These findings lend support to the notion that ventilatory cycling in early development may be a chaotic system. We believe that recent advances in nonlinear dynamics theory and methods have significantly contributed to our understanding of the complex behavior in fetal breathing dynamics.

ACKNOWLEDGMENTS

I wish to thank Drs. Eduard Mulder, Raymond Stark and Carl Schultz for providing the fetal and newborn breathing data that were used in this study. I also wish to thank Drs. Peter Cheng, Gene Dwyer and Dirk Hoyer for their most helpful discussions and encouragement.

REFERENCES

1. Dawes, G. S., H. E. Fox, M. B. Leduc, G. C. Liggins, and R. T. Richards. Respiratory movements and rapid eye movement sleep in the fetal lamb. *J. Physiol.* 220: 119-143, 1972.
2. Stark, R. I., S. S. Daniel, Y.-I. Kim, K. Leung, H. R. Rey, and P. J. Tropper. Patterns of development in fetal breathing activity in the latter third of gestation of the baboon. *Early Hum. Dev.* 32: 31-47, 1993.
3. Mulder, E. J. H., M. Boersma, M. Meeuse, M. Van Der Wal, E. Van De Weerd, and G. H. A. Visser. Patterns of breathing movements in the near-term human fetus: Relationship to behavioural states. *Early Hum. Dev.* 36: 127-135, 1994.
4. Szeto, H. H., P. Y. Cheng, J. A. Decena, D. L. Wu, Y. Cheng, and G. Dwyer. Fractal properties in fetal breathing dynamics. *Am. J. Physiol.* 263: R141-R147, 1992.
5. Szeto, H. H., P. Y. Cheng, J. A. Decena, D. L. Wu, Y. Cheng, and G. Dwyer. Developmental changes in continuity and stability of breathing in the fetal lamb. *Am. J. Physiol.* 262: R452-R458, 1992.
6. Rey, H. R., R. I. Stark, Y. I. Kim, S. S. Daniel, G. MacCarter, and L. S. James. Method for processing of fetal breathing epoch analysis: studies in the primate. *IEEE Engineering in Medicine and Biology* 8: 30-42, 1989.
7. Montroll, E. W. and M. F. Shlesinger. On 1/f noise and distributions with long tails. *Proc. Natl. Acad. Sci. USA* 79: 3380-3383, 1982.
8. Szeto, H. H., Y. S. Zhu, J. G. Umans, G. Dwyer, S. Clare, and J. Amione. Dual action of morphine on fetal breathing movements. *J. Pharmacol. Exp. Ther.* 245: 537-542, 1988.
9. Szeto, H. H., P. Y. Cheng, G. Dwyer, J. A. Decena, D. L. Wu, and Y. Cheng. Morphine-induced stimulation of fetal breathing: role of mu_1-receptors and central muscarinic pathways. *Am. J. Physiol.* 261: R344-R350, 1991.
10. Shinbrot, T., C. Grebogi, E. Ott, and J. A. Yorke. Using small perturbations to control chaos. *Nature* 363: 411-417, 1993.
11. Garfinkel, A., M. L. Spano, W. L. Ditto, and J. N. Weiss. Controlling cardiac chaos. *Science* 257: 1230-1235, 1992.

12

POSSIBLE FRACTAL AND/OR CHAOTIC BREATHING PATTERNS IN RESTING HUMANS

R. L. Hughson, Y. Yamamoto, J. -O. Fortrat, R. Leask, and M. S. Fofana

Department of Kinesiology
University of Waterloo
Waterloo, Ontario N2L 3G1, Canada

1. INTRODUCTION

Déjours[5] was one of the first to comment on the breath-by-breath variations in ventilation ($\dot{V}_E$) and gas exchange (oxygen uptake, $\dot{V}O_2$, and carbon dioxide output, $\dot{V}CO_2$). It was suggested that the spontaneous variations in alveolar ventilation and perfusion of the lungs, as well as variations in venous blood O_2 and CO_2 content, were responsible for these breath-by-breath patterns. He observed that all fundamental respiratory variables varied around the mean value. Since that time, there have been many other observations of breath-by-breath variation in the breathing pattern. Lenfant[13] further characterized the pattern of variation, and Hlastala et al.[9] observed cyclical variations in functional residual capacity (FRC). Attempts to quantify the breath-by-breath variation have ranged from computation of autocorrelation functions by Benchetrit and Bertrand,[1] to analysis of run times[2] to more recent methods of spectral analysis,[11,24] and of chaos theory.[6,15,17-19] The implication of fractal and/or chaotic behaviour in the breathing pattern has wide ranging consequences including, for chaotic systems, the requirement that the system be deterministic. The purpose of this paper is to explore further whether patterns of breathing are consistent with the properties of fractal or chaotic systems.

1.1. Fractal Time Series

Several different types of physiological time series have been identified as having fractal properties. The most commonly studied has been the pattern of heart rate variability. A fractal time series possesses the property of self-similarity. This can be expressed mathematically as:

$$x(ht) = h^H \cdot x(t)$$

Bioengineering Approaches to Pulmonary Physiology and Medicine, edited by Khoo
Plenum Press, New York, 1996

where the equals sign implies that the left and right sides of the equation have the same distribution function. For a self-similar signal, the distribution remains unchanged by the factor h^H (for h>0). Several physiological examples of changes in scale exist in which it is clear that the general form of the time series is preserved even when the total time displayed is changed dramatically such as for heart rate variability as in Saul et al.[21] and Goldberger.[8] This type of scaling behaviour is evident, not only over long time scales, but also in shorter data sets as described by Yamamoto and Hughson.[31]

The method of isolating fractal components from time series by coarse graining spectral analysis has been described in detail by Yamamoto and Hughson.[30] This approach is based on the concept of self-similarity in the fractal components. Thus for any time series signal, the rescaled version of the signal will maintain a high correlation with the original signal for the fractal (self-similar) component, but not for harmonic components. Coarse graining spectral analysis has been demonstrated to return reliable estimates of the Hurst exponent and of the percent fractal in simulated data sets with mixed fractal and harmonic signals.[29,30] Application of this methodology to study of human heart rate variability has shown that changes in state can be tracked on going from supine rest to severe orthostatic challenge associated with near fainting.[3]

Although preliminary observations of the pattern of breathing during exercise suggested that the time series might be fractal,[11] a systematic study has not been undertaken of resting ventilatory patterns. Therefore, it was one purpose of this study to examine the possible fractal nature of the breath-by-breath variations in expired ventilation ($\dot{V}_E$) in humans during quiet seated rest.

1.2. Chaos in Time Series

The primary characteristics of a chaotic system include: the system is governed by a nonlinear deterministic process (that is, it is governed by a set of rules that can, in theory, be completely described); variations in measured values about a so-called set point are bounded by the region of a strange or chaotic attractor; the system is highly sensitive to initial conditions, therefore, adjacent points diverge exponentially. This series of characteristics permit apparently random behaviour to be observed in a system that is in fact tightly controlled by specific interactions. It is the possibility that science might unravel the intricate control mechanisms for a range of physiological systems that has inspired much of the investigation into the possibility of finding chaos.

Whether a process is chaotic or not is subject to potential errors in computation of indicators of chaos. Indeed, several authors have cautioned about the conclusion that chaos exists simply from satisfying a single criterion of chaos.[7,14,25] Further, there are pitfalls within tests that might cause erroneous conclusions. For example, the Lyapunov exponent calculated by Donaldson[6] was with the Wolf algorithm,[28] which has been shown by Theiler et al.[25] to yield false positive exponents when dealing with short duration noisy data. Obviously, human breath-by-breath data sets will be noisy because of multiple factors, including voluntary control, interacting to control respiration, and they will be short because it is difficult to constrain human volunteers within specific conditions for long periods of time to obtain representative data.

In this paper, three different approaches have been taken to explore the possibility that the pattern of breathing is chaotic. First, the rate of exponential divergence of adjacent points was indicated by computation of a positive Lyapunov exponents by two different algorithms. This approach will allow a test of the recent results published by Donaldson,[6] where a positive Lyapunov exponent was taken as evidence of chaos in respiration. Second, the ability to make long-term predictions was explored with the nonlinear prediction algorithm of Sugihara and May.[23] This approach took advantage of the deterministic nature of chaotic systems, where short-term prediction of future events is quite good, while long-term prediction is poor. Third, surrogate data were generated from the original time series and analyzed with the same approaches as described for the original data. This latter technique has proven to be a valuable test of nonlinear deterministic systems versus uncorrelated noise.[25]

2. METHODS

Eight healthy young men were studied during quiet, seated rest while they breathed through a mouthpiece for 60 minutes. All subjects signed a consent form that described the measurements, but did not give specific information that we were studying the pattern of breathing. The environment around the subjects was strictly controlled to be quiet with no distractions or cues to the subject that might influence the pattern of breathing. Subjects were seated in a chair with head and arm rests. They were instructed to keep their eyes open during the test period. The system used to measure ventilation and gas exchange consisted of a volume turbine (Alpha Technologies, VMM-110) and a respiratory mass spectrometer (Marquette, MGA-1100). Data were processed by a computer based system as described in detail elsewhere (10). The turbine was carefully calibrated for low flow rates. Each of tidal volume (V_T), breathing frequency (f_B), minute ventilation ($\dot{V}_E$), inspiratory and expiratory times (T_I and T_E), end-tidal partial pressures of O_2 and CO_2, oxygen uptake ($\dot{V}O_2$) and carbon dioxide output ($\dot{V}CO_2$) were computed for each breath.

2.1. Fractal Analysis

The original time series data were analyzed by coarse graining spectral analysis[30] to extract the fractal and harmonic components. The unequally spaced values were aligned sequentially to obtain equally spaced samples that were taken to have the interval determined from the mean breathing frequency in a manner analogous to that described for evaluation of heart rate variability.[4,31]

The fractal component was evaluated from the slope ($-\beta$) of the relationship between the log of spectral power and the log of frequency.[22,30,31] This is expressed as the spectral exponent, β. To obtain the estimate of β, the unevenly distributed data points (see Fig. 1) were placed in equally spaced bins. In the low frequency region, if no data points were within the bins, values were obtained by interpolation. A linear regression by a least-square procedure was used to calculate the slope. In all cases, the percentage of spectral power found in the fractal components was sufficiently large ($\geq 75\%$) to consider the signal to be fractal and to permit interpretation of the spectral exponent.

2.2. Chaotic Analysis

In order to permit a comparison with the results of Donaldson,[6] the methodology employed in this latter paper was followed closely. The original time series data were resampled at 2 Hz by holding the value of the current breath. Step changes in values were avoided by smoothing the data with a Hanning window. These data were supplied to the Wolf et al.[28] algorithm for determining the positive Lyapunov exponent. As in the original study by Donaldson,[6] we used an embedding dimension of 4 and a phase lag of 4 (i.e. 2 s) so that the data would be directly comparable.

Data were also analyzed with a modified algorithm of Sano and Sawada[5] for determination of the Lyapunov exponent. In this case, the optimal embedding dimension was sought using a modification of the False Neighbours routine.[12] Convergence of the positive exponent was used to select the optimal phase delay. The reason for this was that surrogate data analysis (see below) revealed very different outcomes with the Sano and Sawada compared with the Wolf et al. algorithms. This provided more stable estimates of the Lyapunov exponents.

The nonlinear prediction algorithm of Sugihara and May[23] was tested with these data. Details of this method can be found in Yamamoto and Hughson.[32] Briefly, a vector $X_M(I)$ was constructed in an M-dimensional Euclidean space by embedding delayed samples of the $\dot{V}_E$ time series $x(I)$. The first difference of the $\dot{V}_E$ time series was used in the analysis. A minimal neighbourhood was selected from the second half of the data set such that the vector was contained in the smallest diameter simplex formed from its M+1 closest neighbours from the first half of the data set. The predicted value was obtained from where the original vector had moved within the range of this simplex after the prediction time ($\dot{V}_E$ values), giving exponential weight to its original distances from the relevant neighbours.[23] The

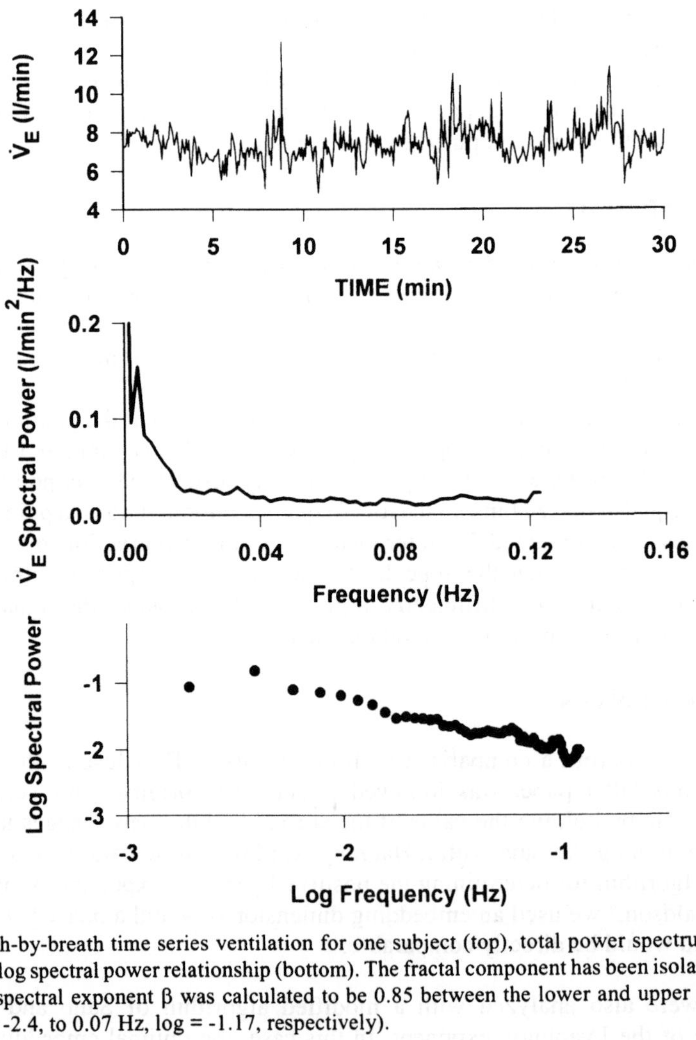

Figure 1. Breath-by-breath time series ventilation for one subject (top), total power spectrum (middle), and log frequency - log spectral power relationship (bottom). The fractal component has been isolated in the bottom panel, and the spectral exponent β was calculated to be 0.85 between the lower and upper frequency limits (0.04 Hz, log = -2.4, to 0.07 Hz, log = -1.17, respectively).

correlation coefficient between the predicted and the original data was computed as an index of the predictive ability.

2.3. Surrogate Data Analysis

Theiler et al.[25] described a method in which the original data are used to generate a surrogate data set that permits the testing of the null hypothesis that the pattern of variation is a result of linearly autocorrelated gaussian noise. The original data were transformed by Fast Fourier Transform (FFT) into the frequency domain. The amplitude components of the FFT analysis were retained, and the phase relationships were randomized prior to using an inverse FFT to generate the surrogate data sets.

3. RESULTS AND DISCUSSION

3.1. Fractal Analysis

The original time series data for $\dot{V}_E$ from each of the 8 subjects is displayed in Fig. 2. Results from coarse graining spectral analysis indicated that 3 subjects had spectral exponents (β) that were in the range of 0.86-1.16, while the remaining 5 subjects had exponents that were less than 0.40 (Table 1). These latter exponents are not notably different from white noise and suggest that the pattern of breathing in these subjects was not fractal. It is important to consider how a fractal pattern of breathing might occur and what its physiological significance might be.

3.2. Lyapunov Exponents

Consistent with the findings of Donaldson,[6] we observed positive Lyapunov exponents in 6 of 7 subjects when calculated with the algorithm of Wolf et al.[28] However, to test the null hypothesis that the positive Lyapunov exponents were simply a function of linearly correlated noise in the time series data, we constructed surrogate data sets as described by Theiler et al.[25] from the inverse FFT with randomized phase. This type of data set retains the

Table 1. Fractal and chaos indicators for expired ventilation ($\dot{V}_E$) during quiet, seated rest. Individual subject values are presented for the spectral exponent (β) of the fractal component obtained from coarse graining spectral analysis, and the Lyapunov exponents calculated with the algorithms of Wolf et al.[28] and Sano and Sawada[20] on the original data and on surrogate data. n.a.= not available because the algorithm failed to converge on this data set. n.c.= not computed

Subject	β for Fractal	Lyapunov (Wolf)	Surrogate (Wolf)	Lyapunov (Sano)	Surrogate (Sano)
BLA	0.35	-0.310	2.485	n.c.	n.c.
GOB	0.21	0.222	2.573	n.c.	n.c.
JOF	1.16	0.131	2.434	1.059	-1.511
KSO	0.20	0.209	2.383	0.730	-1.408
MMC	0.86	0.084	2.353	n.c.	n.c.
TSM	1.11	0.091	2.430	n.c.	n.c.
WEA	0.21	0.153	2.416	n.c.	n.c.
YAM	0.40	n.a.	n.a.	n.a.	n.a.

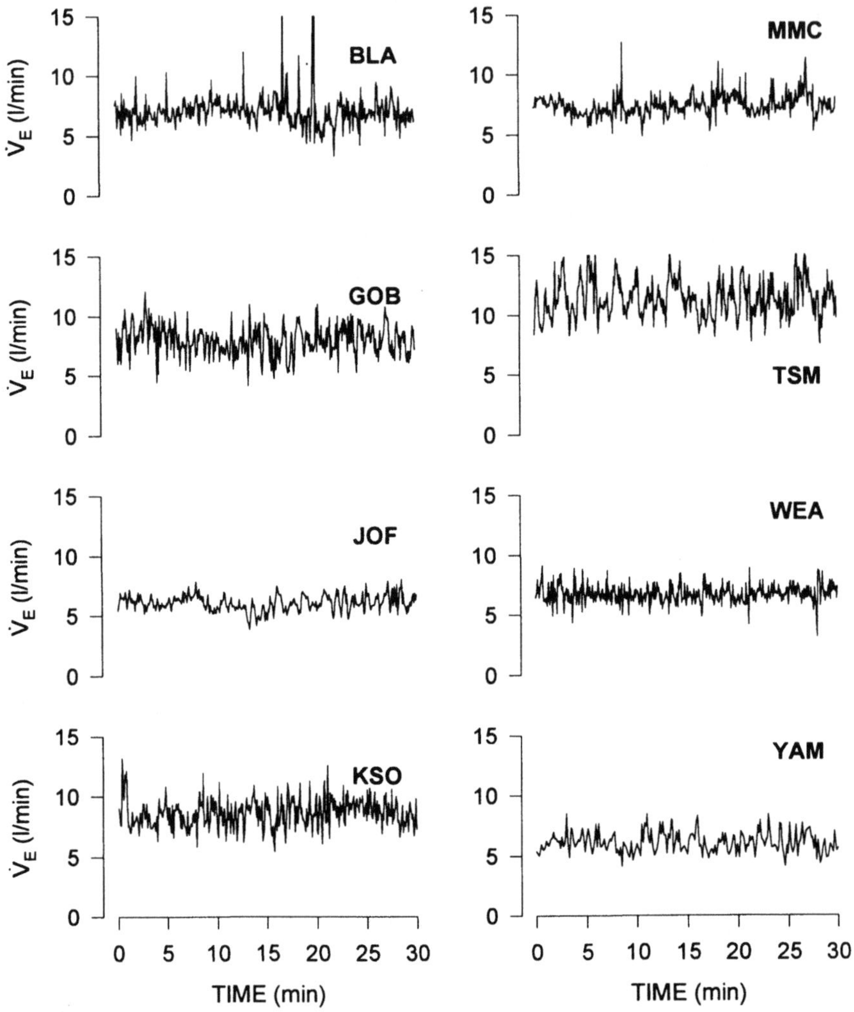

Figure 2. The individual time series for 30 minutes of resting ventilation in all 8 subjects.

same statistical distribution as the original data while preserving only the linearly correlated nature of the data set. The outcome of this analysis with the Wolf et al. algorithm was not consistent with a random data set. Rather than observing negative exponents, the results were in fact more positive (see Table 1). For the linear surrogate data sets, the dynamics are contracting and therefore positive exponents (divergence) should not appear. This finding suggests that the interpretation of the positive Lyapunov exponents for the original time series data must be interpreted with caution for this specific algorithm.

Because of the unexpected findings with the Wolf et al. algorithm, we also tested the $\dot{V}_E$ time series data with the Lyapunov exponent algorithm described by Sano and Sawada.[20] In addition with this analysis, we searched for the optimal embedding dimension and time delay. This step was not taken with the Wolf et al. algorithm because we were attempting to follow as closely as possible the methods described by Donaldson[6] where he selected a fixed embedding dimension of 4 and a time delay of 2 s. It was observed that an embedding dimension of 7 and a time delay of 7 produced consistent, stable results.

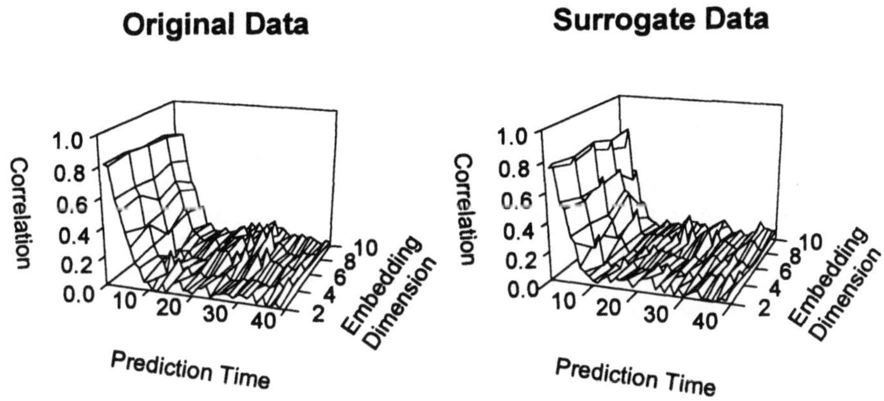

Figure 3. The correlation coefficient is shown as a function of embedding dimension and phase delay for original and surrogate data for a single subject.

The results for the Lyapunov exponents obtained with the original $\dot{V}_E$ data differed between the Sano and Sawada[20] and the Wolf et al.[28] algorithms. In marked contrast with the results for the surrogate data analysis that showed highly positive Lyapunov exponents with the Wolf et al. algorithm, we observed that the exponents were negative with the Sano and Sawada method. This latter finding is consistent with the presence of a nonlinear trend in the original time series data, and stochastic variation in the surrogate data. It is interesting that positive Lyapunov exponents were found with the Sano and Sawado algorithm for two subjects with markedly different power spectral exponent (compare Table 1).

3.3. Non-Linear Prediction

The non-linear prediction method of Sugihara and May[23] was applied over a range of embedding dimensions. The pattern of prediction for the original breath-by-breath data was consistent with deterministic chaos. That is, there was a moderately high correlation for short prediction times, and this decreased rapidly with increasing prediction times. However, there was no difference between the original and the surrogate data for any individual subject ($P>0.05$). An example of the relationship between correlation coefficient, lag time, and embedding dimension is shown in Fig. 3 for each of original data and surrogate.

3.4. On the Presence of Chaos

The results from the present study suggest that chaos might be present in the pattern of normal resting ventilation. However, the answer is not clear because the Wolf et al. algorithm for determining the Lyapunov exponent failed to provide a reliable estimate of the exponents for the surrogate data even though the exponents were within the range anticipated for a chaotic physiological time series. The non-linear prediction method of Sugihara and May[23] also indicated possible chaos in the pattern of correlation between the predicted and original data. Yet, there was no difference between the non-linear prediction for original and surrogate data. In contrast with these findings, the modified Lyapunov exponent algorithm of Sano and Sawada indicated positive Lyapunov exponents for the original data sets analyzed, and negative exponents for the surrogate data. These results suggest that further analysis is warranted with this approach, but caution needs to be applied as it has been pointed out several times that concluding the presence of chaos simply on the basis of results from one test can be misleading.[7,14,25]

If chaos is found in the pattern of breathing, what might it tell us about the way in which ventilation is regulated? The recent series of papers by Sammon, Bruce and colleagues[15-19] has demonstrated exquisitely that chaos can be found in respiration of rats, and that a number of experimental manipulations can evoke specific types of behaviour. If the central respiratory pattern generator is influenced by the combined feedback information from peripheral and central chemoreceptors, from pulmonary stretch receptors, and from other feedback and feed forward inputs, then it might be possible to understand the complex interactions of these factors. A deterministic nature of chaos could mean that the underlying physiology might be described by a relatively simple mathematical relationship.

4. CONCLUSION

The early studies of breathing pattern variability by Lenfant[13] and Hlastala et al.[9] provided the initial information on the application of the Fourier transform to quantification of the nature of respiratory rhythms. They observed fast (2-6 breath), longer (25-50 breath) and slow (150-200 breath) oscillations in breathing pattern. The physiological interpretation of these findings could be that the current pattern of breathing will influence the future pattern of breathing through, for example, the elimination or storage of CO_2 with the impact of arterial PCO_2 on the chemoreceptor output, or perhaps through some short-term memory.[27] In the respiratory system, it can be envisioned that the current value of ventilation or CO_2 output will have effects that last over a number of breaths into the future. Thus, short-term correlations are to be expected. We did find this over a range of up to about 25 breaths.

Perhaps the finding of a range of β values in the present study could be explained by a range of sensitivities of the afferent feedback information. It has been observed that the pattern of breathing shows smaller deviations away from the desired set point while breathing an hypoxic compared to hyperoxic gas mixture.[26] The subjects in this study with β values near, or greater than, 1.0 might have a less tightly regulated pattern of breathing as observed when peripheral chemoreceptor activity was diminished by hyperoxia.

Although the evidence for chaos in the breathing pattern of anesthetized rats is quite convincing,[18,19] the case for chaos in the breathing pattern of resting humans is less so. While the individual breath flow patterns were analyzed in the rat model, breath-by-breath patterns have been analyzed in human subjects. Donaldson[6] concluded that there was chaos in a range of variables measured in humans during resting breathing. However, we have shown by application of the same methods as in this previous study that analysis by only one method could result in a false positive outcome. While there was a positive Lyapunov exponent for $\dot{V}_E$ in this study in 6 of 7 subjects, the surrogate data also had a positive exponent. Based on these data alone, it is not possible to conclude the presence of chaos. Additional analyses of breathing pattern with a different algorithm to calculate the Lyapunov exponents did show that there were positive exponents for the original data, and negative exponents for the surrogate data. Yet, before the conclusion is made that pattern of breathing is chaotic, the results of the non-linear prediction need to be considered as once again, there was no difference between the original and the surrogate data.

Overall, it appears that the pattern of breathing in resting human subjects might have properties consistent with fractal and/or chaotic systems. However, the evidence is not strong. The limitations of attempting to study patterns of breathing in human subjects must be acknowledged. In contrast to the rat model studied by Sammon and colleagues,[18,19] human subjects have the ability to voluntarily over-ride the respiratory controller. Further, even though we attempted to strictly control the environment, the possibility of external cues influencing the pattern of breathing cannot be totally eliminated. Even though the collection of the long data sets required for this type of analysis in humans is fraught with difficulties, the nature of the respiratory control system[15] invites continued exploration of this problem.

ACKNOWLEDGMENTS

Original research included in this paper was supported by the Natural Sciences and Engineering Research Council of Canada, and the Heart and Stroke Foundation of Ontario.

REFERENCES

1. Benchetrit, G. and F. Bertrand. A short term memory in the respiratory centres: statistical analysis. *Resp. Physiol.* 23: 147-158, 1975.
2. Bolton, D. P. G. and J. Marsh. Analysis and interpretation of turning points and run lengths in breath-by-breath ventilatory variables. *J. Physiol. (London)* 351: 451-459, 1984.
3. Butler, G. C., Y. Yamamoto, and R. L. Hughson. Heart rate variability and fractal dimension during orthostatic challenges. *J. Appl. Physiol.* 75: 2602-2612, 1993.
4. DeBoer, R. W., J. M. Karemaker, and J. Strackee. Comparing spectra of a series of point events particularly for heart rate variability data. *IEEE Trans. Biomed. Eng.* 31: 384-387, 1984.
5. Dejours, P., R. Puccinelli, J. Armand, and M. Dicharry. Breath-to-breath variations of pulmonary gas exchange in resting man. *Respir. Physiol.* 1: 265-280, 1966.
6. Donaldson, G. C. The chaotic behaviour of resting human respiration. *Respir. Physiol.* 88: 313-321, 1992.
7. Glass, L. and D. Kaplan. Time series analysis of complex dynamics in physiology and medicine. *Med. Prog. Technol.* 19: 115-128, 1993.
8. Goldberger, A. L., D. R. Rigney, and B. J. West. Chaos and fractals in human physiology. *Sci. Am.* 262: 42-49, 1990.
9. Hlastala, M. P., B. Wranne, and C. J. Lenfant. Cyclical variations in FRC and other respiratory variables in resting man. *J. Appl. Physiol.* 34: 670-676, 1973.
10. Hughson, R. L., D. R. Northey, H. C. Xing, B. H. Dietrich, and J. E. Cochrane. Alignment of ventilation and gas fraction for breath-by-breath respiratory gas exchange calculations in exercise. *Comput. Biomed. Res.* 24: 118-128, 1991.
11. Hughson, R. L. and Y. Yamamoto. On the fractal nature of breath-by-breath variation in ventilation during dynamic exercise. In: *Control of Breathing and Its Modeling Perspectives*, edited by Y. Honda, Y. Miyamoto, K. Konno, and J. Widdicombe. New York: Plenum Press, 1992, p. 255-262.
12. Kennel, M. B., R. Brown, and H. D. I. Abarbanel. Determining embedding dimension for phase-space reconstruction using geometrical construction. *Phys. Rev.* 45: 3403-3410, 1992.
13. Lenfant, C. Time-dependent variations of pulmonary gas exchange in normal man at rest. *J. Appl. Physiol.* 22: 675-684, 1967.
14. Ruelle, D. Deterministic chaos: the science and the fiction. *Proc. R. Soc. Lond. A.* 427: 241-248, 1990.
15. Sammon, M. Symmetry, bifurcations, and chaos in a distributed respiratory control system. *J. Appl. Physiol.* 77: 2481-2495, 1994.
16. Sammon, M. Geometry of respiratory phase switching. *J. Appl. Physiol.* 77: 2468-2480, 1994.
17. Sammon, M., J. R. Romaniuk, and E. N. Bruce. Bifurcations of the respiratory pattern associated with reduced lung volume in the rat. *J. Appl. Physiol.* 75: 887-901, 1993.
18. Sammon, M., J. R. Romaniuk, and E. N. Bruce. Role of deflation-sensitive feedback in control of end-expiratory volume in rats. *J. Appl. Physiol.* 75: 902-911, 1993.
19. Sammon, M., J. R. Romaniuk, and E. N. Bruce. Bifurcations of the respiratory pattern produced with phasic vagal stimulation in the rat. *J. Appl. Physiol.* 75: 912-926, 1993.
20. Sano, M. and Y. Sawada. Measurement of Lyapunov spectrum from a chaotic time series. *Phys. Rev. Lett.* 55: 1082-1085, 1985.
21. Saul, J. P., P. Albrecht, R. D. Berger, and R. J. Cohen. Analysis of long term heart rate variability: methods, 1/f scaling and implications. *Comp. Cardiol.* 14: 419-422, 1988.
22. Schepers, H. E., J. H. G. M. Van Beek, and J. B. Bassingthwaighte. Four methods to estimate the fractal dimension from self-affine signals. *IEEE Eng. Med. Biol.* June: 57-71, 1992.

23. Sugihara, G. and R. M. May. Nonlinear forecasting as a way of distinguishing chaos from measurement error in time series. *Nature* 344: 734-741, 1990.
24. Szeto, H. H., P. Y. Cheng, J. A. Decena, Y. Cheng, D.-L. Wu, and G. Dwyer. Fractal properties in fetal breathing dynamics. *Am. J. Physiol. Regul. Integr. Comp. Physiol.* 263: R141-R147, 1992.
25. Theiler, J., S. Eubank, A. Longtin, B. Galdrikian, and J. D. Farmer. Testing for nonlinearity in time series: the method of surrogate data. *Physica D* 58: 77-94, 1992.
26. Tuck, S. A., Y. Yamamoto, and R. L. Hughson. The effects of hypoxia and hyperoxia on the 1/f nature of breath-by-breath ventilatory variability. In: *Modelling and Control of Ventilation*, edited by S. Semple, L. Adams and B. Whipp. NY: Plenum, 1995, p.297-302.
27. Wagner, P. G. and F. L. Eldridge. Development of short-term potentiation of respiration. *Respir. Physiol.* 83: 129-140, 1991.
28. Wolf, A., J. B. Swift, H. L. Swinney, and J. A. Vastano. Determining Lyapunov exponents from a time series. *Physica D* 16: 285-317, 1985.
29. Yamamoto, Y., J. O. Fortrat, and R. L. Hughson. On the fractal nature of heart rate variability in humans: effects of respiratory sinus arrhythmia. *Am. J. Physiol. Heart Circ. Physiol.* 269:H480-H486, 1995.
30. Yamamoto, Y. and R. L. Hughson. Extracting fractal components from time series. *Physica D* 68: 250-264, 1993.
31. Yamamoto, Y. and R. L. Hughson. On the fractal nature of heart rate variability in humans: effects of data length and β-adrenergic blockade. *Am. J. Physiol.* 266: R40-R49, 1994.
32. Yamamoto, Y., R. L. Hughson, J. R. Sutton, C. S. Houston, A. Cymerman, E. L. Fallen, and M. V. Kamath. Operation Everest II: An indication of deterministic chaos in human heart rate variability at extreme simulated altitude. *Biol. Cybern. 69: 205-212, 1993.*

13

HETEROGENEITY OF PULMONARY PERFUSION CHARACTERIZED BY FRACTALS AND SPATIAL CORRELATIONS

Robb W. Glenny

Division of Pulmonary and Critical Care Medicine
University of Washington School of Medicine
Seattle, Washington 98195

1. INTRODUCTION

New technologies providing high resolution measurements of organ perfusion, have revealed a large degree of spatial heterogeneity. These new observations provide new insights into the mechanisms of blood flow distribution and are revolutionizing our understanding of normal and pathologic physiology. The concepts of heterogeneity, as they apply to blood flow distribution, span all levels of physiology from the organ system to the mitochondrial level. Mechanisms matching heterogeneous perfusion to organ function and cellular needs encompass the same range of scales. New paradigms of organ perfusion and the matching of local blood flow to function are needed to account for these new observations. Fractal analysis, a revolutionary mathematical science, provides insights and tools for constructing such models. Because fractal analysis is not a familiar tool to most physiologic investigators, principal definitions and concepts will be systematically developed. Concurrently, numerical methods to determine whether a structure or process is fractal and to estimate a fractal dimension will be derived.

2. FRACTALS

In 1623, Galileo stated that "[The] universe is written in the language of mathematics and its characters are triangles, circles, and geometrical figures"[1]. In contrast to this perspective, Benoit Mandelbrot claimed that "Clouds are not spheres, mountains are not cones, coastlines are not circles, and bark is not smooth, nor does lightning travel in a straight line"[2]. Mandelbrot has championed the idea that nature does not necessarily have to adhere to man made systems such as Euclidean geometry, but rather may have a language of its own. This language must be simple yet have the potential to create complex structures. Fractals have been introduced by Mandelbrot to characterize structures and processes occurring in nature.

Bioengineering Approaches to Pulmonary Physiology and Medicine, edited by Khoo
Plenum Press, New York, 1996

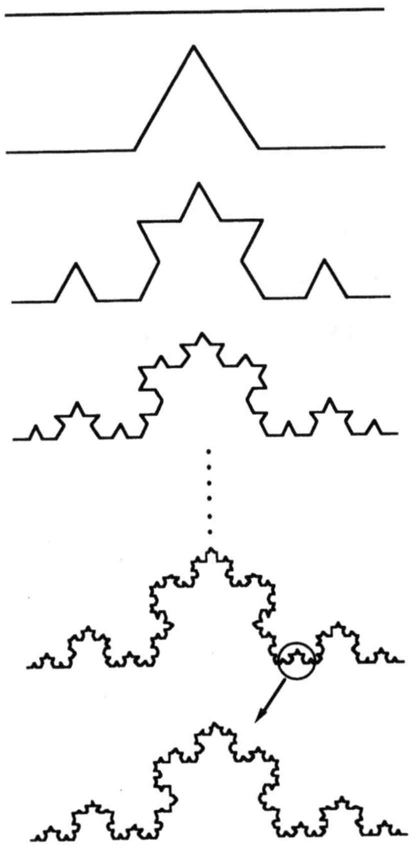

Figure 1. Generation of the Koch curve. The curve is produced by a simple iterative transformation beginning with a straight line (*top*). At each step, the middle third of each line segment is replaced with 2 segments, 1/3 the length of the line, forming part of an equilateral triangle. The completed curve has an infinite number of iterations. Regardless of the magnification of scale, any part of the curve resembles the whole (*bottom*). Reproduced from Glenny and Robertson[3].

A fractal structure or fractal process can be loosely defined as having a characteristic form that remains constant over a magnitude of scales. A *structure* is fractal if its small scale form appears similar to its large scale form. Similarly, a *process* is fractal if a variable as a function of time undergoes characteristic changes which are similar regardless of the time interval over which the observations are made. In the parlance of fractal analysis, this is the quality of *self*-similarity, also termed scale-independence.

The Koch curve (Figure 1), created by the Swedish mathematician Helge von Koch in 1904, is a fractal structure which provides a simple introduction to the concepts of self-similarity and fractal dimensions. This curve is defined by the following iterative transformations. Beginning with a straight line of length l_0 (Figure 1, top line), the middle third of the line is replaced with two segments of length $l_0/3$ forming part of an equilateral triangle (Figure 1, second line). The next iteration repeats the same procedure on each of the four resultant straight line segments. Subsequent generations are formed in an identical fashion, and the completed figure represents the infinite expression of this iterative procedure. The completed Koch curve exemplifies the properties of self-similarity because regardless of the scale used to examine any portion of the Koch curve, it maintains its characteristic form.

Examples of self-similar structures abound in the natural world. A tree maintains a quality of self-similarity independent of the perspective or scale from which it is viewed. The branching angles and proportionate diameters of branches appear to remain constant regardless of whether we are looking at the main trunk or the terminal branches. Clouds are fractal, with each billowing appendage similar in form to its entirety. In fact, without a reference scale, it is not possible to estimate the size of a cloud from a photograph[4]. The classical example of fractal structures are coastlines which appear to maintain the same degree of irregularity regardless of the size or detail of the map studied[5].

2.1. Contour Measuring Method

A second principle of fractal structures and processes is a corollary of self-similarity: since the underlying form of a structure or process remains similar through successive magnifications of scale, it follows that a measured length of its form cannot approach a limit. Remembering the Koch curve, the set of segments which contribute to its length can be expanded indefinitely. Although this property is a necessary requirement for a structure or process to be fractal in a strict mathematical sense, it is possible to discuss fractal properties of natural objects over a limited range of scales.

The Koch curve (Figure 2 *left*) is a good model to formally examine the characteristic of the scale-dependent length of fractal structures. Because of its jaggedness, the apparent length, $L(l)$, of the curve will be dependent on the length, l, of the measuring device chosen. If we use a stick of length l_0, equal to the straight line distance from one end to the other of the curve, none of the protruding structures will be measured and the apparent length of the entire line is l_0. If the measuring stick length, l, is decreased to $(1/3)l_0$ and then $(1/9)l_0$ the apparent contour length increases to $(4/3)l_0$ and $(16/9)l_0$ respectively. Generalizing this process for n iterations and a measuring stick of length $l=(1/3)^n l_0$, the corresponding measured length would be $(4/3)^n l_0$. Therefore, as l becomes infinitely small, or as $n \rightarrow \infty$, the apparent contour length of the Koch curve becomes infinite. As increasing magnification reveals more detail, the overall appearance of the new segment examined remains similar to that of the previous segment.

Mandelbrot derived the fractal dimension to characterize the complexity of fractal figures[2]. The ratio of the contour lengths can be related to the ruler lengths by the fractal dimension, D:

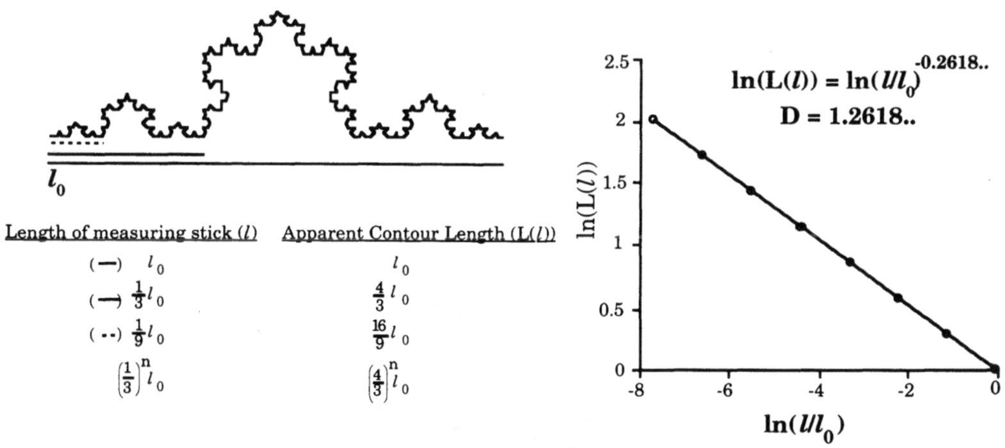

Figure 2. The apparent contour length of the Koch curve is dependent on the length of the measuring stick. *Left.* The finer the scale (greater magnification) the greater the apparent length of the curve. *Right.* Fractal (log-log) plot of apparent length, $L(l)$, of the Koch Curve as a function the length of the measuring device, l relative to l_0. The line through the points represents the least squares linear regression fit . The relationship appears linear with a slope of -0.2618.. and thus a fractal dimension, D = 1.2618. Reproduced from Glenny and Robertson[6].

$$\frac{L(l)}{L(l_0)} = [\frac{l}{l_0}]^{1-D} \tag{1}$$

For the Koch curve, in which $L(l)/L(l_0) = 4/3$ and $l/l_0 = 1/3$ for each iteration, the fractal dimension, $D = \ln(4)/\ln(3) = 1.2618...$ Unlike the Euclidean dimension, the dimension of a fractal structure is not usually an integer, but a fractal structure will always have a dimension which is equal to or less than the Euclidean dimension of space in which the structure is defined.

Taking the logarithm of both sides of Equation (1) and rearranging the terms yields

$$\ln L(l) = (1 - D) \cdot \ln[\frac{l}{l_0}] + \ln L(l_0) \tag{2}$$

In this form, a log-log plot of $L(l)$ versus l/l_0 produces a line with a slope of (1-D) and an intercept of $\ln(L(l_0))$. Fig. 2 (*right)* shows such a plot for the Koch curve, demonstrating a slope of -0.261.. or a D = 1.261... Equation 2 thus provides a working definition of a fractal process or structure. A process or object may be fractal if the logarithm of the measured value is linearly related to the logarithm of the scale of measurement. The fractal dimension, D, is 1.0 minus the slope of this linear relationship.

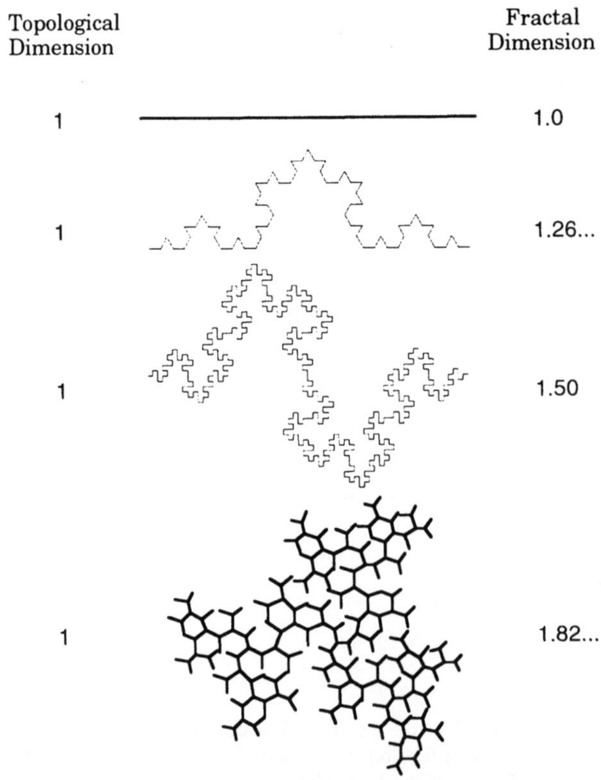

Figure 3. Lines of topological dimension 1 with different fractal dimensions. The fractal dimension is bounded by the topological dimension and the Euclidean dimension (1 and 2 in this case). The greater the irregularity of the line, or the more space it fills, the greater the fractal dimension. Reproduced from Glenny and Robertson[3].

An intuitive grasp of the meaning of a fractal dimension can be obtained from examination of some different fractal figures. A straight line has properties of self-similarity, in that at higher and higher resolutions it continues to show its same straight shape. It has a topological, fractal and Euclidean dimension of one. For different fractal line algorithms generated in two space, a topological dimension of one and a Euclidean dimension of two are maintained, but the more complex line figures have progressively increasing fractal dimensions. When the line becomes so complex that it nearly fills the plane, the D_s approaches 2.0. Figure 3 illustrates some fractal curves which are iteratively produced by different rules, yielding different fractal dimensions. It can be seen that even at the limited level of iteration illustrated in Figure 3, the curves with higher fractal dimensions are more space-filling. The fractal dimension therefore serves as a measure of complexity that is independent of the scale of magnification. As we will later show, this measurement of complexity can used to characterize physiologic structures and processes which have fractal properties.

2.2. Fractal Characterization of Spatial Heterogeneity Using RD Analysis

While the small scale variability in organ flow has been described as random, the branching structure of vascular anatomy suggests that regional flow is also best described by fractal measures. Heterogeneity of regional blood flow in an organ can be characterized by measuring the relative dispersion (RD=100·standard deviation/mean) of the regional flows when the organ is divided into a number of pieces. The observed relative dispersion is a sum of the spatial variation and the fluctuation of local flows over time[7,8]. When the distribution of flows is measured by a single rapid injection of a deposited flow marker, there is little contribution from the temporal fluctuations to the total heterogeneity.

When the heterogeneity of organ blood flow is characterized by this approach, the calculated spatial RD, abbreviated as RD_s, is dependent on the size of the sampled pieces[7]. If the blood flow in each of four pieces of an organ is measured, one can obtain the mean, standard deviation, and hence relative dispersion of flow in the organ. If these same pieces are progressively subdivided, then for 16, 64, 256, or more regions, the mean remains constant but the estimate of the standard deviation and relative dispersion increase (Figure 4). Even after appropriate corrections for experimental error, the largest estimate of RD_s will be obtained from the finest subdivisions of the organ[7].

The heterogeneity of organ blood flow can be characterized independently of scale by employing fractal analysis[3,7,8]. The fractal equation describing the relative dispersion of flows for a given spatial resolution (piece size) is given by rephrasing Equation 1 using RD_s as a function of a volume of size v:

$$\frac{RD_s(V)}{RD_S(V_0)} = [\frac{V}{V_0}]^{1-D_S} \tag{3}$$

Here $RD_s(v)$ is the measured relative dispersion when the organ is partitioned into regions of volume v, $RD_s(v_0)$ is the RD_s found for an arbitrarily chosen piece size and D_s is the derived spatial fractal dimension. Multiplying both sides of Equation 3 by RD_s and taking the logarithms, we obtain

$$\ln RD_S(V) = (1 - D_S) \cdot \ln[\frac{V}{V_0}] + \ln RD_S(v_0) \tag{4}$$

If the slope of log $RD_s(v)$ vs $\log(v/v_0)$ is constant over a range of partitions, the system is said to behave fractally within that range. The greater the rate of increase in observable heterogeneity with an increase in resolution, the greater is the fractal dimension. The fractal

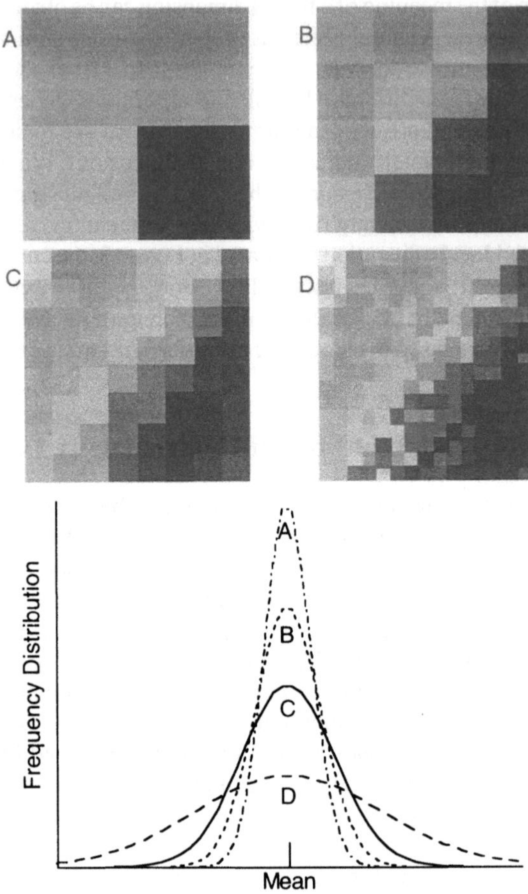

Figure 4. Relative dispersion increases as piece size decreases. With increasing resolution from (decreasing piece size) from A to D, the mean of the distribution is constant but the standard deviation increases.

dimension therefore serves as a measure of the scale independent irregularity, roughness, or variation of a system[7]. An advantage of this analytical approach is that it provides an estimate of D_s from easily obtained measurements of the RD_s of regional organ flow during successive subdivisions of the tissue pieces down to v_0.

Although regional blood flow to an organ is distributed in three dimensional space, when characterized as a relative dispersion, flow heterogeneity is 1-dimensional[3, 7, 8]. In RD analysis, a D_s of 1.0 indicates totally correlated magnitudes of flow between neighboring regions of the organ in that the flow is the same everywhere, while a D_s of +1.5 indicates that the magnitude of flow is not correlated or randomly distributed among neighboring pieces of the organ. D_s's above 1.5 indicate inversely or negatively correlated flows.

Pulmonary blood flow distribution can be characterized by fractal methods[3]. A composite fractal plot of the $RD_s(v)$ for six supine dogs is presented in Figure 5. The fractal dimension, D_s, for these animals ranged from 1.07 to 1.12 with an average of 1.09. The data fit the fractal model well with an average correlation coefficient, $r = 0.98$. It is interesting to note that the observed measures of RD_s appear to oscillate about the linear regression line. As discussed earlier, this may be due to the fact that the lung pieces (measuring stick) are not the proper shape or orientation for our measurement. The appropriate sectioning of the

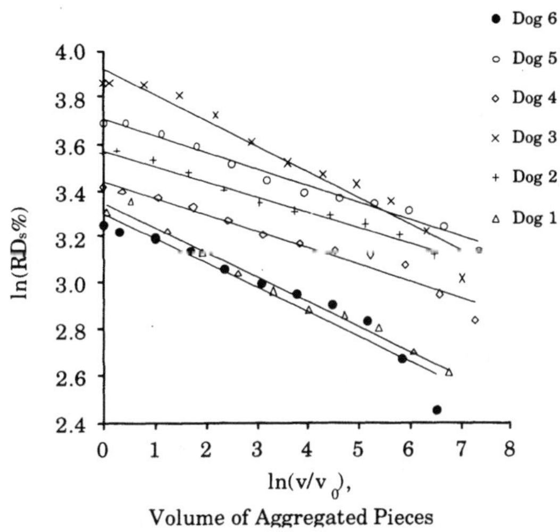

Figure 5. Relative dispersion of regional pulmonary blood flows in 6 dogs plotted as a function of the volume of the aggregated lung pieces. The smallest regions (v_0) are "voxels" from a planar gamma camera in which the voxels are 1.5 x 1.5 x 11.5 mm or 24 mm^3. Reproduced from Glenny and Robertson[3].

organ would be along the vascular tree, with regions of common perfusion being grouped together. The data points shown in Figure 5 do not include the first four subdivisions of the lungs, as those measurements are disproportionately smaller.

The heterogeneity of myocardial and skeletal muscle blood flow has been characterized by fractal analysis as well[3, 9, 10]. The distribution of radio-labeled microspheres to the heart was analyzed by progressively subdividing the heart into finer pieces. In this original application of the RD approach, regional flows were normalized to mass rather than to volume. The spatial heterogeneity of cardiac blood flow in baboons, sheep, and rabbits, as characterized by the fractal dimension, is shown in Table 1, along with the fractal dimensions for pulmonary blood flow in dogs. Although the number of pieces in the cardiac data sets are relatively small, the fractal dimensions are significantly different between some of the species and organs. This tells us that the spatial distribution of flow is different among these organs and suggests that this is necessary for their different functions or due to dissimilar morphogenesis.

This comparison demonstrates an advantage of fractal analysis in that comparisons of measurements can be made between experiments, species and laboratories, regardless of units or scales of measure. The heterogeneity of blood flow in the hearts of baboons and sheep measured by one technique can be compared to the heterogeneity of blood flow in dog lungs measured in another lab using very different methods.

Blood flow hetergogeneity can be fractal only over a limited range. If smaller and smaller pieces are used to measure flow to a region of tissue, eventually the flows will become

Table 1. Fractal dimensions of blood flow heterogeneity to different organs[3, 9, 10]

Animal	No. of animals	D_s
Baboon hearts	10	1.21 ± 0.04
Sheep hearts	11	1.17 ± 0.06
Rabbit hearts	6	1.25 ± 0.07
Dog lungs	6	1.09 ± 0.02
Rabbit muscle	46	1.37 ± 0.06
Sheep muscle	5	1.35 ± 0.08
Cat muscle	8	1.40 ± 0.07

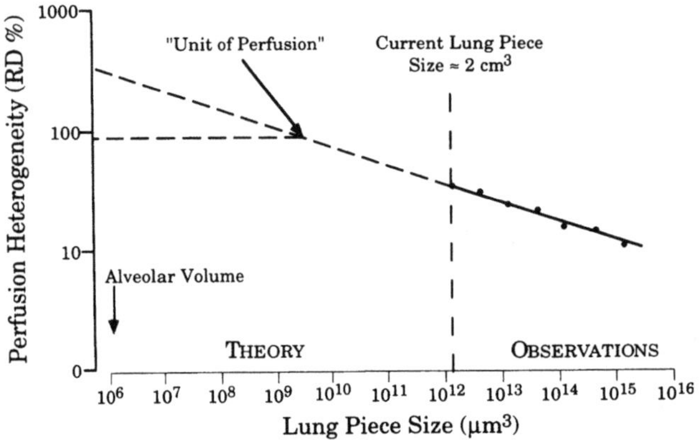

Figure 6. Fractal plot showing perfusion heterogeneity as a function of piece size. Experimental data demonstrate a linear relationship across the range of piece sizes obtained with current methods. Theoretically, perfusion heterogeneity could continue down to the alveolar level or it may plateau at the "unit of perfusion". Recent data using higher resolution methods suggest that flow heterogeneity continues to increase in a fractal manner down to the gas exchanging level.

more similar as the anatomic limit of a capillary is reached[8]. As long as heterogeneity is fractal, the log of RD_s will remain linear with respect to the log of the volume of pieces. However, as the functional unit of perfusion is approached with smaller piece sizes the relative dispersion will stabilize, causing a plateau in the fractal plot[8] (Figure 6). Theoretically, fractal analysis could identify the size of the functional unit of flow in a lung by finding the piece size where there is an inflection in the slope of the fractal plot[7]. The method we used to measure regional blood flow distributions was able to examine pieces of lung which were 24 mm^3. No inflection point could be detected, suggesting that the unit of uniform perfusion is smaller than 24 mm^3.

To explore blood flow distribution at higher resolutions, a fluorescent cryomicrotome is being developed that allows the spatial distribution of organ blood flows to be measured at the capillary level. The system rapidly collects data from organs containing up to four

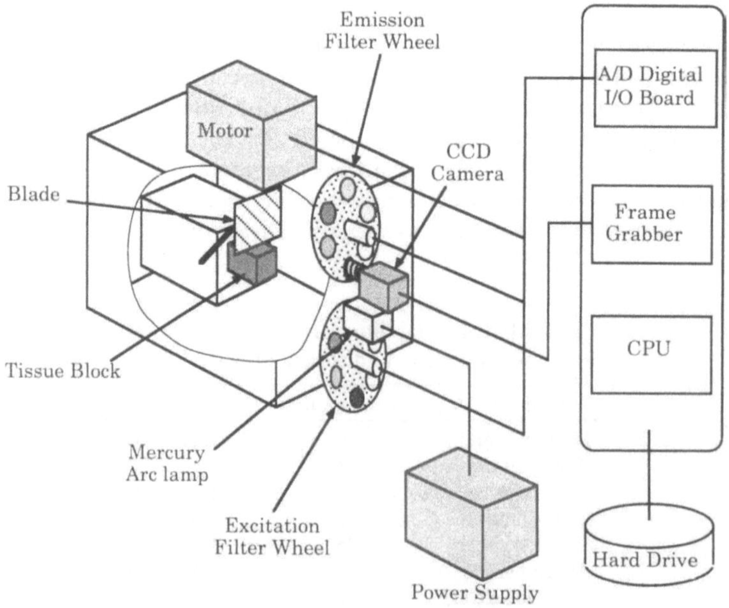

Figure 7. Schematic of fluorescent cryomicrotome.

different colors of fluorescent microspheres. The instrument (Figure 7) consists of a charge-coupled device (CCD) video camera, a microcomputer with I/O capabilities, a separate hard disk drive, mercury arc lamp, an excitation filter-changer wheel, an emission filter-changer wheel, and a cryostatic-microtome. Fluorescence images are acquired using a digital camera (2040 x 2040 pixel array). A mercury arc lamp illuminates the frozen sample. Two motorized filter wheels containing excitation and emission filters are mounted in front of the camera and their positions are controlled by the computer through photomicrosensors. An American Optical Cryo-Cut microtome is outfitted with an asynchronous stepper motor to serial section frozen organs. Computer control of the microtome motor, emission filter wheel, xenon flash triggering, and image capture and display is accomplished through a virtual instrument written in the LabView 2 (National Instruments) programming language.

The cryomicrotome serially sections through the frozen organ at the selected slice thickness. Following each slice, five digital images of the tissue surface (*en face*) are obtained. Four of the images are acquired with appropriate excitation and emission filters to isolated each of the individual fluorescent colors. The X, Y, and Z (slice) locations of each microsphere are determined and saved in a text file. The fifth image of the lung cross-section produces a three-dimensional binary map defining the spatial location of lung parenchyma.

In preliminary studies[11], four ~30-gm mice were anesthetized with intraperitoneal injections of pentobarbital. After sonicating and vortexing the microspheres, approximately 30,000 red microspheres suspended in 300 μl of saline were injected into a tail vein of each mouse. The trachea of each mouse was exposed and a loose ligature placed around it. The lungs were processed as above and the X, Y, and Z locations of each microsphere determined. The lungs were sampled in a random fashion at a preselected sample volume (2×10^{-4} mm^3). The number of microspheres in each sampled volume was determined and a frequency distribution obtained from the pilot mouse experiments (Figure 8).

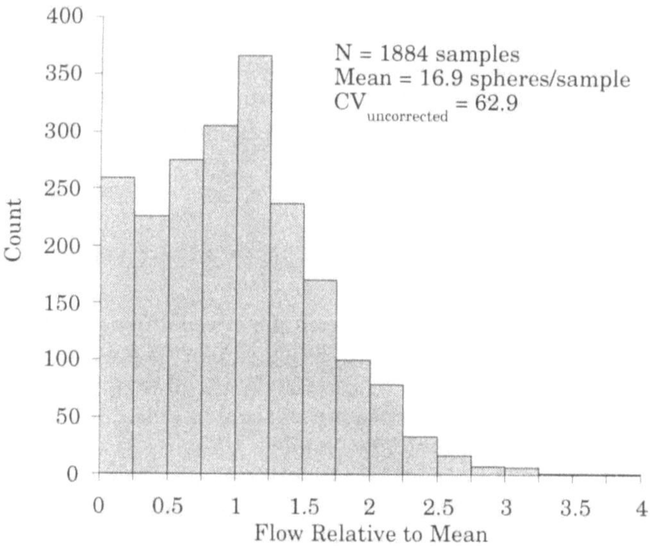

Figure 8. Frequency distribution of number of microspheres counted per sampled volume of lung. 31,694 microspheres were injected with an average of 16.9 microspheres/sample volume. The sample volume was ~ 2×10^{-4} mm^3. The distribution is similar to those seen in dogs and pigs in that it is heterogeneous and skewed to the right. Flows derived from number of microspheres/piece were normalized to a mean flow of 1 for this histogram.

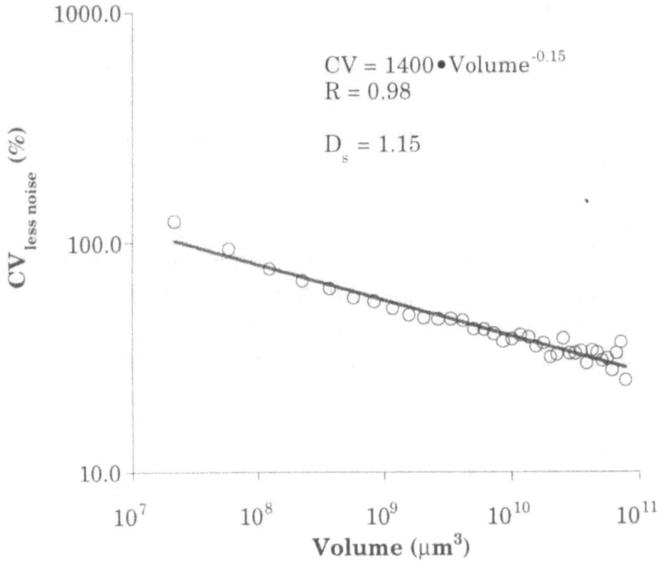

Figure 9. Plot of perfusion heterogeneity as a function of sampled volume of lung. The relationship remains linear down to a volume of ~ 10^7 μm^3. A mouse alveolar volume is ~10^6 μm^3.

This distribution is similar to pulmonary blood flow in larger animals, such as dogs. The distribution is heterogeneous and skewed to the right. The sampling volume used in this distribution, totaled 1884 pieces, considerably more than usually obtained from a 25-kg dog, in which we cut the lung into ~ 1400 (2 cm^3) pieces.

A fractal plot of perfusion heterogeneity as a function of the sampled volume of lung can be constructed from this high resolution data (Figure 9). The data are obtained by random sampling of the lung at increasing sample volumes, determining the number of microspheres in each sampled volume, and calculating a coefficient of variation for each sample size. The coefficient of variation is corrected for noise introduced by the small number of microspheres in the smaller sample volumes. Of particular interest is the observation that the relationship remains linear as the piece size decreases down to the smallest size. This is the first observation demonstrating that perfusion continues to become increasingly heterogeneous down to the acinar level.

3. SPATIAL CORRELATION OF REGIONAL BLOOD FLOW

In apparent contrast to the perfusion heterogeneity, pulmonary blood flow is spatially correlated with neighboring regions of lung having similar magnitudes of flow and distant pieces exhibiting negative correlation. When regional blood flow is determined to ~ 2 cm^3 volumes of lung in dogs, high flow regions are adjacent to other high flow regions and low flow regions are adjacent to other low flow regions[12] (Figure 10).

This relationship can be quantified using the correlation coefficient between flows to pairs of lung pieces. Let $f(i,j,k)$ be a process defined over the 3-dimensional space. The spatial correlation, r(d), is the correlation between the values of f at one position and a second position displaced from the first by a distance d. For example, the cubes in Figure 11 represent pieces of an organ distributed in space. The magnitude of blood flow to each piece is known. The spatial correlation between adjacent pieces, r(1), is therefore determined by creating all possible pairs of cubes that have a distance of one between their centers and calculating the correlation coefficient for all of these pairs. The spatial correlation between pieces that are

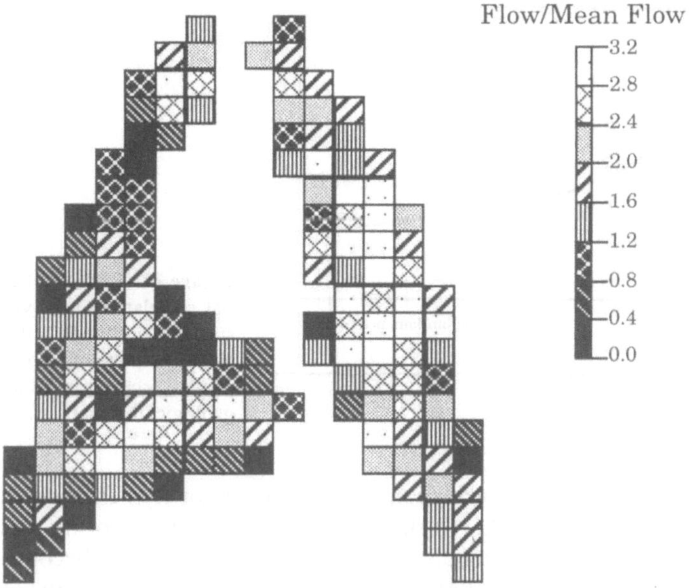

Figure 10. Distribution of regional perfusion within an isogravitational plane. The flows have been binned into 8 levels to facilitate the graphical presentation. Regional perfusion is spatially correlated with high flow regions near other areas of high flow and low flow regions near other areas of low flow. Reproduced from Glenny [12].

separated by a distance of two, r(2), is determined by creating all possible pairs of cubes that have a distance of two between their centers. The correlation coefficient has a range $-1.0 \leq r(d) \leq 1.0$ in which $r(d) = 1.0$ indicates a perfect positive correlation between all pairs, $r(d) = -1.0$ means a perfect negative correlation between pairs, and $r(d) = 0.0$ indicates a random association between all pairs.

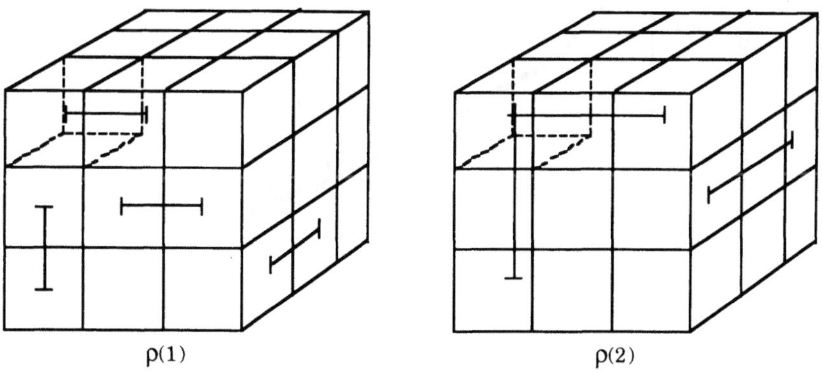

Figure 11. Concept of spatial correlation. The cubes represent pieces of an organ distributed in space. The magnitude of blood flow to each piece is known. The spatial correlation between adjacent pieces, r(1), is determined by creating all possible pairs of cubes that have a distance of one between their centers and calculating the correlation coefficient for all of these pairs. The spatial correlation between pieces that are separated by a distance of two, r(2), is determined by creating all possible pairs of cubes that have a distance of two between their centers.

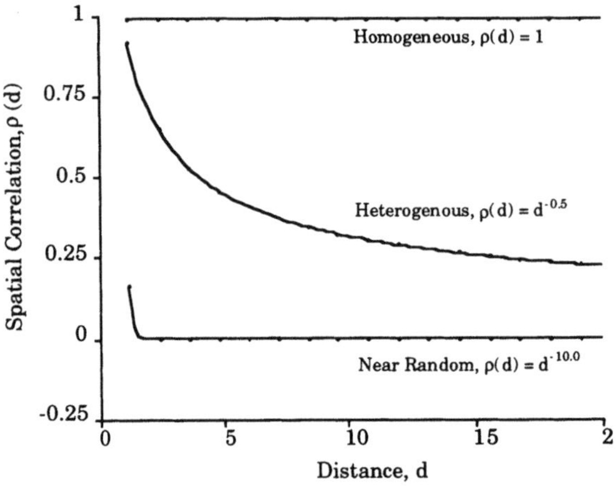

Figure 12. Expected behavior for the spatial correlation in a system with different spatial distributions of an observed variable. The power law relationship is an empirical prediction from agricultural and geologic observations.

Intuitively, the spatial correlation of regional pulmonary perfusion should behave in the following manner (Figure 12). It should be greatest for neighboring pieces when $d = 1$. It should decrease with increasing distance between pieces. As flow becomes more uniform throughout the organ, the correlation should approach 1.0. And the more heterogeneous the flow is, the faster the correlation should drop off with distance. Observations from agriculture and geology suggest the empirical formula that the spatial correlation decays as a power law of distance:

$$\rho(d) = C\ d^{-\alpha} \tag{5}$$

When the spatial correlation of perfusion is determined for regional pulmonary perfusion, a strong positive correlation is noted for neighboring pieces (Figure 13). On

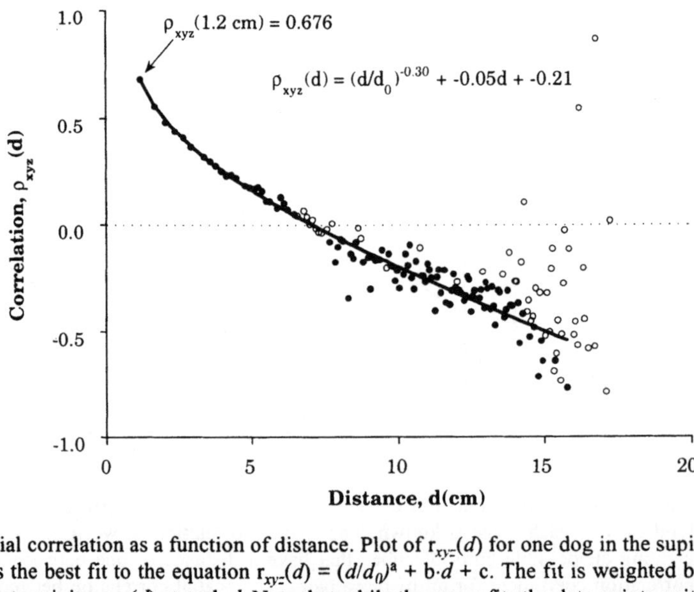

Figure 13. Spatial correlation as a function of distance. Plot of $r_{xyz}(d)$ for one dog in the supine position. The curve represents the best fit to the equation $r_{xyz}(d) = (d/d_0)^a + b \cdot d + c$. The fit is weighted by the number of paired regions determining $r_{xyz}(d)$ at each d. Note that while the curve fits the data points quite well when d is small, there is a large amount of scatter in the data points for larger d. The filled circles indicate those points at which the spatial correlation is significantly different from 0.0 ($p < 0.05$). Reproduced from Glenny et al.[13].

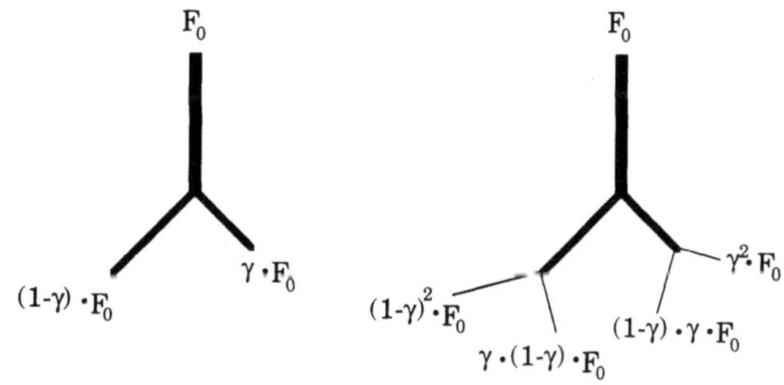

Figure 14. Dichotomously branching fractal model. *Left*: Basic element in which fractions g and 1-g of total flow F_0 are distributed to daughter branches. *Right*: Flow at terminal branches in network of 2 generations. Reproduced from Glenny and Robertson[15].

average, r(*d*) = 0.72, with d = 1.2 cm. As expected, the spatial correlation decreased with distance between pieces. Surprisingly though, the spatial correlation eventually became negative in all cases. This observation was not predicted by the empirical power laws observed in other natural processes and is likely due to the fact that blood flow is conserved within an organ[12]. Because a fixed amount of blood is distributed to a limited volume, high flow regions can exists only at the expense of flow to low flow regions. Regions of similar flow magnitude tend to be near each other because of the shared heritage of the vascular tree. No other examples of negative spatial correlation have been reported, suggesting that it may be unique to organ blood flow.

The spatial correlation of pulmonary perfusion is likely due the branching pattern of the vascular tree. To test this hypothesis, blood flow can be modeled by a dichotomously branching tree in which the fraction of flow from parent to daughter branches is g and 1-g at each bifurcation (Figure 14)[14]. The flow asymmetry parameter, g, is randomly chosen for each bifurcation from a normal distribution with a mean of 0.5 with a standard deviation of s. This model produces flow distributions similar to those observed in experimental animals, with heterogeneous flows that are skewed to the right (Figure 15).

This 2-dimensional model can be extended to 3-dimensional space by requiring branches to branch along one of three orthogonal directions to assure a space filling structure (Figure 16) [13]. Modeled blood flow values can be calculated to each terminal branch and the spatial correlation of flow determined for the 3-dimensional distribution (Figure 17). The

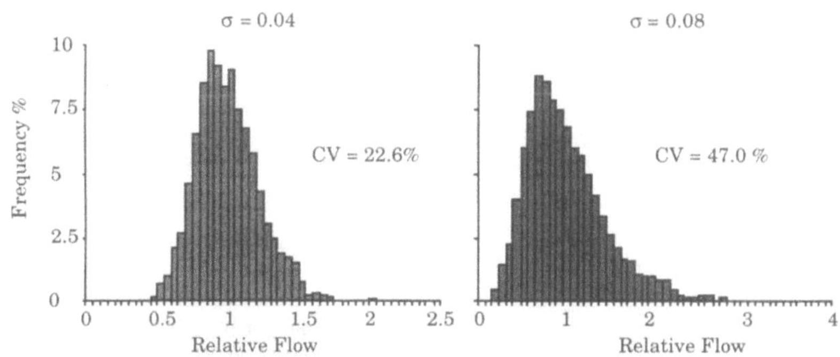

Figure 15. Heterogeneity of perfusion created by vascular branching model with different values of s. As s increases, the heterogeneity of perfusion increases. This degree of heterogeneity is comparable to previously reported distributions using lung pieces of similar scale. Note the scale differences in the abscissa (relative flow) between the two distributions. Reproduced from Glenny et al[13].

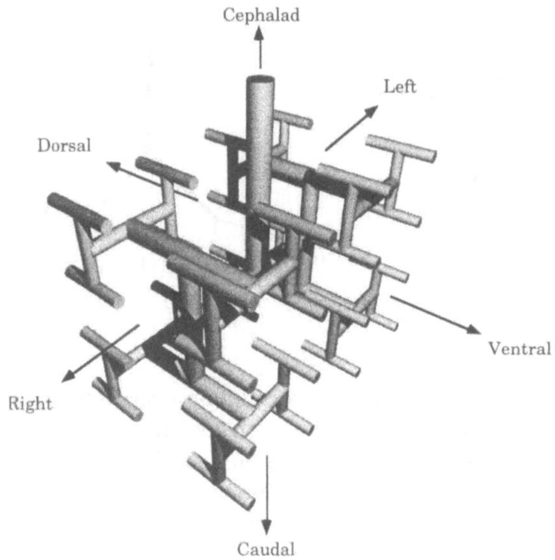

Figure 16. Three-dimensional branching model. Dichotomous branching model is extended into three dimensions by designating branching directions at each bifurcation along one of three orthogonal axes. The branching directions can be arbitrarily designated as dorsal/ventral, caudal/cephalad, and right/left. Reproduced from Glenny and Robertson[13].

spatial distributions are similar to those of experimental animals, with highly correlated flow to neighboring regions and negatively correlated flows at distances[13].

3. CONCLUSION

The traditional analytic tools of scientists are measures of means and variances with statistical approaches based on the assumption of random error. Until recently, observed pulmonary perfusion heterogeneity has been interpreted within the context of the gravitational model as random noise. The fractal revolution has brought the realization that this "error" can be analyzed as a fundamental property of the biologic system. Fractal analysis permits characterization of processes or structures that are not easily represented by traditional analytic tools. By providing a geometric framework for the description of apparently irregular patterns, fractal analysis is able to capture both the richness of physiologic structure and its function in a single model. The robust descriptive properties of fractal analysis to

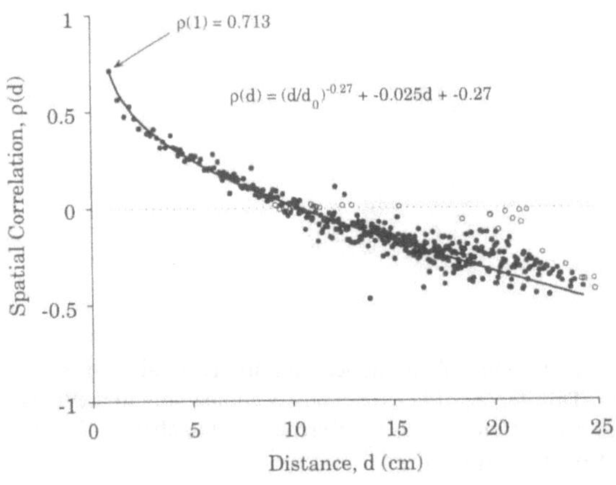

Figure 17. Spatial correlation as a function of distance. Plot of $r_{xyz}(d)$ for a modeled data set The curve represents the best fit to the equation $r_{xyz}(d) = (d/d_0)^a + b \cdot d + c$. The spatial correlation of the modeled data is very similar to the experimental data (see Figure 16). The filled circles indicate those points at which the spatial correlation is significantly different from 0.0 ($p < 0.05$). Reproduced from Glenny and Robertson[13].

biologic variability suggest that it may signal the development of a new paradigm, compelling the attention of investigators from diverse areas of scientific inquiry.

REFERENCES

1. Voss RF: Fractals in nature: From characterization to simulation. In: *The Science of Fractal Images*, edited by H.-O. Peitgen and D. Saupe. New York: Springer Verlag, 1988, p. 21-70.
2. Mandelbrot BB. *The Fractal Geometry of Nature*. San Francisco: W. H. Freeman, 1983.
3. Glenny, R.W., and H.T. Robertson. Fractal properties of pulmonary blood flow: characterization of spatial heterogeneity. *J. Appl. Physiol.* 69: 532-545; 1990.
4. Feder, J. *Fractals*. New York: Plenum Press, 1988.
5. Mandelbrot, B. How long is the coast of Britain? Statistical self-similarity and fractal dimension. *Science* 156: 636-638, 1967.
6. Glenny, R.W., H.T. Robertson, S. Yamashiro, and J.B. Bassingthwaighte. Applications of fractal analysis to physiology. *J. Appl. Physiol.* 70: 2351-2367, 1991.
7. Bassingthwaighte, J.B. Physiologic heterogeneity: fractals link determinism and randomness in structures and functions. *News Physiol Sci.* 5-1, 1988.
8. Bassingthwaighte, J.B., and J.H.G.M. van Beek. Lightning and the heart: fractal behavior in cardiac function. *Proc. IEEE* 76: 693-699, 1988.
9. Bassingthwaighte, J.B.,R.B. King, and S.A. Roger. Fractal nature of regional myocardial blood flow heterogeneity. *Circ. Res.* 65: 578-590, 1989.
10. Iversen, P.O., and G. Nicolaysen. Fractals describe blood flow heterogeneity within skeletal muscle and within myocardium. *Am. J. Physiol.* 268: H112-H116, 1995.
11. Glenny, R., S. Bernard, C. Barlow, J. Kelly, and H.T. Robertson. Spatial distribution of pulmonary blood flow at a microscopic scale of resolution. *Am. J. Respir. Crit. Care Med.* 151: A518, 1995.
12. Glenny, R.W. Spatial correlation of regional pulmonary perfusion. *J. Appl. Physiol.* 72: 2378-2386, 1992.
13. Glenny, R., and H.T. Robertson. A computer simulation of pulmonary perfusion in three dimensions. *J. Appl. Physiol.* 79: 357-369, 1995.
14. van Beek, J.H.G.M., S.A. Roger, and J.B. Bassingthwaighte. Regional myocardial flow heterogeneity explained with fractal networks. *Am. J. Physiol.* 257: H1670-H1680, 1989.
15. Glenny, R.W., and H.T. Robertson. Fractal modeling of pulmonary blood flow heterogeneity. *J. Appl. Physiol.* 70: 1024-1030, 1991.

14

THE TEMPORAL DYNAMICS OF ACUTE INDUCED BRONCHOCONSTRICTION

Jason H. T. Bates

Meakins–Christie Laboratories and
Departmant of Biomedical Engineering
McGill University Montreal
Quebec, Canada, H2X 2P2

1. INTRODUCTION

The mechanical properties of the lungs are crucial determinants of our ability to breathe effectively. There are a number of important diseases in which these mechanical properties change markedly. These properties can also be quickly altered by contraction of the smooth muscle which is wrapped around the conducting airways. This process is known as bronchoconstriction and is a hallmark feature of asthma, a common condition characterized by an abnormal bronchoconstrictor response. That is, when an asthmatic subject is exposed to stimuli that cause the airway smooth muscle to contract, the resulting contraction is much more extreme than in a normal individual[1].

The study of asthma, and bronchial hyperresponsiveness in general, is frequently done by the application of bronchoactive agents to the lungs in animals. This produces many of the symptoms of an acute asthma attack. The extent of the mechanical changes produced is taken to be a measure of the animal's degree of bronchial responsiveness. An issue of great concern is thus to elucidate those factors that determine bronchial responsiveness in animals, the hope being that this will lead to an understanding of the greatly increased responsiveness seen in asthmatic human subjects.

The conventional approach to studying the respiratory mechanical effects of a bronchoactive drug is that of constructing a dose-response curve[2-4]. That is, one measures respiratory system or lung impedance as a function of the dose of the drug of interest, for a range of doses. However, this approach neglects an entire major facet of the bronchoconstriction process, namely the temporal dynamics of the response to a given dose. Recent work from the author's laboratory[5-8] has shown that significant physiological information is present in the bronchoconstriction time course that is completely missed in the dose-response curve. This work has involved the application of methods in recursive parameter estimation and time/frequency analysis to data collected from animals under anesthesia. These methods are reviewed in this chapter, along with some of the physiological insights they have helped to produce.

Bioengineering Approaches to Pulmonary Physiology and Medicine, edited by Khoo
Plenum Press, New York, 1996

2. ASSESSMENT OF RESPIRATORY MECHANICS

The assessment of respiratory mechanics is essentially an exercise in inverse modeling. That is, one applies certain inputs to the system, measures the corresponding outputs, and then tries to relate the inputs and outputs to each other in terms of a mathematical model of the system. The particular model used depends critically both on the nature of the applied inputs and on which outputs can be measured[9,10]. The usefulness of the endeavor depends on the extent to which the components of the model can be taken to reflect important physiological counterparts.

Most of our detailed knowledge of respiratory mechanics comes from experiments involving animals, which have the great advantage over those involving humans of allowing much greater access to key measurement sites. The most important of these is at the proximal end of the trachea, henceforth referred to as the airway opening. This is accessed most satisfactorily (from an experimental point of view) via a tracheostomy where the trachea is sectioned below the larynx and brought out through an incision in the neck. Most respiratory mechanics estimation scenarios consider the flow ($\dot{V}$) at the airway opening to be the system input. The pressure (P) required to generate this flow is the corresponding output. P is equal to the pressure at the airway opening (Pao) if the subject under study is being mechanically ventilated by a gas source connected to the airway opening. In this case, the dynamic relationships between P and $\dot{V}$ reflect the mechanical properties of the entire respiratory system.

However, if the subject is breathing spontaneously then P is produced entirely by the respiratory muscles and Pao is zero (atmospheric). In this case one can obtain the pressure across the lungs (so-called transpulmonary pressure, Ptp) from the pleural pressure. This pressure is most conveniently approximated by the pressure in the esophagus (Pes). Pes can be measured via a balloon-tipped catheter placed in the esophagus and connected to a pressure transducer[11]. The dynamic relationships between Pes and $\dot{V}$ now reflect the mechanical properties of the lung alone (i.e. without any contribution from the thorax).

Another respiratory measurement of great importance in animal studies is that provided by the alveolar capsule technique which can be applied in animals whose lungs have been exposed by surgically opening the chest[12,13]. The capsule is a small plastic conduit (i.d. typically 4 mm) with a flange on one end. The flange is glued to the lung surface with cyanoacrylate glue so that the capsule isolates a small portion of the pleural surface within its chamber. This surface is then carefully punctured to a depth of about 2 mm several times with a needle, thereby exposing the sub-pleural alveoli to the capsule chamber. Finally, a miniature piezoresistive pressure transducer is lodged in the capsule chamber so that it can continuously record alveolar pressure (P_A). Several alveolar sites can be monitored simultaneously in this way using multiple capsules.

3. A VISCOELASTIC MODEL OF THE LUNG

The simplest and most widely invoked model of respiratory mechanics is the single-compartment model which represents the alveolar regions of the lung as a single elastic compartment served by a single flow-resistive airway. Figure 1 shows how this model can be conceptualized. The single alveolar compartment is represented by a chamber consisting of two opposing cylindrical containers, one fitting inside the other. We will assume an airtight fit with no friction between the two containers and no significant mass to either container. A conduit leading into the compartment through one of the containers represents the conducting airways of the lung, and has resistance Raw to flow. Each of the two

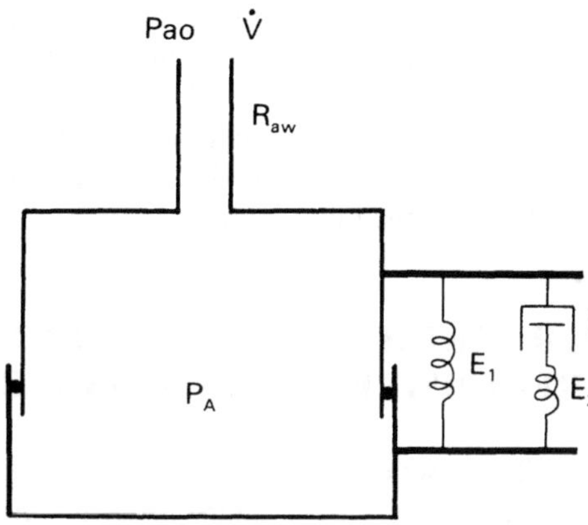

Figure 1. The homogeneous viscoelastic model of the lung. Raw is the flow resistance of the single airway, while the three components of the Kelvin body (E_1, E_2 and R_2) account for the viscoelastic properties of the lung tissue.

containers has a rigid bar connected to it. The bars are parallel to each other and are separated by a distance proportional to the volume of air in the compartment. The elastic recoil of the lung tissues is represented by a Hookean spring (elastance E_1) connected between the two rigid bars. As gas flows into the compartment through the airway, the two bars move apart and so stretch the spring. This produces a pressure inside the compartment just as inflating the lung produces an elastic recoil pressure within the alveoli.

With only the spring E_1 in place, the tissues of the lung are modeled as a purely elastic material. However, the lung tissues are actually viscoelastic[14,15]. This means that they can dissipate energy when stretched as well as just storing it. Furthermore, the relative amounts of energy stored and dissipated depend greatly on the frequency at which lung volume is oscillated. The gross features of this behavior can be conveniently represented by a collection of three elements - two springs and a dashpot (linear friction element) - as shown in Fig. 1. These three elements together constitute a Kelvin body, or linear viscoelastic solid. The series spring-dashpot pair (E_2 and R_2) constitute a Maxwell element[14].

Suppose $\dot{V}$ is quasi-sinusoidal and is provided by a mechanical ventilator. This is approximately the case during normal mechanical ventilation when most of the power in the $\dot{V}$ signal is at the fundamental ventilation frequency. If one assumes that the various resistive and elastic components of the model are constant and do not depend on flow or volume, then the equation of motion of the model is

$$Pao = (Raw + Rt)\dot{V} + EtV \tag{1}$$

where Rt and Et are the effective resistance and elastance of the tissues, respectively. Rt and Et depend on the frequency at which $\dot{V}$ is oscillated in and out of the lungs, and are functions of the three quantities E_1, E_2 and R_2.

If we also have a measure of P_A (provided by an alveolar capsule) then we can partition the lung's mechanical properties into those due to the conducting airways and those due to the tissues. That is,

$$Pao - P_A = Raw\,\dot{V} \tag{2}$$

and

$$P_A = Rt\dot{V} + EtV \tag{3}$$

This model assumes that the lung is homogeneous, which is a good approximation in the normal dog lung at normal breathing frequencies[15]. However, it becomes less valid as frequencies rise above a few hertz or when the lung becomes bronchoconstricted[12,13]. Nevertheless, to the extent that the lung does behave homogeneously, the model provides a very convenient basis upon which to interpret lung mechanics and the changes that can be induced by various interventions.

4. TRACKING CHANGES IN LUNG MECHANICS USING RECURSIVE LEAST SQUARES

The viscoelastic model shown in Fig. 1 can be fit to measurements of Pao, P_A and $\dot{V}$ by adjusting the parameters in its governing equations (Eqs. 1-3) until the pressures predicted by the model match measured pressures as closely as possible. Fortunately, the parameters are all linearly related to pressure in these equations, so parameter estimation can be performed by multiple linear regression. That is, if **Y** is a (column) vector of values of the dependent variable, **A** is a vector of model parameters, and **X** is a matrix of dependent variables then the system of equations produced by N measurements of the variables is

$$Y = XA + W \tag{4}$$

where **W** is a vector of errors which includes all deviations between model predictions of **Y** and its measured values. In the case of Eq. 1, Eq. 4 has the following specific realization:

$$\begin{bmatrix} Pao_1 \\ Pao_2 \\ Pao_3 \\ \cdot \\ \cdot \\ \cdot \\ Pao_N \end{bmatrix} = \begin{bmatrix} \dot{V}_1 & V_1 \\ \dot{V}_2 & V_2 \\ \dot{V}_3 & V_3 \\ \cdot & \cdot \\ \cdot & \cdot \\ \cdot & \cdot \\ \dot{V}_N & V_N \end{bmatrix} \begin{bmatrix} R \\ E \end{bmatrix} + \begin{bmatrix} w_1 \\ w_2 \\ w_3 \\ \cdot \\ \cdot \\ \cdot \\ w_N \end{bmatrix} \tag{5}$$

where R is Raw + Rt and E is Et. Equation 3 can be dealt with in the same manner, in which case R is just Rt.

The least-squares estimate, $\hat{\mathbf{A}}$, of the parameter vector is[16]

$$\hat{A} = [X^T X]^{-1} X^T Y \tag{6}$$

Equation 6 gives the parameter vector that minimizes the sum of the squared deviations between the measured pressure measurements and those predicted by the model, and so provides a single "best fit" value for R and E. However, this procedure is only useful if the respiratory system (or lung) is mechanically stable over the time during which measurements are made. If the mechanical properties of the system change significantly during the measurement period, then one would expect that the values of R and E necessary to describe the data should also change. Such changing parameter values can be tracked by implementing the multiple linear regression algorithm in recursive form. This allows the parameter values to be continuously updated at each time step.

Equation 6 provides the least-squares estimate of **A** given N measurements of the variables. If N+1 measurements were made (that is, the original N plus the next one in time), one could calculated the least squares **A** for all N+1 measurements simply by recalculating Eq. 6, except this time running the array indices up to N+1 instead of just N. However, this is computationally very inefficient if one has already gone to the trouble of calculating $\hat{A}$ based on the first N measurements. Let the estimate of **A** based on the first N measurements be $\hat{\mathbf{A}}_N$, while that based on all N+1 measurements is $\hat{\mathbf{A}}_{N+1}$. The recursive least squares (RLS) algorithm allows one to express $\hat{\mathbf{A}}_{N+1}$ in terms of $\hat{\mathbf{A}}_N$ plus a correction term proportional to the difference between the N+1th measurement of the dependent variable (y_{N+1}) and its prediction from the model using $\hat{\mathbf{A}}_N$, thus[17,19]

$$\hat{A}_{N+1} = \hat{A}_N + P_{N+1}X_{N+1}(y_{N+1} - X_{N+1}{}^T\hat{A}_N)/(\rho + X_{N+1}{}^T P_N X_{N+1}) \tag{7}$$

where

$$P_{N+1} = (1/\rho)[P_N - P_N X_{N+1} X_{N+1}{}^T P_N/(\rho + x_{N+1}{}^T P_N X_{N+1})] \tag{8}$$

Equation 8 keeps track of the covariance matrix **P**. $\mathbf{X}_{N+1}$ is the vector of independent variables obtained at the N+1th time-step (i.e. it would form the N+1th row of the **X** matrix in Eq. 4).

The real utility of the RLS algorithm embodied in Eqs. 7 and 8 is provided by the quantity ρ which is known as the forgetting factor and which takes a value between 0 and 1. When ρ is 1 the RLS algorithm works as a purely recursive form of conventional least-squares. However, when ρ is less than 1 the algorithm becomes the recursive form of weighted least-squares, with the data weights decreasing exponentially back in time from the current data point. The time-constant of the exponential is

$$\tau = -\delta t/\ln(\rho) \tag{9}$$

where δt is the data sampling interval and ln(...) denotes the natural logarithm of the bracketed quantity. This effectively imbues the RLS algorithm with a finite memory so that the current parameter estimates (i.e. the members of $\hat{\mathbf{A}}_{N+1}$) are determined most strongly be recent data while older data have progressively less influence. As ρ decreases, so does the time-constant of the data window.

In order to implement the RLS algorithm one must have initial estimates of $\hat{\mathbf{A}}$ and **P** as starting points for Eqs. 7 and 8. The most convenient way to do this is to set the elements of $\hat{\mathbf{A}}$ and those elements of **P** not on the leading diagonal all to zero. The elements of **P** on the leading diagonal are then set to some extremely large value. This has the effect of assigning negligible confidence to the initial estimates of **A** so that their values are quickly changed to the correct values in the light of incoming data.

A final issue that must be dealt with when implementing the RLS algorithm using Eqs. 1 and 3 concerns volume drift. That is, V is obtained by numerically integrating $\dot{V}$, and this is invariably accompanied by a low-frequency baseline drift in V that arises from a variety of sources such as flow transducer asymmetry, non-unity expiratory exchange ratio, and differences in inspired and expired gas temperatures and humidities. These are collectively virtually impossible to eliminate, and so it is necessary to take steps to artificially force a stable baseline in V. We have found that satisfactory results can be obtained by subtracting from $\dot{V}$ its recursively calculated, exponentially weighted running mean prior to integration (this is equivalent to passing it through a recursive first-order high-pass digital filter). Lauzon and Bates[19] tested the RLS algorithm with this kind of drift correction on computer-generated respiratory data and found that they could adequately follow the most rapid changes in mechanics one could expect to achieve by injecting a large dose of bronchial agonist.

5. TIME-COURSE OF PULMONARY MECHANICS DURING ACUTE BRONCHOCONSTRICTION IN DOGS

We applied the RLS algorithm with finite memory to tracheal pressure (Ptr-$\dot{V}$)data collected from anesthetized, paralyzed dogs during conventional mechanical ventilation following a bolus injection of histamine into the vena cava[5]. The bronchoconstriction elicited

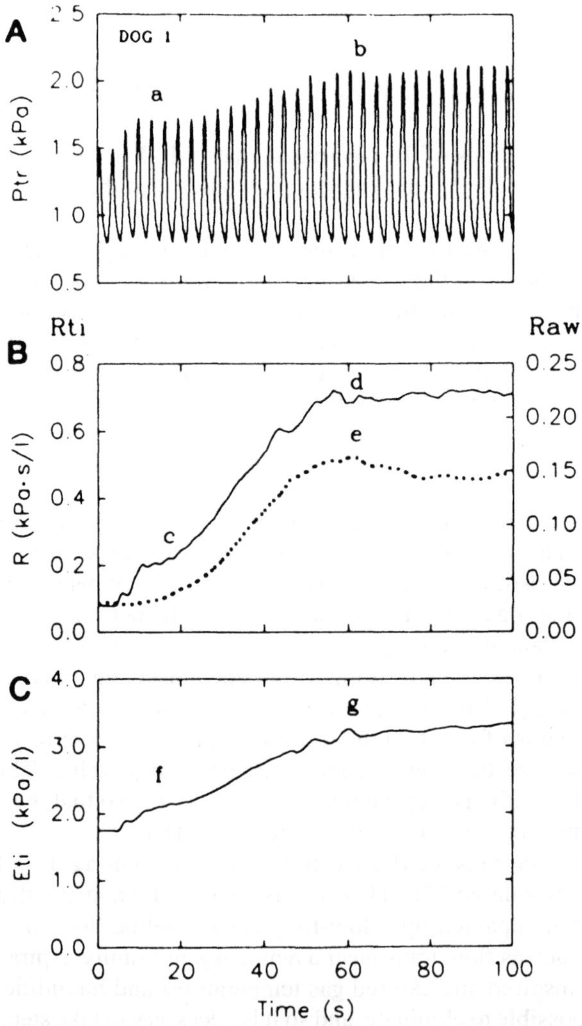

Figure 2. Time-courses of (A) tracheal pressure (Ptr), (B) tissue resistance (Rti - solid line) and airway resistance (Raw - dotted line), and (C) tissue elastance (Eti). These signals were obtained from an anesthetized dog during regular mechanical ventilation following i.v. injection of a histamine bolus. Adapted from Ref.5.

by the histamine as it arrived in the lungs from the right heart was manifest as a change in the Ptr profile occurring each breath. At the same time, we also measured P_A with alveolar capsules. We fit Eq. 3 to the P_A-$\dot{V}$ data using the RLS algorithm with a memory time-constant of 1 s. This gave continuous estimates of Rt and Et throughout the data collection period. We also obtained a continuous estimate of Raw via Eq. 2. Fig. 2 shows the Ptr, Raw, Rt and Et obtained in a typical dog during the 100 s immediately following histamine injection. These data show that the responses in Ptr, Rt and Et were bi-phasic, with an initial response occurring about 10 s after injection (indicated by a, c and f in Figs. 2A, B and C, respectively) with a second larger response starting at about 30 s (indicated by points b, d and g in Fig. 2). In contrast, Raw exhibited only a single response (indicated by e in Fig. 2B) in phase with the second responses of the other quantities.

We interpreted these results as reflecting the nature of the histamine delivery to the lungs[5]. Specifically, after a short delay due to the transit time through the heart and the pulmonary artery, the histamine reached the pulmonary circulation where it stimulated smooth muscle in the very periphery of the lung (including that in the alveolar ducts and peripheral airways). This caused a general stiffening of the lung tissues which was manifest as an increase in Ptr, Rt and Et. It did not, however, affect Raw to a noticeable extent because the very peripheral airways do not contribute significantly to Raw. However, the histamine then travelled back into the heart and out into the peripheral circulation which includes the bronchial supply that pervades the airway tree. This allowed the histamine to act on the major portion of the airway smooth musculature and so induce a massive constriction of the airways which produced the single phase increase in Raw.

Of particular interest in these results is the fact that the major change in lung tissue mechanics (i.e. the second phases in Rt and Et) appears to occur at the same time as the change in Raw. As the change in Raw was ascribed to central airway constriction, this suggests that most of the tissue changes occurred secondary to the airway constriction and not as a result of some active mechanism within the tissues themselves. Whether or not the tissues themselves have some intrinsic ability to respond to bronchial agonists has been something of a controversy ever since the alveolar capsule technique showed an apparent tissue response to agonist. Some workers have suggested that most of the action is in the tissues[20,21], while others have presented evidence that tissue changes follow purely from an active airway response[22]. Our results suggest that most of the observed tissue response is due to effects produced by contracting airways. One way this might happen is if tissue distortion is induced by narrowing of the embedded airways. Another mechanism is through ventilation inhomogeneity. That is, if the constriction of the airways is not uniform throughout the lungs then the apparent resistance and elastance of the lung will be increased in part purely as a result of the inhomogeneity, by the mechanism first expounded by Otis *et al*[23]. Given that the alveolar capsules always look into alveoli on the top of the lung (for technical reasons) it is likely that they give a biased sampling following bronchoconstriction when inhomogeneity is very likely to develop.

Although the evidence presented so far suggests that most changes in tissue mechanics are secondary to the constriction of the central conducting airways, there seems to be no doubt that some active elements out in the periphery of the lung do act to change lung tissue mechanics when stimulated. This is evidenced by the existence of the first response phase in Rt and Et (Fig. 2B and C) which occurred prior to any effect on Raw and which was presumed to result from the action of histamine on the lung periphery via the pulmonary circulation. Additional evidence for the existence of a peripheral response was obtained in dogs maintained at constant lung volume while bronchoconstriction developed following i.v. injection of histamine[6]. Lung mechanics were assessed during this period by imposing very small amplitude volume oscillations at 6 Hz at the trachea with a piston pump while Ptr was measured. This produced an oscillating Ptr signal (similar to that shown in Fig. 4)

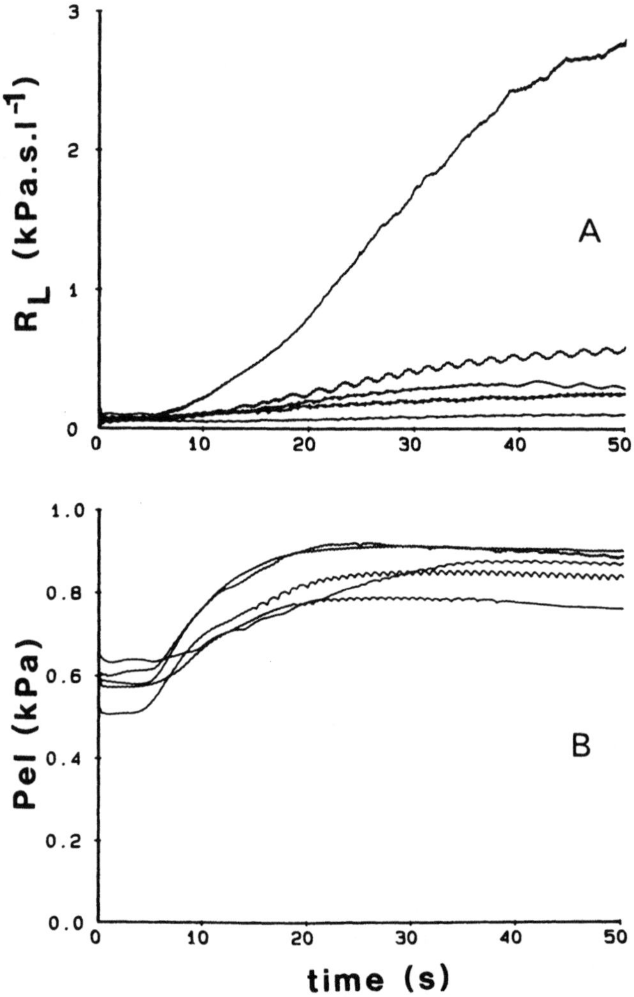

Figure 3. Time-courses of (A) lung resistance at 6 Hz (R_L - approximating airway resistance) and (B) lung elastic recoil pressure (Pel) (from Ref.6). The five curves in each graph were obtained from 5 different anesthetized dogs given an i.v. bolus of histamine at 0 time. Lung volume was kept constant throughout the 50 s measurement period.

that increased both in its mean value and in the amplitude of its oscillations, after an initial transit delay required for the histamine to reached the lungs. The mean Ptr, denoted elastic recoil pressure (Pel in Fig. 3B) was extracted from the total Ptr signal by passing a running mean of 1 s window length over the data. Subtracting Pel from Ptr left the 6 Hz oscillations whose amplitude continued to grow throughout the 50 s measurement period. These oscillations were established as being almost entirely due to airway resistance[6], which enabled Raw to be followed continuously by fitting the model

$$\mathrm{Ptr} = \mathrm{Raw}\dot{V} + K \tag{10}$$

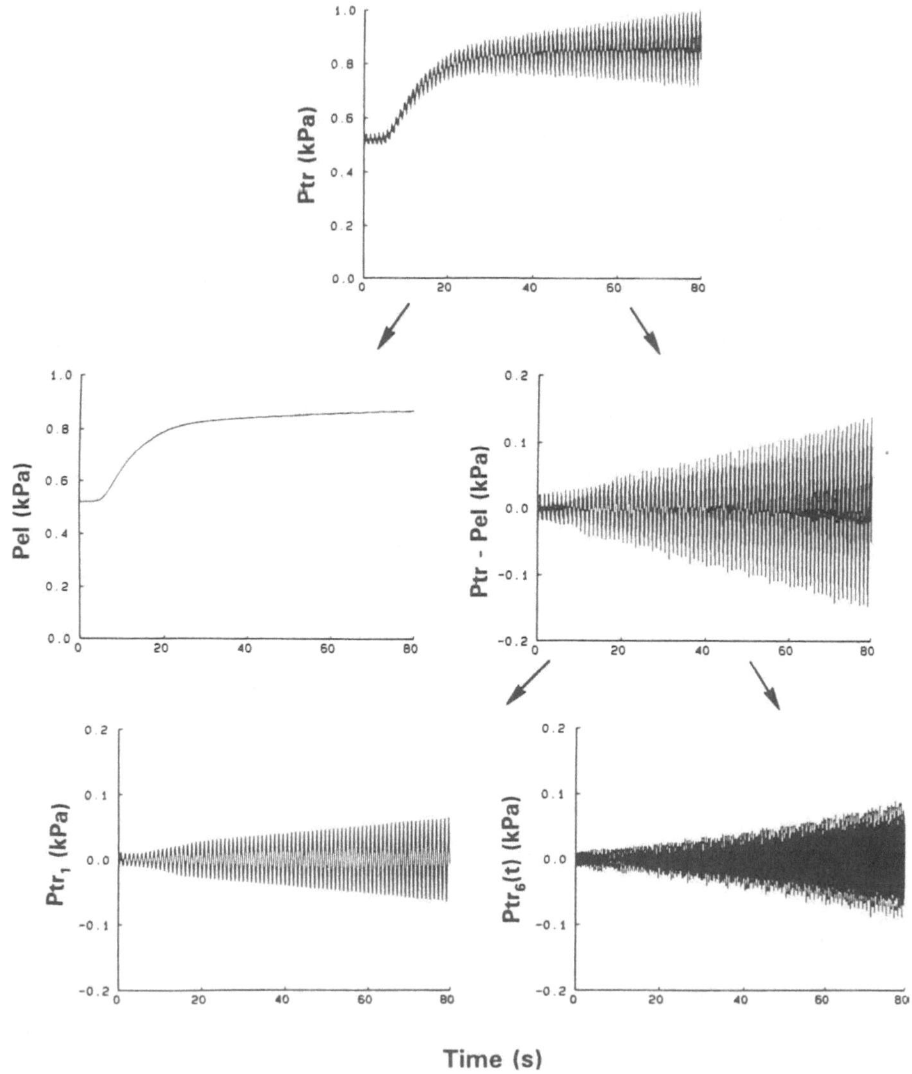

Figure 4. Time-course of tracheal pressure (Ptr) and its decomposition into individual frequency components (from Ref.7), obtained from an anesthetized dog at constant lung volume following a bolus i.v. administration of histamine at 0 time. Elastic recoil pressure (Pel) was obtained as the running mean of Ptr. The remainder (Ptr - Pel) constituted the oscillations alone. These were digitally band-pass filtered into their 1 Hz (Ptr_1) and 6 Hz (Ptr_6) frequency components.

to the 6 Hz oscillations in Ptr and $\dot{V}$ using the RLS algorithm with a very short memory time-constant of 0.5 s. The Pel signals obtained in five dogs were remarkably similar and peaked at about 25 s after injection, remaining fairly constant thereafter (Fig. 3B). In contrast, Raw increased monotonically throughout the 50 s measurement period and showed great variability among the dogs (Fig. 3A). The differences between Pel and Raw dynamics were again interpreted in terms of the nature of the circulation. Specifically, the early increase in Pel was due to the first passage of histamine through the pulmonary circulation where it acted upon peripheral contractile elements and so stiffened the lung tissue. The subsequent passage of the histamine through the bronchial circulation then caused the central airways to contract and so produced the later steady rise in Raw.

An interesting observation in these experiments is that the time-course of Pel exhibited only a single early peak (Fig. 3B). This contrasts with the two-phase response seen in dynamic Et in the earlier experiments during conventional mechanical ventilation (Fig. 2C). Insofar as both Pel and Et are assumed to reflect the properties of the lung tissues, this posed something of a dilemma. Its resolution came with a further set of experiments in which the evolution of bronchoconstriction was again followed at constant lung volume, this time however with two oscillation frequencies used simultaneously in the $\dot{V}$ applied to the tracheal opening[7]. The two frequencies were 6 Hz as before and 1 Hz. Fig. 4 shows a typical Ptr recording obtained during an 80 s measurement period at the beginning of which a bolus of histamine was injected into the vena cava. The steps involved in separating Ptr into its running mean (Pel), its 1 Hz and its 6 Hz components are also shown in Fig. 4. The same process was applied to$\dot{V}$ so that the dynamic behavior of the lung at both 1 Hz and 6 Hz could be followed simultaneously. The model fit to the 6 Hz data was the same as before (Eq. 10). The model fit to the 1 Hz data was Eq. 1. This enabled us to follow four parameters continuously over the 80 s measurement period: Pel, Raw, Rt at 1 Hz and Et at 1 Hz. Typical profiles of Rt and Et are shown in Figs. 5D and C as R_{L1}-R_{L6} and E_{L1}, respectively, and demonstrate two important features. First, they both exhibit early rapid increases synchronous with those in Pel (Fig. 3A). However, rather than leveling off as Pel did, they then both continued to increase steadily in a manner much more reminiscent of Raw (estimated as R_{L6}

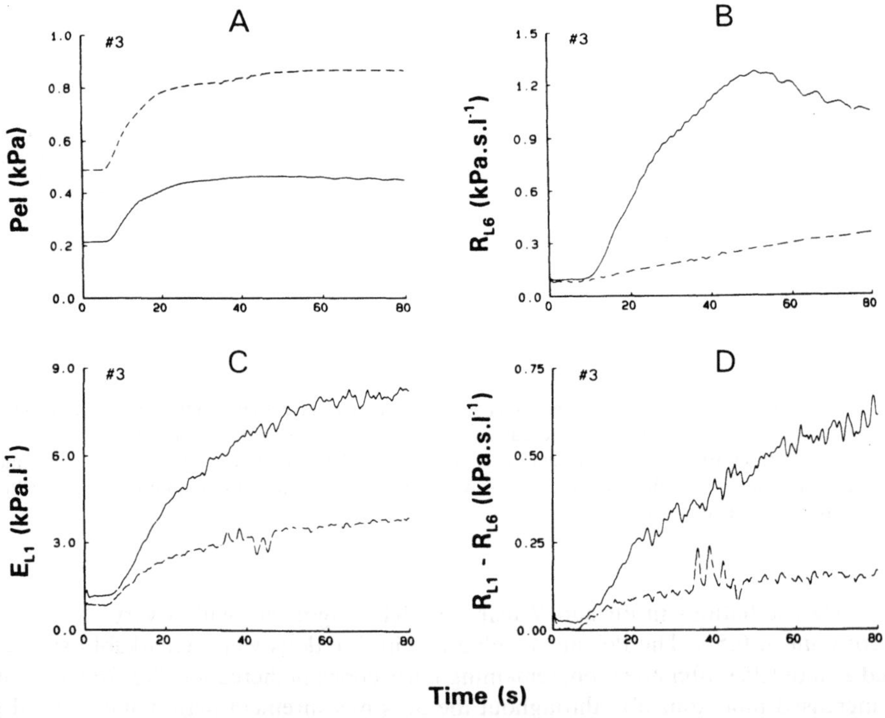

Figure 5. Time-courses of (A) lung elastic recoil pressure (Pel), (B) lung resistance at 6 Hz (R_{L6} -approximating airway resistance), (C) lung elastance at 1 Hz (E_{L1}) and (D) lung tissue resistance at 1 Hz (approximated by R_{L1} - R_{L6}). These data were obtained from an anesthetized dog at constant lung volume following a bolus i.v. administration of histamine. Adapted from Ref.7.

in Fig. 3B). In other words, dynamic measures of lung tissue mechanics, such as Et and Rt at 1 Hz, show both early and later phases just as did the initial observations of Lauzon *et al*[5] (Figs. 2B and C). It is only a static measure of tissue mechanics (i.e. Pel) that exhibits just the first phase. Since it seems reasonable that tissue distortion should affect both Pel and dynamic tissue properties, this suggests that the first phases in Et and Rt are produced by tissue distortion while the second phases are produced by developing ventilation inhomogeneity throughout the lungs.

Lauzon *et al*[24] also examined the complete time-course of the acute bronchoconstrictive response to histamine in dogs. This response takes about 15 or 20 min to return to baseline following histamine injection and has the peaked shape and drawn out tail reminiscent of many biological responses to impulsive pharmacologic stimuli. Whereas the rapid rise of the response reflects the dynamics of drug delivery to the lungs and the nature of smooth muscle contraction as described above, the descending tail contains information about the rate of drug clearance either by circulatory washout or degradation at the site of action.

6. TIME-COURSE OF REGIONAL LUNG IMPEDANCE DURING ACUTE BRONCHOCONSTRICTION IN DOGS

The alveolar capsule technique described above has provided key experimental evidence in various animal species about the nature of the mechanical heterogeneity throughout the lung[12,13]. By using several capsules simultaneously at different sites on the lung surface, one can assess how uniform P_A is under a given set of conditions and following various interventions. However, this technique does not allow one to determine if any observed heterogeneity in alveolar pressure is due to a change in the local resistive properties of the lung, the local elastic properties, or both. In order to make such a determination one would need to have local alveolar flow as well as P_A. In order to overcome this limitation, we have recently developed a technique to measure regional lung impedance on a very small scale in animals[25]. The technique involves applying forced oscillations in flow to the lungs through an alveolar capsule and provides measures of alveolar input impedance, Z_A. The device we use to apply flows through an alveolar capsule is depicted in Fig. 6. We call this device an alveolar capsule oscillator (ACO). It consists of a small load-speaker (approx. 2 cm diameter) encapsulated on both sides within rigid plexiglass chambers. The back chamber is completely closed and a miniature pressure transducer monitors its pressure changes as the speaker cone moves. This gives the volume displacement of the cone via Boyle's law. The front chamber connects to the alveolar capsule. Movement of the speaker cone also causes some gas compression within this chamber, while the rest of the volume displacement of the cone goes into flow through the capsule and into the lung. By dynamically calibrating this device we can determine precisely how much of the speaker cone displacement produces gas compression and how much produces flow into the lung[25].

We have studied open-chest dogs during acutely induced bronchoconstriction using two ACO's simultaneously at different sites[8,25,26]. The protocol used was essentially identical to that described above for the application of flow oscillations at the tracheal opening at fixed lung volume. This time, however, we applied a short flow burst from each ACO every 2 seconds over the 80 s measurement period following i.v. injection of histamine. The bursts consisted of composites of sinewaves having frequencies spanning the range 26 to 200 Hz. Fourier analysis of the P_A-alveolar flow signal pairs then produced 40 equally spaced measures of Z_A for each ACO. Each Z_A was fit to a simple model consisting of a sub-pleural elastic compartment connected to the ACO via a short flow-resistive pathway and leading

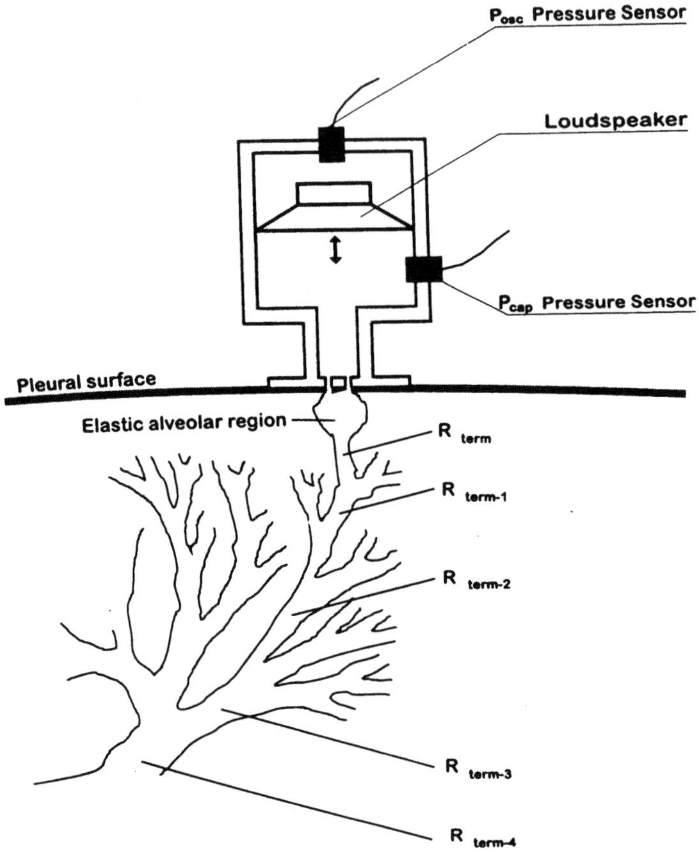

Figure 6. Alveolar capsule oscillator. The gas volume displaced by movement of the loudspeaker cone is compressed in the back chamber, producing the pressure signal Posc. Part of this volume goes into gas compression in the chamber in front of the cone and is measured by the capsule pressure Posc, while the remainder flows into the elastic alveolar region of the lung. This region is connected to the central airways of the lung by a series of ever widening airways beginning with the terminal bronchiole (resistance R_{term}) and proceeding to its parent (resistance R_{term-1}), and so on.

out to the rest of the lung via another pathway. The elastance of the compartment (E_A) and the resistance of the pathway leading to the rest of the lung (R_A) were taken as the parameters of physiological interest and represent, respectively, a very localized parenchymal elastance and peripheral airway resistance. Based on the magnitudes of E_A and R_A, we have argued that they correspond to a volume of the peripheral lung containing a very small number of acini[27].

A key result that has come out of these studies is that the response of the lung periphery to induced bronchoconstriction is extremely inhomogeneous in every respect, both temporally and spatially[8]. Fig. 7A shows the R_A obtained from two ACO's on each of 6 dogs plotted against each other, while Fig. 7B shows the corresponding E_A. A homogeneous response of the lung to bronchial agonist would be represented by each curve following the line of identity at 45° to each axis. The actual trajectories are so widely scattered, however, that almost no correspondence between the two ACO's on a given dog is apparent. In a second study using the ACO technique, we investigated how regional Z_A evolves when the lungs are bronchoconstricted at different inflation volumes[25]. As expected, the increases in R_A were

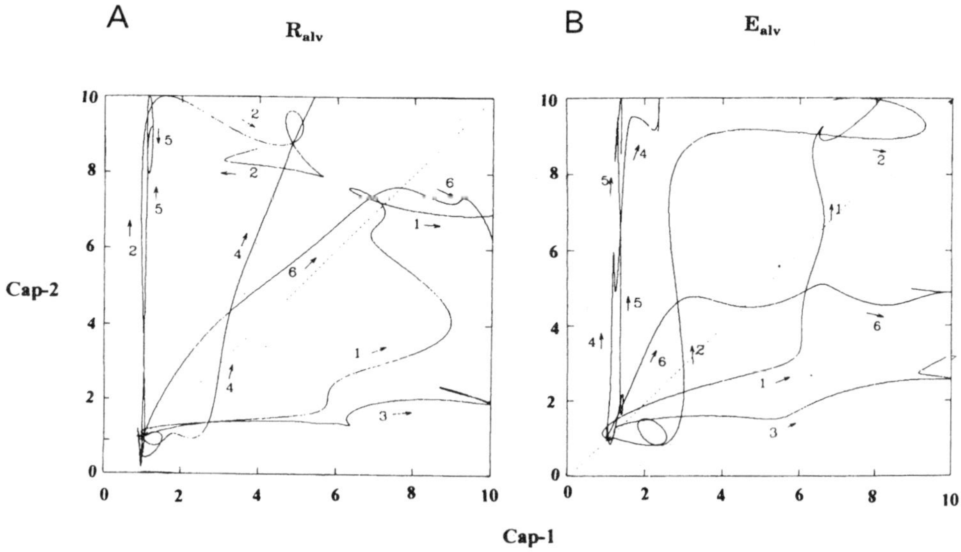

Figure 7. E_A and R_A obtained from pairs of alveolar capsule oscillators placed on the lungs of 6 open-chest dogs (from Ref.8).

much greater, on average, at the lower lung volumes as the decreased forces of parenchymal interdependence allowed a greater degree of airway smooth muscle shortening. However, the results also showed that the degree of regional heterogeneity increased markedly as lung volume decreased.

7. SUMMARY

Recent work from the author's laboratory has shown that the time-course of acute induced bronchoconstriction contains a great deal of important physiological information that is not present in the conventional dose-response representation of a bronchial agonist's effect on the lung. In particular, the temporal aspects of circulatory delivery of an agonist can be discerned. The relationships between airway and lung tissue responses to i.v. administered agonists has provided evidence that the constriction of the airways induces both a distorting effect on the parenchyma and an inhomogeneous delivery of airflow throughout the lungs. Both phenomena make it appear as if the tissues themselves are responding directly to the agonist. Finally, alveolar capsule studies, and more recently the alveolar capsule oscillator, have begun to elucidate the nature of the regional inhomogeneity that develops as bronchoconstriction proceeds.

REFERENCES

1. Macklem, P.T. Bronchial hyporesponsiveness. *Chest.* 87:1585-1595, 1985.
2. Itkin, I.H., S.C. Anand, M. Yau, and G. Middlebrook. Quantitative inhalation challenge in allergic asthma. *J. Allergy.* 34:97-106, 1963

3. Felarca, A.B., amd I.H. Itkin. Studies with the quantitative inhalation challenge technique. I. Curve of dose response to acetyl-beta-methylcholine in patients with asthma of known and unknown origin, hay fever subjects, and nonatopic volunteers. *J. Allergy.* 37:223-235, 1966.
4. Ludwig, M.S., P.V. Romero, and J.H.T. Bates. A Comparison of the dose-response behavior of canie airways and parenchyma. *J. Appl. Physiol.* 67:1220-1225, 1989.
5. Lauzon, A.-M., G. Dechman, and J.H.T. Bates. Time course of respiratory mechanics during histamine challenge in the dog. *J. Appl. Physiol.* 73:2643-2647, 1992.
6. Bates J.H.T., and R. Peslin. Acute pulmonary response to i.v. histamine and fixed lung volume in dogs. *J. Appl. Physiol.* 75:405-411, 1993.
7. Bates J.H.T., A-M Lauzon, G.S. Dechman, G.N. Maksym and T.F. Shuessler. Temporal dynamics of pulmonary response to intravenous histamine in dogs: effects of dose and lung volume. *J. Appl. Physiol.* 76:616-626, 1994.
8. Mishima, M., Z. Balassy and J.H.T. Bates. Acute pulmonary response to i.v. histamine using forced oscillations through alveolar capsules in dogs. *J. Appl. Physiol.* 77:2140-2148, 1994.
9. Similowski, T., and J.H.T. Bates. Two-compartment modelling of respiratory system mechanics at low frequencies: gas redistribution or tissue rheology? *Eur. Respir. J.* 4:353-358, 1991.
10. Lutchen, K.R., Z. Hantos, and A.C. Jackson. Importance of low-frequency impedance data for reliably quantifying parallel inhomogeneities of respiratory mechanics. *IEEE Trans. Biomed. Eng.* 35:472-481, 1988.
11. Baydur, A., P.K. Behrakis, W.A. Zin, M. Jaeger, and J. Milic-Emili. A simple method for assessing the validity of the esophageal balloon technique. *Am. Rev. Respir. Dis.* 126:788-791, 1982.
12. Fredberg, J.J., D.H. Keefe, G.M. Glass, R.G. Castile, and I.D. Franz. Alveolar pressure nonhomogeneity during small-amplitude high-frequency oscillation. *J. Appl. Physiol.* 57:788-800, 1984.
13. Fredberg, J.J., R.H. Ingram, R.G. Castile, G.M. Glass and J.M. Drazen. Nonhomogeneity of lung response to inhaled histamine assessed with alveolar capsules. *J. Appl. Physiol.* 58:1914-1922, 1985.
14. Fung, Y.C.B. *Biomechanics. Mechanical properties of living tissues.* New York: Springer-Verlag, 1981.
15. Sato, J., B.L.K. Davey, F. Shardonofsky and J.H.T. Bates. Low-frequency respiratory system resistance in the normal dog during mathematical ventilation. *J. Appl. Physiol.* 70:1536-1543, 1991.
16. Draper, N.R., and H. Smith. *Applied Regression Analysis.* New York: Wiley, 1966, ch. 2.
17. Hsia, T.C. *System Identification.* Lexington, MA: D.C. Heath, 1977.
18. Avanzolini, G., P. Barbini, A. Cappello, and G. Cevenini. Real-time tracking of parameters of lung mechanics: Emphasis on algorithm tuning. *J. Biomed. Eng.* 12:489-495, 1990.
19. Lauzon, A.-M. and J.H.T. Bates. Estimation of time-varying respiratory mechanical parameters by recursive least squares. *J. Appl. Physiol.* 71:1159-1165, 1991.
20. Ludwig, M.S., I. Dreshaj, J. Solway, A. Munoz, and R.H. Ingram, Jr. Partitioning of pulmonary resistance during constriction in the dog: effects of volume history. *J. Appl. Physiol.* 62:807-815, 1987.
21. Karıya, S.T., L.M. Thompson, E.P. Ingenito, and R.H. Ingram, Jr. Effects of lung volume, volume history, and methacholine on lung tissue viscance. *J. Appl. Physiol.* 66:977-982, 1989.
22. Mitzner, W., S. Blosser, D. Yager, and E. Wagner. Effect of bronchial smooth muscle contraction on lung compliance. *J. Appl. Physiol.* 72:158-167, 1992.
23. Otis, A.B., C.B. McKerrow, R.A. Bartlett, J. Mead, M.B. McIlroy, N.J. Selverstone and E.P. Radford. Mechanical factors in the distribution of pulmonary ventilation. *J. Appl. Physiol.* 8:427-443, 1956.
24. Lauzon, A.-M., G. Dechman, J.G. Martin, and J.H.T. Bates. The complete time course of bronchoconstriction by i.v. bolus histamine in the dog. *Respir. Physiol.* 99:127-138, 1995.
25. Davey, B.L.K. and J.H.T. Bates. Regional lung impedance from forced oscillations through alveolar capsules. *Resp. Physiol.* 91:165-182, 1993.
26. Balassy, Z., M. Mishima, and J.H.T. Bates. Changes in regional lung impedance following i.v. histamine bolus in dogs: effects of lung volume. *J. Appl. Physiol.* 78:875-880, 1995.
27. Bates, J.H.T., M. Mishima, and Z. Balassy. Measuring the mechanical properties of the lung in vivo with spatial resolution at the acinar level. *Physiol. Meas.* 16:151-159, 1995.

15

UNDERSTANDING PULMONARY MECHANICS USING THE FORCED OSCILLATIONS TECHNIQUE

Emphasis on Breathing Frequencies

Kenneth R. Lutchen and Béla Suki

Department of Biomedical Engineering
Boston University
Boston, Massechusetts 02215

1. INTRODUCTION

The lung is a marvelously over-designed mechanical ventilation system. It is capable of sustaining substantial injury or alteration while still maintaining life-sustaining blood gas levels. Nevertheless, these alterations often lead to compromised lung function and breathing discomfort. Consider two basic questions: What is the relative compromise in airway versus tissue properties during asthma?; and How do airways and tissues contribute to alterations in breathing discomfort (dyspnea) and breathing function (i.e., to adequately ventilate)? Answers to these questions are required for designing and monitoring effective treatment protocols. Unfortunately, obtaining answers is fraught with ambiguities as they will be highly dependent on the breathing frequency and amplitude, on the alterations in the mechanical constituents comprising the airways and tissues (eg., airway geometry and wall properties and the tissue fiber system), and on the topological properties of the interdependent airway-tissue matrix. What is needed is a measurement approach which can provide specific and reliable insight on the mechanical properties and important mechanisms that contribute to breathing. Preferably, the method should be amenable to clinical applications and assessment.

In 1956, Dubois et. al.[18] introduced a simple, minimally invasive forced oscillation approach to measure the mechanical impedance of the lung or respiratory system. However, up until eight years ago the impact of the approach was confined to gross clinical categorization with little physiological or anatomical resolution. Substantial advances in technology, signal processing, and modeling have recently resulted in the potential to use of forced oscillations to extract far more specific insight on lung mechanical alterations during disease than any other existing measurement of pulmonary mechanics. The overall goal of this chapter is to summarize these advances.

Bioengineering Approaches to Pulmonary Physiology and Medicine, edited by Khoo
Plenum Press, New York, 1996

An overview of the forced oscillations technique will be provided first. The emphasis will be on the critical impact that frequency range has on the specific mechanisms that effect impedance data. The primary focus of the chapter will be on impedance data surrounding breathing frequencies (0.1-3 Hz). Recent methodology now reveals that this data can indeed be reliably acquired in humans, and the data are richly sensitive to airway and tissue mechanisms that influence breathing. These mechanisms include tissue structure and function, airway-tissue coupling, lung inhomogeneities and lung nonlinearities. A new technique, termed the "Optimal Ventilator Waveform (OVW)" [63,64] will be described and shown to allow reliable separation of airway and tissue properties in mild-to-moderately constricted lungs. During more severe constriction, lung inhomogeneities and nonlinearities can confound data interpretation. High fidelity modeling and novel experimental approaches are presented that may help resolve if and to what extent these phenomena can influence lung mechanics during physiological breathing conditions.

2. REVIEW OF THE FORCED OSCILLATION APPROACH

2.1. Basic Forms of Impedance

There are two common methods for applying forced oscillations to measure a mechanical impedance. With input impedance (Z_{in}) forced oscillations are imposed at the airway opening and one measures the complex ratio of airway opening pressure to flow ($Z_{in} = P_{ao}/\dot{V}_{ao}$) [17,44,60,73,82,,85,109,110,114]. With transfer impedance (Z_{tr}), forced oscillations are typically imposed around the chest wall (P_{cw}) and one measures $P_{cw}/\dot{V}_{ao}$ [60,62,84,85,93,113]. In either case, an impedance spectrum containing the magnitude and phase or real and imaginary parts of impedance vs frequency is estimated. A simple conceptualization is that the respiratory system is comprised of airways and tissues separated by a volume of compressible alveolar gas. The mechanical structures within the airways and tissues produce effective resistance, inertance and compliance for the airways and tissues which can

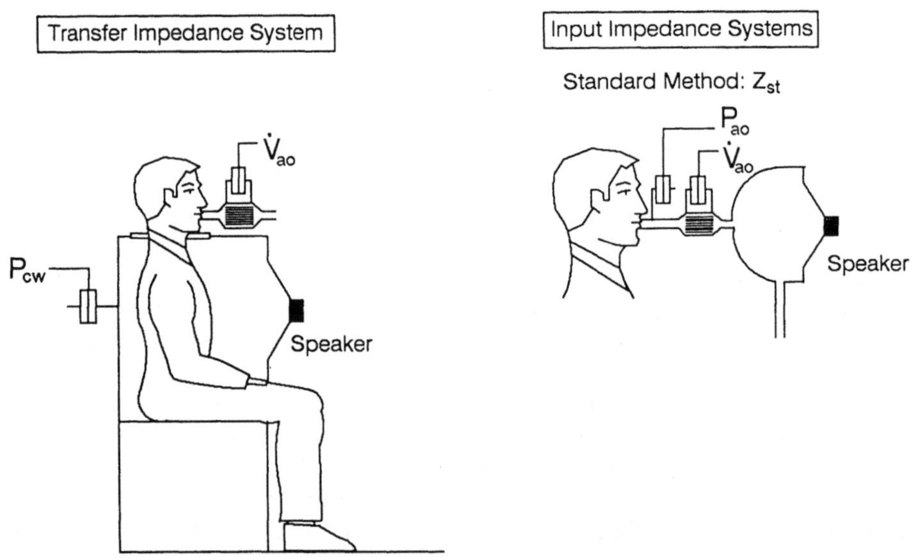

Figure 1. Schematic of measurement systems for transfer and input impedances. Adopted from Lutchen et. al. [62].

influence the overall impedance spectrum. One can see, then, the potential for these measurements to provide a noninvasive means for probing changes in these mechanical properties, and perhaps infering anatomic locations from which these changes originate. Moreover the fact that the technique requires no patient cooperation suggests an attractive potential pulmonary function test, particularly for infants and ventilator dependent patients. There are, however, some obvious disadvantages. First, the measurements will be influenced by all respiratory structures, including the upper airways and cheeks[41,82,19,114], chest wall, and lungs, only the latter of which are of interest. Second, physiological and clinical interpretation will be model and frequency range dependent [42,44,53,55,60,62,84].

2.2. Input Impedance in Earlier Studies

Traditionally, data surrounding the breathing frequencies (<3 Hz) were not acquired because of interference from the subject's own breathing. At the advent of cost-effective computers, Michaelson et al.[73] in 1975 proposed the Fourier approach to measure impedance at several frequencies above 2 Hz simultaneously. Here, a small-amplitude pseudorandom noise signal is provided by a loud-speaker while the patient breathes. The recordings are high-pass filtered to remove interference from the subjects breathing and subsequently processed as described in several studies and outlined in the Appendix A[16,68,73]. Since the Michaelson study, a plethora of Zin (and Ztr) studies were performed in which the bandwidth of the forcing signal was restricted to 2-32 Hz or less [11,78,80,82,83,85,92,93,109,110,113,114]. For Zin, these data were generally non-informative. As is seen in Fig. 2, in healthy subjects the real part is constant and the imaginary part is monotonically increasing; behavior well described

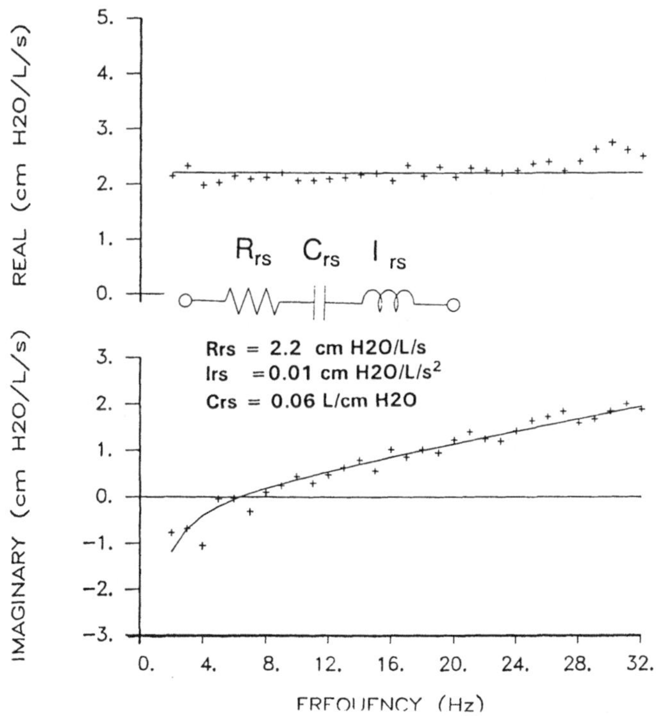

Figure 2. Typical healthy human input impedance data (+) from 2-32 Hz and corresponding model fit (solid line) using a simple 3-element series R-I-C model.

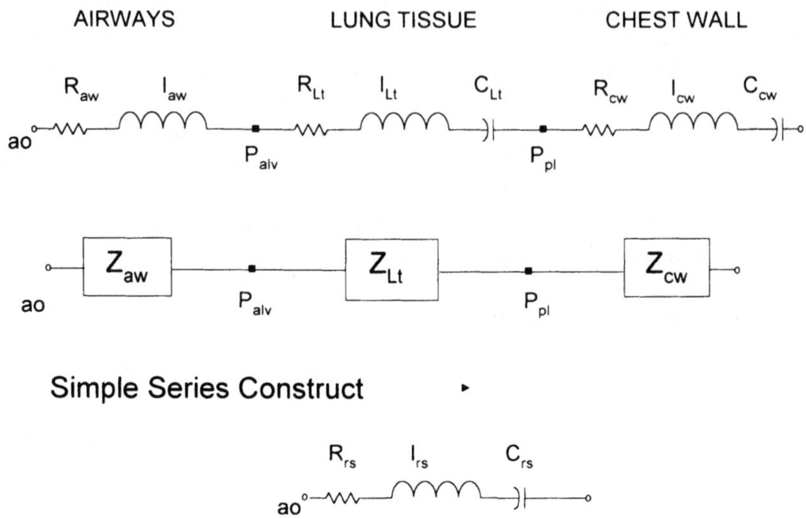

Figure 3. Lumped element model conceptualization of respiratory mechanical system for application to input impedance data from 2-32 Hz. The series combination of airway, tissue and chest-wall properties reduce to a single series R-I-C model.

by a simple series combination of a respiratory resistance, inertance, and compliance (R_{rs}, I_{rs}, C_{rs}). How can the complex respiratory system produce such behavior for Z_{in}. If from 2-32 Hz the impact of alveolar gas compressibility is negligible, the respiratory system can be conceptualized as indicated in Fig. 3. Thus, individual effective resistances of homogeneous airway, lung tissue, and chest wall tissue combine serially to produce one effective R_{rs}-I_{rs}-C_{rs} system. Hence, the data and mechanical properties are anatomically and functionally non-specific. Indeed, only severe overall lung obstruction of constriction will noticeably alter Z_{in} features from 2-32 Hz or even 2-64 Hz in humans [56,109,110].

2.3. Transfer Impedance

Unlike Z_{in}, Z_{tr} data over a slightly extended frequency range is not easily described by such a simple model. Figure 4 shows recent Z_{tr} mean ± s.d. bounds from 2-72 Hz for normal and COPD patients. In both subjects the real part shows a clear frequency dependent decrease. Most encouraging, several aspects of the Z_{tr} spectra appear highly distinct for COPD vs healthy subjects, particularly when comparing the real parts. The high frequency measurement limit on Z_{tr} is primarily set by the physical dimensions of the body-box within which the pressure oscillations are generated (Fig. 1). The prevailing assumption is that there is uniform forcing around the thoracic cage. As the frequency increases, standing pressure waves can develop which will violate this assumption and confound data interpretation [62]. The data of Fig. 4 are well described by the 6-element model of Fig. 5 (cf. Ref.60). Of course the same respiratory system is present during Z_{in} and Z_{tr} measurements. However, Z_{in} and Z_{tr} emphasize the airway (Z_{aw}), tissue (Z_T) and gas compression (Z_g) components of the 6-element system of Fig. 5 differently [60,85]. In particular

$$Z_{in} = Z_{aw} + \frac{Z_T Z_g}{Z_T + Z_g} \qquad Z_{tr} = Z_{aw} + Z_T + \frac{Z_{aw} Z_T}{Z_g} \tag{1}$$

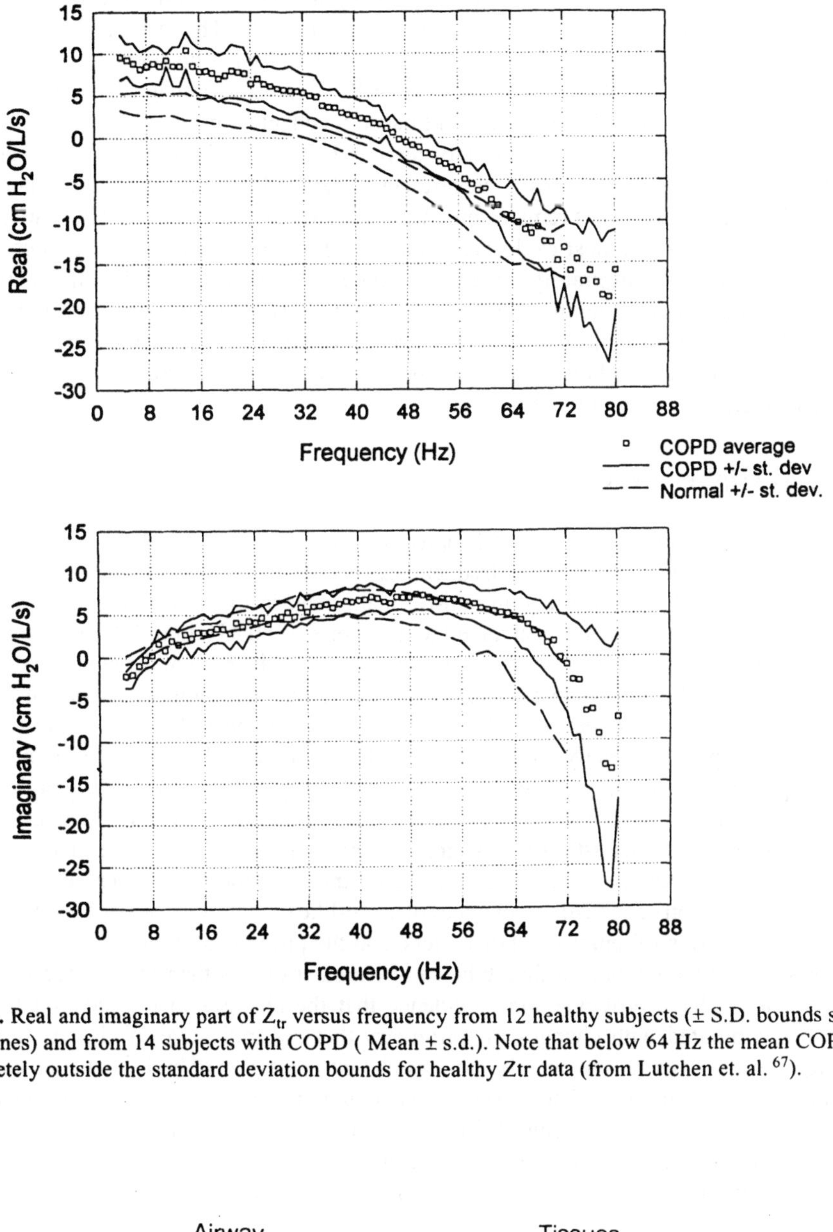

Figure 4. Real and imaginary part of Z_{tr} versus frequency from 12 healthy subjects (± S.D. bounds shown in dashed lines) and from 14 subjects with COPD (Mean ± s.d.). Note that below 64 Hz the mean COPD ± s.d. is completely outside the standard deviation bounds for healthy Ztr data (from Lutchen et. al. [67]).

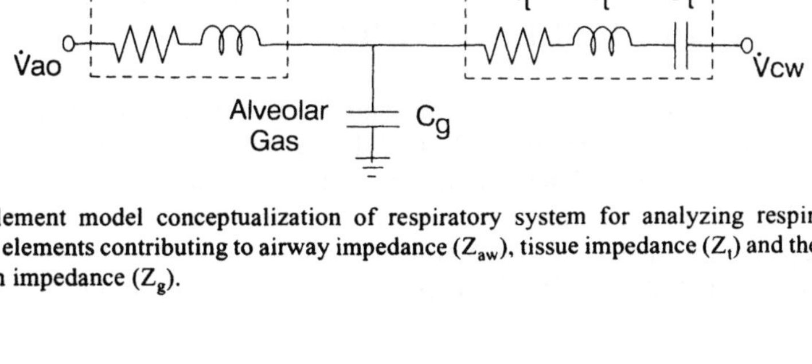

Figure 5. Six-element model conceptualization of respiratory system for analyzing respiratory Z_{tr} data. Grouped are the elements contributing to airway impedance (Z_{aw}), tissue impedance (Z_t) and the shunt alveolar gas compression impedance (Z_g).

Lutchen and Jackson[60], have shown that a human sized C_g can significantly influence Z_{tr} from 2-32 or 2-64 Hz while having a negligible influence on Z_{in}. The result is that Z_{tr} displays a frequency dependent decrease in the real part which can no longer be adequately modeled by a simple series R-I-C model. Moreover, Lutchen et. al.[59,60,62] has shown that if Z_{tr} data from 2-64 Hz is acquired (eg., as in Fig. 5) the airway and tissue mechanical properties estimated from the 6-element model applied to Z_{tr} data are statistically reliable. Finally, Z_{tr} data is less sensitive to upper airway and cheek structures[56]. Nevertheless, detailed mechanistic interpretation of Z_{tr} relative to breathing mechanics remains elusive. Because only data above 2 Hz is analyzed, the tissue resistance reflects almost exclusively chest wall and not parenchymal tissues (see next section below). Also, the respiratory tissue compliance (C_t) is the serial combination of parenchymal and chest wall tissues, and the reliability of the estimated value is quite poor with data only above 2-4 Hz[60] (i.e., Ct primarily influences Z_{in} or Z_{tr} well below 2 Hz). Consequently, Z_{tr} is best suited as a noninvasive method for estimating R_{aw} and I_{aw} and for providing spectral features sensitive to alterations in the airway system.

2.4. Z_{in} at Higher Frequencies

An exciting and potentially useful approach for extracting more detailed information from Zin is to extend the data to much higher frequencies[9,13,17,19,21,29,41]. Jackson et al.[41] have shown that around 160 Hz the Z_{in} data display an anti-resonant peak in the real part. This peak shifts in frequency when the resident gas is changed to a mixture of 80%Helium-20% Oxygen (HeOx). Since the frequency shift is proportional to the square root of the ratio of densities of air to HeOx, it is concluded that the anti-resonance represents the first acoustic anti-resonance of the airway system. At even higher frequencies additional resonances appear. As such, these Z_{in} spectral features are indigenous strictly to the airways. The technical challenge for high frequency measurements are maintaining reliable transducer frequency response (including high common-mode rejection) in the face of decreasing signal-to-noise ratio as frequency increases. At these higher frequencies, the acoustic relation between pressure and flow in the airways becomes distributed (i.e., the airways behave as an acoustic transmission line as described in Ref. 10. Generally, the location of acoustic resonances of a tree-like branching tube system depends on the geometric distribution of lengths and diameters, and the properties of the airway walls (i.e., the speed of sound in a compliant walled tube can be greater or less than that of free space [12,28,99] Chalker et al.[9,13] have provided some evidence that the changes in the location of acoustic resonances correlated with changes in spirometry that are indicative of airway disease. The challenge ahead is to extract an unambiguous model interpretation of high frequency Zin that can be related to airway structure. The most promising approach to date is to apply the anatomically consistent branching airway systems of Horsfield et. al. to Z_{in}[29,38,39,55]. Nevertheless, work remains on confirming the uniqueness of such an approach on an individual subject basis and on a reliable approach to incorporate the upper airway structures in awake, nonintubated patients.

3. MOTIVATION FOR MEASURING LUNG IMPEDANCE AROUND PHYSIOLOGICAL FREQUENCIES

Why should we be interested in measurement of mechanical impedances over the physiological frequency range (0.1 - 2 Hz)? The answer lies in the fact that this is the operating frequency range of the respiratory system and there is a need to establish the primary mechanical properties and mechanisms that effect breathing in subjects with healthy and sick lungs. Unfortunately, because of the combined problems of removing interference from the subjects breathing from the applied forcing waveform, and the need to use an

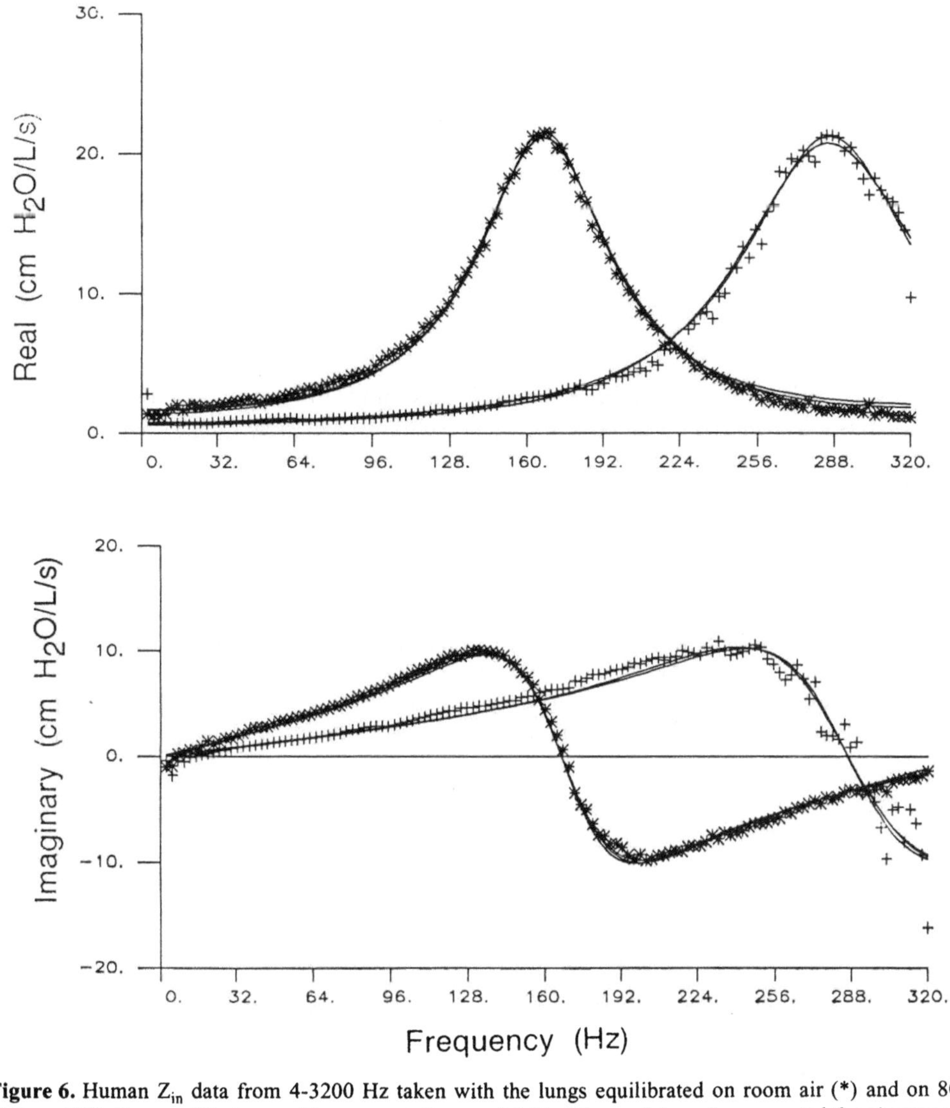

Figure 6. Human Z_{in} data from 4-3200 Hz taken with the lungs equilibrated on room air (*) and on 80% Helium-20% Oxygen (+) along with corresponding model fits. The model used represented the airways as a bundle of parallel rigid-tubes and their associated distributed acoustic transmission line impedance all leading to an alveolus-tissue element. From Ref. 61.

esophageal balloon to estimate changes in pleural pressure, there is quite a sparsity of these data in humans. Nevertheless, some recent studies have provided substantial motivation for acquiring these data on a much larger scale, particularly in humans. Here we overview these studies and then later return to introduce a more recent technique that will allow us to acquire these data reliably and *in situ*.

3.1. Alveolar Capsule Studies

The alveolar capsule technique was introduced in the mid-1980s[22,23]. Here a pressure transducer is fixed directly to a small puncture made through the pleural surface thereby measuring the pressure in the terminal alveoli (P_{alv}). During forced oscillations at the airway opening, one could then separate the total lung impedance (Z_L) into its airway (Z_{aw}) and tissue (Z_T) components as follows:

$$Z_L = \frac{P_{ao}}{\dot{V}_{ao}} = Z_{aw} + Z_T \tag{2}$$

with

$$Z_{aw} = \frac{P_{ao} - P_{alv}}{\dot{V}_{ao}}, \qquad Z_T = \frac{P_{alv}}{\dot{V}_{ao}} \tag{3}$$

It is critical to note that the separation uses input flow rather than the flow directly into the tissue subtended by the alveolar pressure transducer. The lung, of course, consists of many parallel pathways (Fig. 7) and the sampling is performed on a single pathway. Nevertheless, one can show[23] that if the input flow is distributed uniformly (i.e., if the lungs were mechanically homogeneous), the Z_{aw} and Z_T obtained from a single capsule are equivalent to those for the whole lung.

By virtue of this technique, in many animal studies and across several species[9,20,31-33,40,49-51,64,74,90,91,95,96], the paradigm of Fig. 8 has become widely accepted for the healthy lung with regard to the contributions of airway resistance (R_{aw}) and tissue resistance (R_T) to total lung resistance (R_L). The R_{aw} is constant with frequency (although, over this frequency range positive frequency dependence of R_{aw} can occur if the peak flow rates increase sufficiently to cause a significant turbulent component). The R_T decreases quasi-hyperboli-

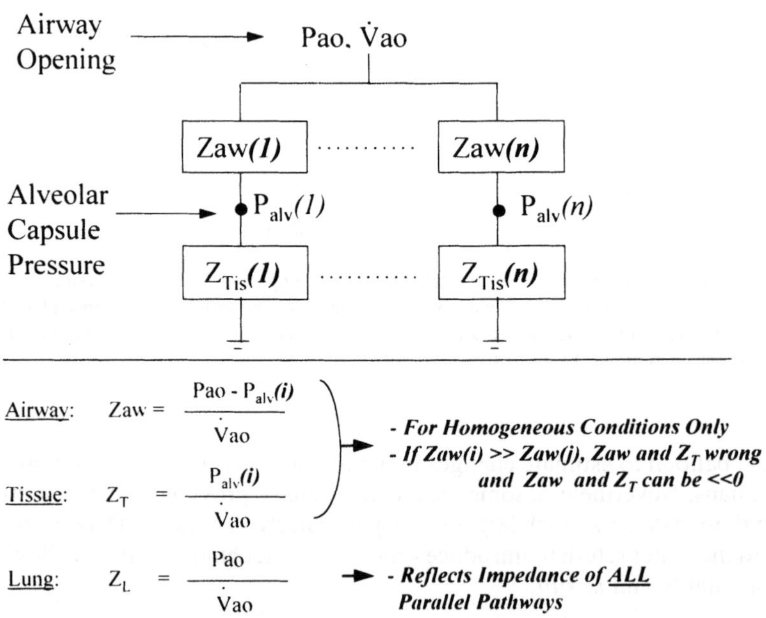

Figure 7. Schematic depiction of alveolar capsule concept in which an alveolar capsule pressure transducer is placed on one of the "n" possible parallel pathways in the model. The separation of input lung impedance (Z_L) into its airway and tissue components from the data is as shown, and is strictly valid only if the impedance is distributed homogeneously among the parallel pathways.

cally with frequency, asymptotically approaching zero near 2-4 Hz. This decrease is a consequence of the lung tissue being viscoelastic and void of any purely Newtonian resistance[7,20,64,27]. The R_L is the sum of these two components. This paradigm reveals a perspective that has only recently been appreciated; namely that during normal (< 0.5 Hz) breathing the preponderance of total lung resistance is due to the tissues and not the airways.

3.2. Partitioning of Airway and Tissue Properties

A powerful implication of the above paradigm is that separation of R_L into its airway and tissue components should be feasible without the use of the alveolar capsules, i.e., from Z_L alone. With regard to the latter, the R_L behavior is well described by the model indicated in Fig.8 in which a homogeneous airways compartment containing an R_{aw} and airway inertance (I_{aw}) lead to a viscoelastic tissue compartment described by a so called constant-phase tissue model. This tissue model arises naturally from viscoelastic tissues that display a stress relaxation response in which pressure decays as a power-law in time[31-36]. A mathematical development and potential molecular basis have been presented for the model[104]. There are two tissue properties, tissue damping (G) and tissue elastance (H). Tissue resistance decreases quasi-hyperbolically as $R_T = G/\omega^\alpha$ and tissue elastance increases slightly with frequency as $E_T = H(\omega/\omega^\alpha)$. The degree of frequency dependence is governed by α which is related to tissue hysteresivity or $\eta = G/H$[20] (Fig. 8). By fitting this model to input lung impedance data one can separately estimate the airways and tissue properties. Indeed for the healthy lung, two studies have confirmed that the separation is consistent with that

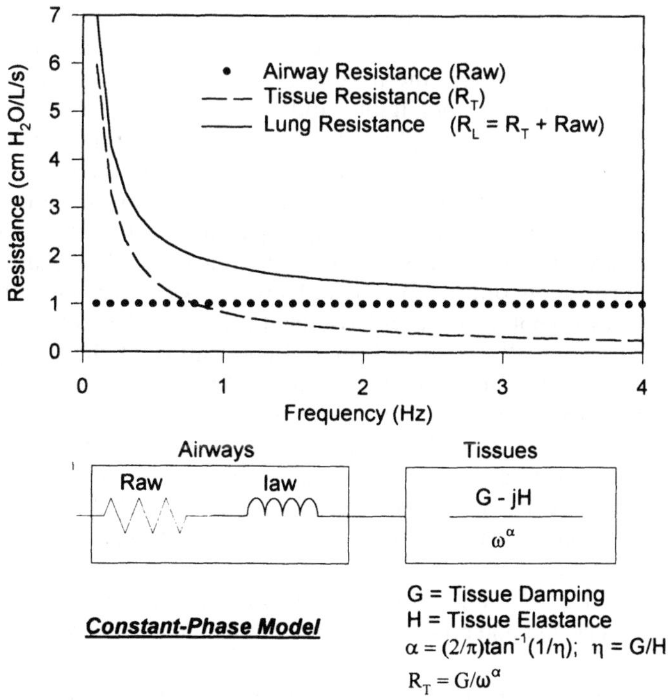

Figure 8. Schematic of airway versus tissue contributions to total lung resistance (R_L) over the physiological breathing frequencies. Also show is a simple model of homogeneous airways with airway resistance (R_{aw}) and inertance (I_{aw}) leading to a viscoelastic constant-phase model of tissues containing tissue damping (G) and tissue elastance (H) properties. Figure adopted from Ref. 65.

measured with alveolar capsules[64,87]. In fact, in the Lutchen et al. study[64] the separation from Z_L were reasonably consistent with those from alveolar capsules even after inducing bronchoconstriction in dogs. The implications are substantial with respect to eventual human studies for which the use of alveolar capsules is impossible.

3.3. Impact of Additional Mechanisms during Lung Disease (Tissue Constriction, Nonlinearities, and Inhomogeneities)

Several studies using either alveolar capsules or application of the constant-phase model to Z_L suggest that after administering a bronchoconstrictor (eg., histamine or methacholine) there is a substantial increase in R_T as well as R_{aw} [33,40,49-51,64,76,89-91] and the increase in R_T is equal or greater than that of R_{aw}. The implication is that the lung constriction substantially involves the tissues and the airways. Moreover, after administration of a bronchoconstrictor or during induced pulmonary edema, lung tissue resistance and elastance below 0.5 Hz show an enhanced inverse dependence on tidal volume, V_T [1,2,3,4,5,36,57,64]. This means there is a nonlinear component to dynamic tissue mechanics.

With respect to the nonlinearity, one must now be wary of using small amplitude broadband forcing to measure Z_L. First the Z_L measured with small amplitude forcing may not reflect that associated with a physiological tidal excursion. Second, if the input is broadband, the nonlinearities can cause harmonic interactions that distort the measured impedance at each frequency [100,102,105,115]. The nonlinearities can distort basic measures of the reliability of a Z_L measurement (eg., coherences) as well (see Appendix A). Ideas on how to address these concerns are described below.

With regard to the inference that there is increased tissue constriction, we must consider that during lung disease, the lungs are expected to become mechanically more inhomogeneous. According to Fig. 7 the alveolar capsule technique ideally works only when mechanical inhomogeneities are negligible. Indeed, if a capsule subtends an unusually obstructed pathway the measured "Z_{aw}" and "Z_T" can no longer be interpreted as airway and tissue impedance, respectively. This occurs because the input flow is no longer in phase with the flow within the capsule pathway and there is pendulluft among all parallel pathways[23,87] One might then ask if inhomogeneities would equally distort the use of Z_L to separate airway and tissue properties via the constant-phase model approach (Fig. 8). Figure 9 contrasts two kinds of airway constriction, one in which parallel pathways undergo identical constriction and one in which they do not. Both cases have identical tissue elements in the pathways. In the former case, the entire R_L is shifted up and application of the homogeneous single pathway model will result in an accurate estimation of the parallel combination of airway and tissue properties over the whole lung. However, in the latter case, inhomogeneities develop in the airway-tissue parallel mechanical time constants (i.e, between the effective R_i/H_i of each pathway) similar to that described by Otis et. al[81]. This causes an additional frequency dependence in R_L and, while not shown, E_L too. If the homogeneous single pathway model were applied to the latter case, the resulting estimate of tissue G (and perhaps H) would be distorted as these are the only parameters that can adjust to the frequency dependence in R_L and E_L[32,87]. We will return to this issue and present some approaches for evaluating the presence and potential influence of inhomogeneities on assessment of airway and tissue mechanics.

In summary, we have motivated the need to measure Z_L from 0.1-3 Hz as it contains important information on the relative impact of airway and tissue properties and structure on breathing mechanics. Before discussing current approaches for reliably identifying and quantifying this physiological information, we now review a potential method for wide-spread acquisition of these data in humans.

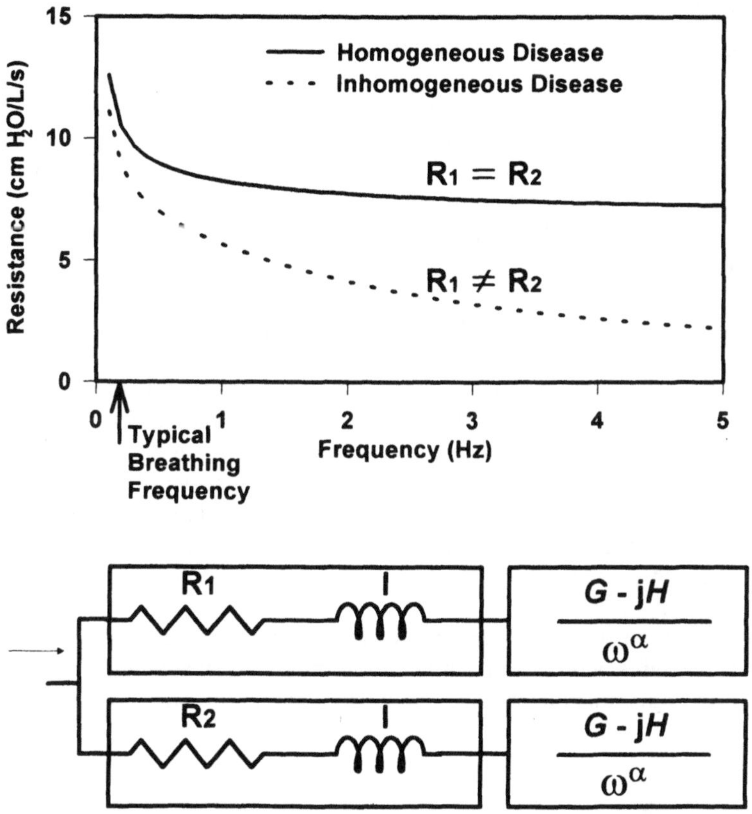

Figure 9. Alterations in lung resistance vs frequency when the airway resistances (R_1 and R_2) increase homogeneously (solid line) vs inhomogeneously (dashed). Note the added frequency dependence in R_L for the inhomogeneous case (i.e., a frequency dependence unrelated to that due to the constant-phase tissues).

4. TECHNIQUES FOR MEASURING Z_L AT LOW FREQUENCIES: THE OPTIMAL VENTILATOR WAVEFORM APPROACH

4.1. Past Approaches

There are two obvious practical limitations with alveolar capsules. First, since only 4-5 capsules maximum and typically two are used, there is necessarily huge under sampling of the alveolar pressures. Second, in humans the use of alveolar capsules is not possible. Three kinds of approaches have been used for partitioning airway and tissue properties in humans in the past. One approach is to use plethysmography with an esophageal balloon in place[69,70,71]. The esophageal pressure (P_{es}) and airway opening flow provides R_L while the R_{aw} is estimated plethysmographically. However, this separation is at one frequency and typically at the panting frequency near 1.5 Hz required for reliable plethysmography. At this frequency the influence of R_T is well below what it would most likely be at a lower, more physiological breathing frequency (Fig. 8). Barnas et. al. advocate applying forcing at the mouth sweeping with discrete sine waves from 0.1-4 Hz (1-4) of typical tidal volume magnitude. This works fine with subjects with stable lung conditions, but would be far to time consuming for probing the dynamics of a bronchial challenge. Moreover, it is unclear what the tidal volumes (V_T) should be at higher frequencies. In the past they have maintained the same V_T at all frequencies which then results in non-physiologically large peak flows at the higher frequencies. This overly exaggerates the peak flow dependence that can occur in airway resistance[5,4]. An alternative is to use small amplitude pseudorandom noise containing energy from 0.125-5 Hz as was done in a pioneering study on healthy humans by Hantos et. al.[30] in 1986. However, due to the nonlinearities such data may not relate to the mechanical properties that influence normal tidal excursions, and harmonic distortion can occur on the Z_L. Also, this technique requires that the subject remain apneic for approximately 30 sec at FRC with the glottis open while the forcing is applied. This is clearly impractical for most clinical conditions.

4.2. Optimal Ventilation Waveform Approach

The ideal waveform will produce reliable Z_L estimates from 0.1 - 4 Hz in healthy and diseased patients in a practical and efficient manner. One candidate is standard ventilator waveform (SVW) of a step flow on inspiration and a passive expiration. This waveform should contain energy at frequencies beyond the fundamental ventilation frequency. However, Lutchen et al.[58] showed that the amount of energy beyond the fundamental depends on the subject's lung rather than the ventilator itself. Indeed, the signal-to-noise ratio at higher frequencies in healthy humans is inadequate. Moreover, the waveform is subject to nonlinear harmonic distortions[58,100,102,103,105].

In 1994 Lutchen et al.[63] presented a design for a forcing signal that is broadband, but simultaneously sustains ventilation while minimizing the harmonic interactions due to nonlinearities. The spectral concept is to use approximately the same fundamental energy as

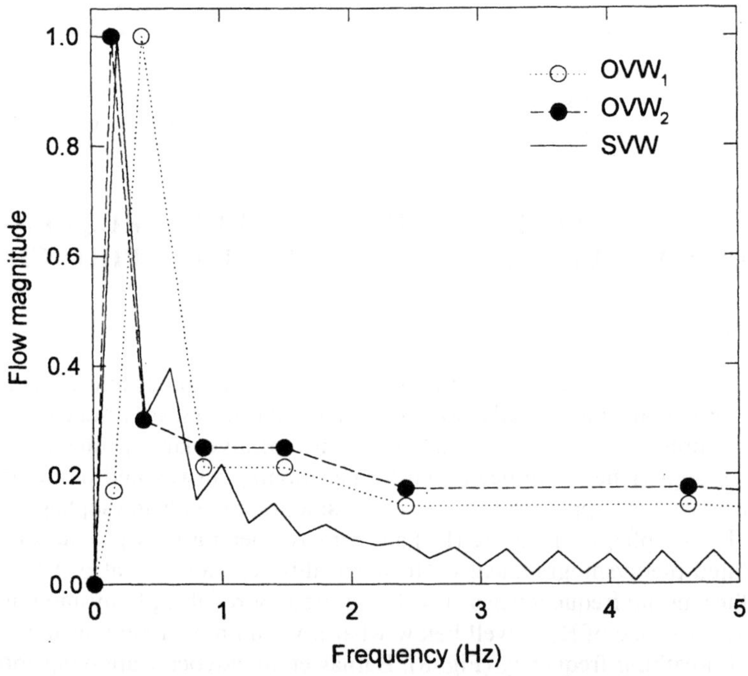

Figure 10. Comparison of flow magnitude spectrum for a standard human step-ventilation pattern (SVW) and for a two types of optimal ventilation waveforms. The OVW_1 places the primary energy around 0.4 Hz, but modulates it with lower frequency energy at 0.156 Hz while OVW_2 has the primary energy at 0.156 Hz. The impact on the ventilation patterns is shown in Fig. 12.

a SVW for producing a given V_T and breathing rate, and then increase the energy at the other frequencies. The waveform uses fewer frequencies than are contained in the SVW. Moreover, the frequencies selected are of a non-sum non-difference (NSND) design of order "n" as developed by Suki and Lutchen[100]. Here none of the frequencies are integer multiples of one another and none can be created by the sums or differences of combinations of "n" others. This means that the output pressure examined at the input frequencies only will not be subject to harmonic distortion (lower frequencies distorting higher ones) and harmonic crosstalk (higher frequencies distorting lower ones) will be significantly reduced. The phases of each frequency are then optimized to produce a waveform which maximizes volume delivered while minimizing peak-to-peak airway opening pressure.

The optimization procedure is with respect to a target Z_L one intends to measure as shown in Fig. 11. A set of amplitudes and phases defines a $\dot{V}(j\omega)$ which, when multiplied by the $Z_L(j\omega)$ produces the corresponding $P_{ao}(j\omega)$. After an inverse Fourier transform (FFT^{-1}), the time domain $\dot{V}(t)$ and p(t) waveforms are obtained. A global optimization procedure is used[14] to identify the set of phases which minimize a performance index which simultaneously minimizes the peak-to-peak airway opening pressure while maximizing the tidal volume delivered. Generally, one can select from several candidate waveforms that satisfy the condition of being clinically practical. Note that the Z_L for which a waveform is desired can either be that simulated from a model which represents the system to be probed, or an actual data set from that subject type taken by some related technique. The latter allows for a sequential approach in which the optimization is repeated as the Z_L becomes more reliable. A given waveform is fairly robust within a range of lung disease for a particular species[63]. An interesting variation of the waveform is to adjust the magnitude spectra so that the primary ventilation frequency is comfortable for a subject (say 0.4 Hz) but modulated by lower amplitude energy at lower frequencies (Fig. 10B). This produces a Z_L spectrum from 0.1-4 Hz while maintaining primary ventilation at 0.4 Hz. Recently, Rotger et al employed a similar

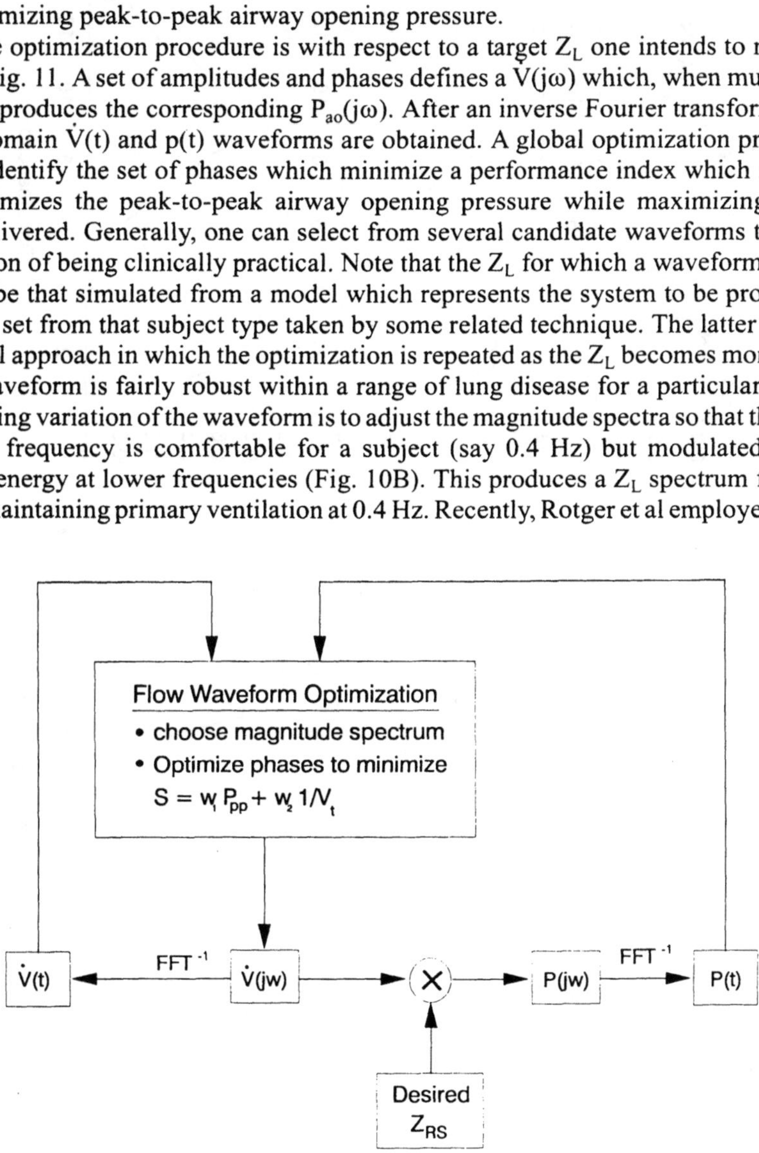

Figure 11. Summary of algorithm for generating an optimal ventilation waveform (OVW). Here $\dot{V}(j\omega)$ and $P(j\omega)$ are the frequency domain representation of flow and pressure associated with an impedance (Z_{rs}) to which the waveform is intended to be applied. The time domain representations ((t) and P(t) are derived by applying the inverse Fast Fourier Transform, FFT^{-1}). From Ref.63.

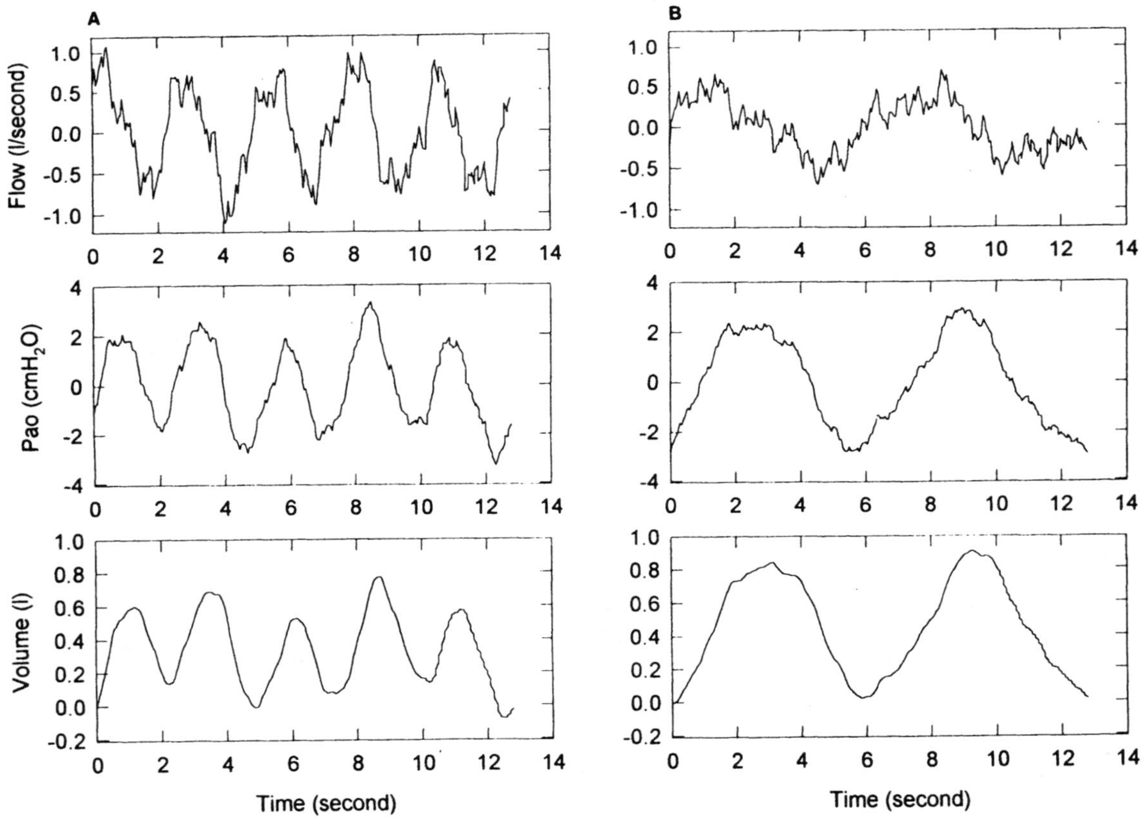

Figure 12. Examples of OVW flow and corresponding pressure and volume waveforms applied to human subjects .In one case (OVW_1) the highest energy is at the third frequency (.41 Hz) with lower-frequency modulations at 0.07 and 0.156 Hz) as well as higher frequency energy. In another case (OVW_2), the highest energy exists at the breathing rate (i.e., .156 Hz). In both cases the waveform is periodic over 12.8 sec (i.e., one OVW cycle contains either 6 or 2 physiological breaths, for OVW_1 and OVW_2 respectively). The flow magnitude spectra for these waveforms were shown in Fig. 10.

OVW approach, but in combination with the SVW which significantly corrupted the NSND nature of the waveform[94]. An example of these waveforms are shown in Fig. 12 and the quality of impedance measurements in Fig 13. One can see that the measured Z is smooth and quite sensitive to the obstructive state of the lung, even for closed chested measurements with the esophageal balloon technique. In more recent studies the OVW approach has been used to compare open to closed chested Z_L in dogs[106,106,107]. Even in healthy lungs there were small, but significant differences in the estimates airway and tissue properties with the closed chested lungs containing lower R_{aw} (35%) and G (13%) and higher H (15%). More startling is that the closed chested lung is far more sensitive to a given level of bronchoconstrictor[107]. Thus, the advantage of the OVW to measure Z_L in situ is more than just one of practicality (i.e, not requiring open chest conditions), but provides data far more relevant to the natural physiological mechanical and blood chemistry conditions in the lung. Finally, several other laboratories have recently employed the OVW approach for clinical evaluation or physiological investigation[86,94].

5. AIRWAY VERSUS TISSUE RESPONSE: QUANTIFYING THE ROLE OF INHOMOGENEITIES AND NONLINEARITIES ON OSCILLATORY LUNG MECHANICS

A key challenge is to determine the nature of the so-called tissue response during bronchoconstriction. Is it due to a separate contractile response within the parenchyma (a tissue response), to airway-tissue interdependence (a combined airway and tissue response), or is it an artifact due to airway inhomogeneities (an airway response alone)? Also, to what extend are nonlinearities distorting the estimated responses?

5.1. Inhomogeneities

A recent experimental study strongly suggest that inhomogeneous airway constriction is the dominant alteration in the rat lung[66]. Here Z_L was measured with the lungs equilibrated on room air and then again on a mixture of Neon and oxygen (NeOx). The NeOx mixture has a significantly higher viscosity than air. It was reasoned that if inhomogeneities were negligible, only the airway properties (R_{aw} and I_{aw}) would be different on NeOx and by an amount consistent with the ratio of gas viscosities. However, if there were significant inhomogeneities, the distribution of parallel time constants would be different for each gas such that their contribution to the frequency dependence in R_L and E_L would differ. The result would be a different tissue G and H for each gas. Indeed, as summarized in Fig. 14, during control conditions the tissue properties were identical on both gases while the Raw's scaled with viscosity. However, during constriction the estimated G was now significantly dependent on the resident gas viscosity while the H remained unaltered. This is experimental evidence that during lung constriction inhomogeneities do contribute to an overestimation of tissue resistance and that the increase in overall lung tissue stiffness is relatively inconsequential.

Several modeling approaches[32,55,64,87] have examined this issue and all have indicated that under conditions of inhomogeneous airway constriction, there can be an artifactual increase in the estimated tissue damping. One of the more powerful modeling approaches is to invoke a structurally based model of the airways consistent with the Horsfield anatomic casting studies[21,37,38,39,42,55,65]. The model contains a fixed number of airway generations, each with a designated length and diameter. The branching pattern is asymmetric. The impedance of a single generation is governed by the acoustic transmission line approach outlined in Benade[10]. At the

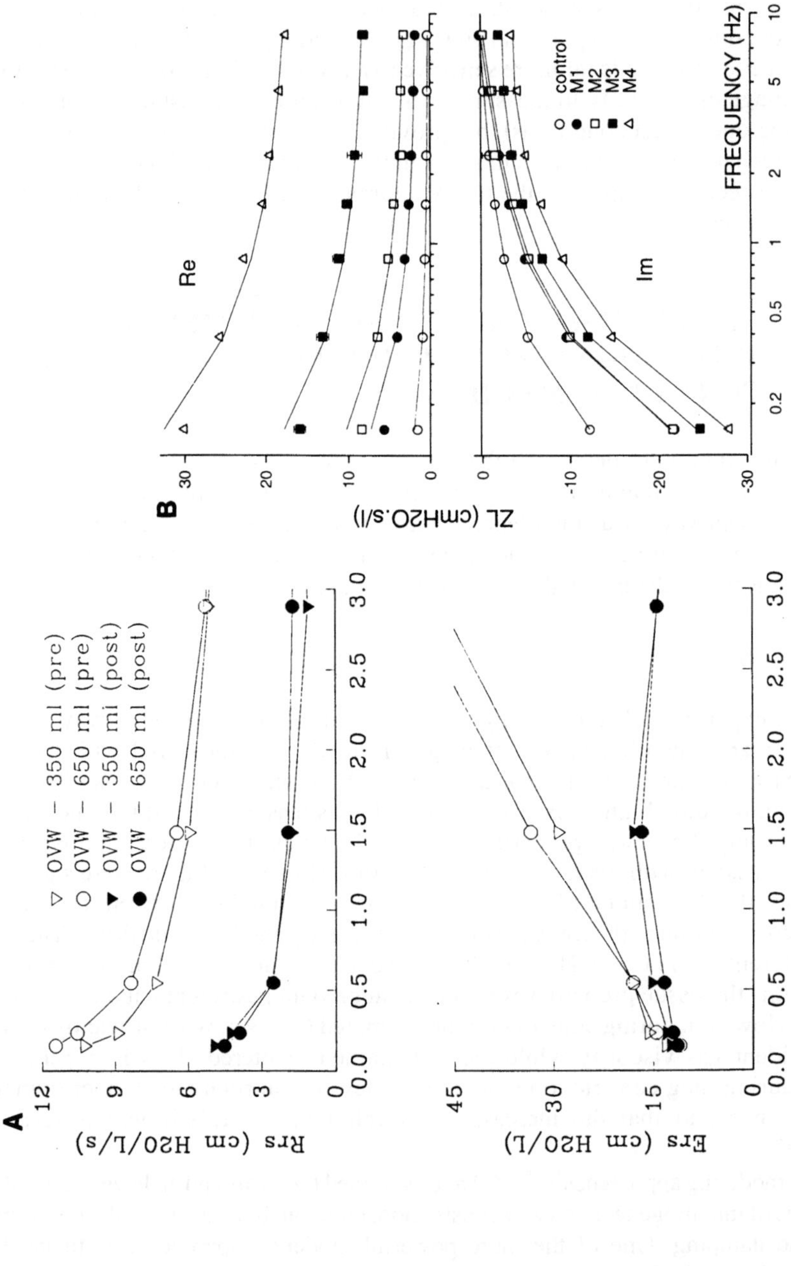

Figure 13. Example impedance spectra estimated while applying OVW. (A) Human asthmatic Z_{rs} pre and post bronchodilator while ventilated at different tidal volumes (from Ref. 63); and (B) Dog lung Z_L during control and after four increasing doses of methacholine all measured in situ via an esophageal balloon (from Ref.107). Note log frequency scale for (B).

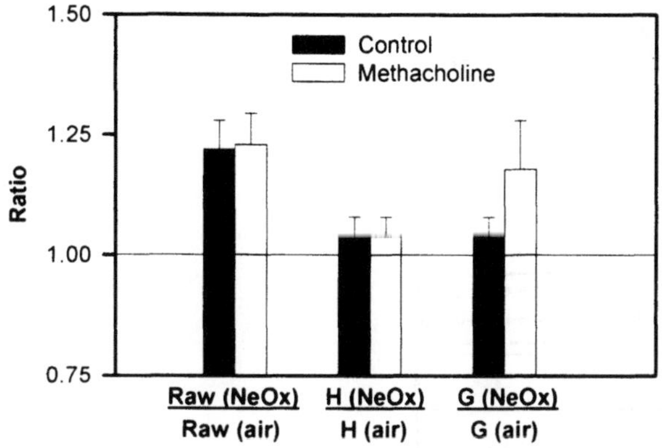

Figure 14. Ratio of airway and tissue properties (mean ± s.d.) estimated in the model of Fig. 8 from seven rat lungs with the lungs equilibrated on room air versus a mixture of NeOx and air. Results shown for control and post methacholine challenge conditions. The NeOx mixture was approximately 20% more viscous than the air mixture. Note that during control the tissue properties were independent of the resident gas but that during methacholine the estimated tissue damping parameter (G) depended on the gas (from Lutchen et. al. [66])

terminal airways one can attach an alveolar-tissue element which accounts for the gas compression in the alveolus and the constant-phase tissues of the surrounding parenchymal wall. Techniques have been developed to take advantage of the self-consistency in the model for calculating the impedance of the entire network[21]. We have recently modified the model to allow for inhomogeneous airway constriction[65]. Figure 16 shows simulated R_L and E_L spectra for a healthy lung compared to a condition where 80% of the peripheral airways undergo a 40% or 60% diameter constriction. Also shown are the corresponding homogeneous constant phase model (Fig. 8) fits to these data. First, we note that as constriction increases, the frequency dependence of R_L and E_L increases. Indeed, with a 60% diameter constriction there is a huge frequency

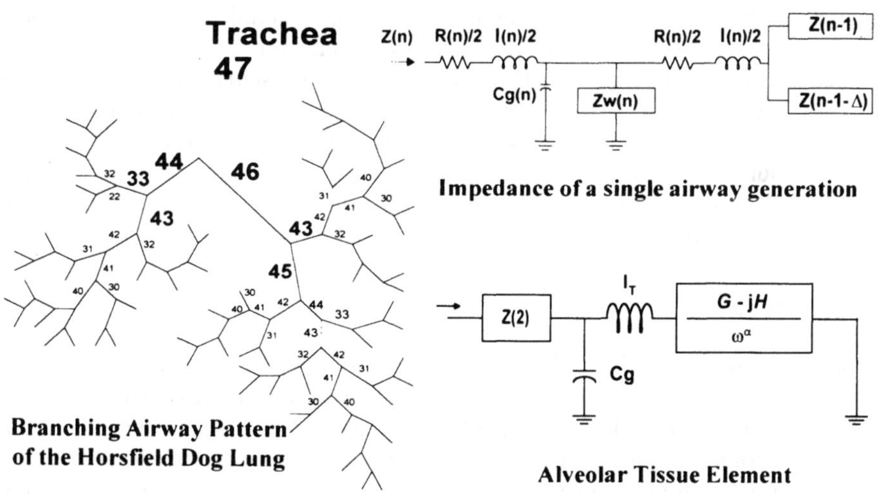

Figure 15. Schematic depiction of Horsfield based structural lung model. Shown are the concept of the asymmetric branching airway system; the impedance model for a single generation of the airways; and the alveolar -tissue element attached to the terminal airways.

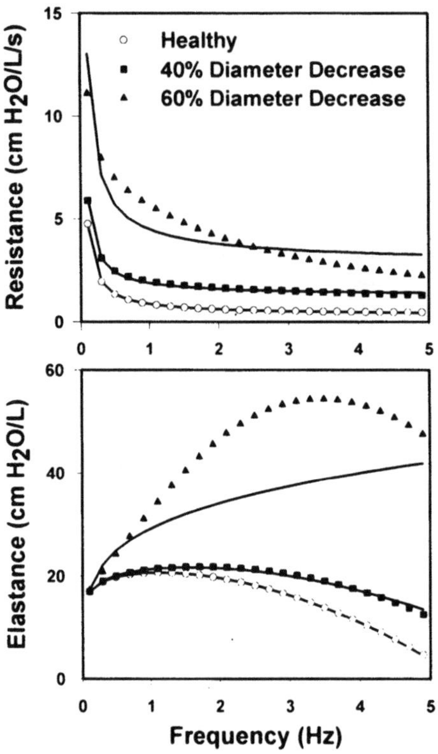

Figure 16. Simulated R_L and E_L spectra from the Horsfield model of Fig. 15 for healthy conditions and at increasing diameter reductions applied to 80% of the peripheral airways in the model (i.e., airways below generation 22 which were originally 2 mm in diameter). Also shown are the homogeneous constant-phase model fits (Fig. 8).

dependence of E_L such that if data were acquired at only a single frequency around 0.5 Hz, one would infer a large increases in tissue elastance when, in fact, there has been none. Second, we also see that a milder constriction (40% diameter reduction), the homogeneous model still fits the data well but during severe constriction it does not. Indeed, at the lower level of constriction the estimated tissue properties were nearly identical to the true tissue properties assigned to the Horsfield model, but at the higher constriction level (60%), the additional frequency dependence prevented a good fit and resulted in a substantial overestimation of tissue G. The estimated H remained equal to the true H. Generally, H is governed by the lowest frequency included in the data fit and in this case, the E_L at 0.1 Hz remained unaffected by constriction. The implications of this kind of modeling study are: 1) milder inhomogeneous constriction will not substantially distort the partitioning of lung airway and tissue properties via application of the homogeneous constant phase model to Z_L; and 2) more severe inhomogeneous constriction will distort the separation, and may be identified by a clear degeneration in the quality of fit when using this model.

5.2. Nonlinearities

Several studies[1,2,3,4,5,36,57,64] show that during pulmonary disease in addition to increased frequency dependence of R_L and E_L there is also increased (and an inverse) amplitude dependence. Thus R_L and E_L tend to decrease as tidal volume increases indicating an increasing amount of nonlinearity. This raises an important question: What is the relation between these nonlinear mechanisms and the assessment of frequency dependence based on estimates of an apparent linear impedance via the OVW-NSND waveforms? We have recently addressed this problem using a nonlinear block-structure approach[105]. Recall, that with an OVW-NSND waveform the output time-domain pressure contains information related to the linear processes, including inhomogeneity, as well as nonlinear processes, including harmonic interactions. The apparent linear impedance is calculated only at the NSND frequencies, thus avoiding the latter. However, using parameter estimation techniques in the time domain, we can identify linear and nonlinear processes simultaneously. We found that the best structure for the lungs (pre and post histamine) is a Wiener-type model for the lung tissues in series with a linear airway compartment. The Wiener model is a cascade connection of a linear dynamic block (i.e., the constant phase model) with a nonlinear, zero-memory block (which was a third-order polynomial model for distorting the pressure waveform). We showed that the primary cause for increased amplitude dependence after constriction was not due to a change in the underlying nonlinear mechanisms. Instead, it results from exacerbation of the nonlinearity through an increase in the magnitude of the linear tissue impedance. This nonlinear model, however, cannot account for the increased frequency dependence due to the development of inhomogeneities after constriction. Nevertheless, we can further enhance the model by substituting for homogeneous airway compartment with a model embodying a continuous distribution of parallel pathways, each terminating on an identical Wiener nonlinear tissue model. An important practical consequence is that with such a model one may be able to distinguish the influence of parallel inhomogeneities from tissue changes following an agonist induced challenge. The reason is that parallel time constant inequalities constitute an inherently linear phenomona whereas real tissue changes should always be accompanied by increased nonlinearities[105]. The latter can be captured by simultaneously fitting the impedance in the frequency and time domains.

6. SUMMARY AND FUTURE DIRECTIONS

We have shown that forced oscillations from low to high frequencies provide a noninvasive means for probing the mechanical properties and behavior of the respiratory and lung systems during healthy and diseased conditions. The extent of insight is strongly a function of the frequency range over which the data is acquired. Figure 17 summarizes this point. Unfortunately, the traditional and easiest to obtain data (from 2-32 Hz) is the least informative, only permitting estimation of overall lung or respiratory resistance, inertance, or compliance and/or the presence of severe obstruction. Transfer impedance data out to higher frequencies (2-64 Hz) show much promise with respect to clinically distinguishing patients with lung disease and in quantifying airway resistance (Fig. 4). Very high frequency Z_{in} data may provide detail on the serial distribution of airway diameters, lengths, and wall properties (Fig. 6). However, with respect to mechanical phenomena that influence breathing and which are invariably altered during disease, the very low frequency range (<2 Hz) is simply rich in information. Here there is an impact of tissue structure and function, airway-tissue coupling, nonlinear vs linear mechanisms, and airway or tissue inhomogeneities, all of which are considered altered during disease. The challenge to the biomedical engineer is to design protocols and modeling analyses to distinguish how each mechanism impacts a physiological breath for a particular lung condition. Only recently have experimental and modeling techniques been developed to measure and interpret these data. The use of the OVW with NSND design has potential clinical utility for acquiring reliable Z_L data with minimal nonlinear distortion. The use of different resident gases and structural models[8,65] show much promise for distinguishing mechanical inhomogeneities from alterations in the viscoelastic properties of the lung tissues.

The next major hurdle is to acquire a data base from human subjects spanning a large variety of type and severity of lung disease, including asthmatics pre and post bronchodilator, lung transplant and reduction therapy patients, and restrictive lung disease. The goal will be not only to quantify differences in the Z_L of these groups, but in interpreting the structural

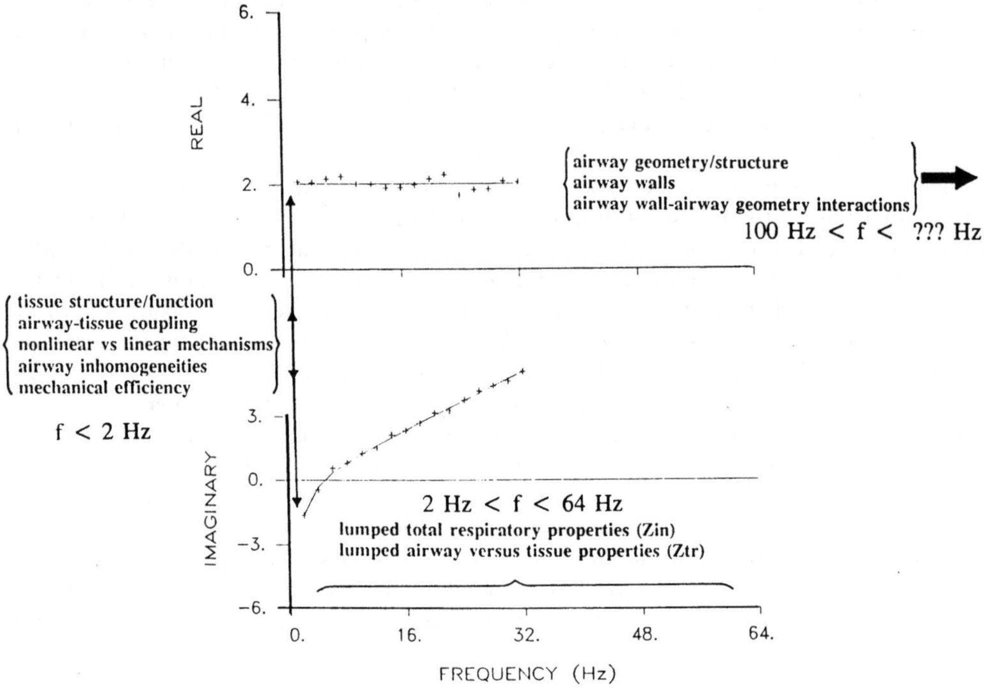

Figure 17. Summary of mechanistic information available from impedance data as a function of frequency range superimposed on typical data from the most commonly reported range.

and functional origins of these differences. If we can localize the pathology to be primarily in tissue (inflammation), inhomogeneous peripheral airway constriction, or homogeneous central airway constriction the clinician can chose between systemic vs topical administration of steroids or bronchodilator.

APPENDIX A: SIGNAL PROCESSING ISSUES IN THE ESTIMATION OF IMPEDANCE

The mechanical impedance (Z) of the respiratory system or the lungs is usually determined from measured pressure (P) and flow ($\dot{V}$) data records using Fourier analysis. The quality of the measurement is then assessed by computing the coherence function (γ^2). When the input signal is a small-amplitude random noise, the original approach presented by Michaelson et al.[73] can be invoked. More recently, however, investigators have realized the advantages provided by the pseudorandom noise (PRN) inputs since they are less sensitive to noise and require shorter data records. In this brief appendix, we describe some recent advances in the signal processing of Z and γ^2 using PRN inputs.

The conventional way of estimating Z as a function of frequency (f) is to calculate the ratio of the cross-power spectrum of P and to the autopower spectrum of

$$Z(f) = \frac{G_{P\dot{V}}(f)}{G_{\dot{V}\dot{V}}(f)} \tag{A-1}$$

and γ^2 is obtained as

$$\lambda^2(f) = \frac{|G_{P\dot{V}}(f)|^2}{G_{\dot{V}\dot{V}}(f)\ G_{PP}(f)} \tag{A-2}$$

where G_{PP} is the autopower spectrum of P. The power spectrum G_{XY} can be calculated from N repetition of the PRN input signal as

$$G_{XY}(f) = \frac{1}{N} \sum_{k=1}^{N} X_k(f)\ Y_k(f)^* \tag{A-3}$$

where X and Y are the fast Fourier transforms (FFT) of P and/or . If we assume that the data acquisition was synchronized with the PRN stimulus (i.e., the number of points in a cycle of the PRN input is an integer power of 2) then additional windowing of the time record is not necessary since the rectangular window is self-windowing (i.e., the zeros of the FFT of the rectangular window fall exactly on the frequencies where the spectra are evaluated)[16]. Nevertheless, when substantial breathing components are also present, use of a non-rectangular window (e.g., Hamming) or digital filtering before the FFT can improve the data. In order to obtain reasonable estimates of the spectra, one needs to include in Eq. A-3 at least a few complete periods. For low frequency measurements where the period of the input is long, there may be only 1 or 2 complete cycles available. In this case the number of data blocks in Eq. A-3 can be increased by using overlapping windows. Additionally, the low frequency amplitudes can be elevated which increases the signal-to-noise ratio and the quality of Z at low frequencies. Generally, the PRN spectral content, the phase angle distribution and the specific frequencies included in a PRN signal are often optimized to provide the best estimate of Z for the specific application at hand[15,63,100,103]. If nonlinearities are present, the spectra in Eq. A-3 will also contain harmonic distortion and cross-talk terms and, as a consequence, the estimates of Z and γ^2 will be biased[100]. Since harmonic distortion and cross-talk are deterministic averaging does not help reducing their influence. For strongly nonlinear systems or using large amplitude inputs this bias can amount to such an extent that the characteristic features of Z disappear[101,102]. Unfortunately, in contrast to random inputs, the γ^2 using PRN inputs will not detect the presence of nonlinearities. A sure sign of nonlinearities is when Z appears very "noisy" look-like, however, the standard deviation of this noisy behavior is very small and γ^2 is high. Maki[68] and Daroczy et al.[15] proposed non-integer multiple PRN inputs which eliminate harmonic distortion while Suki and Lutchen[100] developed a class of PRN inputs called Non Sum Non Difference (NSND) PRN inputs that minimize cross-talk at the frequencies of the input. The NSND inputs provide smooth spectral estimates of Z even using physiological tidal volumes (e.g., OVW inputs[63]) but at the expense of a sparser frequency resolution.

APPENDIX B: MODEL VALIDATION: PARAMETER SENSITIVITY ANALYSIS

Model interpretation of impedance data routinely involves fitting a proposed model to the data. A variety of optimization schemes exist to perform this parameter estimation problem[6,14,59]. However, acquiring a set of parameters is only part of a larger problem associated with model validation. The estimation of parameters is a model identification step, i.e., what are the model parameters that best match a given impedance data based on some performance criteria? Of course the key questions to be addressed in this step are: Is the match to the data acceptable?; and Are the parameter estimates physically sensible? However, before wasting time on the second step, it is imperative to first answer two other critical questions: How unique are the parameter values, i.e., how much can we vary them from their optimal estimates without significantly altering the quality of the model match to the (noisy) data? And can we use this model to predict an alternative measurement? i.e., is the identified model specific only to the data it was fit to? This appendix will briefly address the first question which requires a sensitivity analysis. A more thorough analysis can be found in the literature[53,59,60,93].

Model parameters are estimated by finding the vector of parameters, $\underline{\theta}$, that minimize some performance index (P.I.) function, eg.:

$$P.I.(\underline{\theta}) = \sum_{k=1}^{n} [w_i(Y_{d_i} - Y_{m_i}(\underline{\theta}))] \tag{B-1}$$

where Y_d represents the data, $Y_m(\underline{\theta})$ the model predicted value for a given set of $\underline{\theta}$, and w_i a weighting factor which can account for relative accuracy of the Y_d (eg., a maximum likelihood approach) or, perhaps, for differences in the units of several different physical variables to be fit simultaneously[59]. The optimal parameter vector, $\underline{\theta}^*$, minimizes P.I.. Generally, P.I.($\underline{\theta}^*$) is not zero because of measurement noise and of an imperfect model (i.e., the model does not capture all mechanisms contributing to the data). The sensitivity analysis question is to find all $\underline{\theta} \neq \underline{\theta}^*$ such that

$$P.I.(\underline{\theta}) - P.I.(\underline{\theta}^*) < \epsilon \tag{B-2}$$

The choice of ε can be such that one is answering a non-statistical sensitivity analysis question. For example, one can set ε to a be a certain percentage increase in P.I. ($\underline{\theta}^*$). Alternatively, one can assume that the residuals and measurement noise are distributed as a white Gaussian process and chose ε so that confidence interval for $\underline{\theta}^*$ is determined at some specified probability level α. In this case

$$\epsilon = 2\sigma^2 p F_{1-\alpha}(p, n-p) \tag{B-3}$$

where p is the number of parameters, n is the number of data points fit, F relates to the F-distribution and σ^2 is the variance of the measurement noise. The σ^2 can be estimated as

$$s^2 = \frac{P.I.(\underline{\theta}^*)}{N-p} \tag{B-4}$$

For either choice of ε, eq. B-2 is generally nonlinear in θ. There are two approaches to solving eq. B-2. First, one can perform a Taylor series expansion on P.I.(θ^*) for changes in the parameters $\delta\underline{\theta}$ about $\underline{\theta}^*$. In this case one can show that the solution is

$$\delta\underline{\theta}^T H \delta\underline{\theta} = \epsilon \tag{B-5}$$

where

$$H = 2S^T W S \quad \text{with} \quad S_{ij} = \frac{\partial Y_m(f_i, \underline{\theta})}{\partial \theta_j} \tag{B-6}$$

with W a matrix of the weights from eq. B-1. Alternatively, one can use a numerical Monte-Carlo approach to solve for all $\delta\underline{\theta}$ such that[60,93]:

$$P.I.(\underline{\theta}^* + \delta\underline{\theta}) - P.I.(\underline{\theta}) < \in \qquad \text{(B-7)}$$

In either case, the solution represents the allowable simultaneous changes in the estimated parameters such that the quality of fit is not compromised by more than the ε selected. This is more powerful information than determining an individual parameters standard deviation or confidence interval. It is very common in structural or mechanistic based models for an individual confidence interval to be quite small while the solution of eq. B-7 might indicate extremely large changes are allowed if several parameters can changes simultaneously[60,53,52]. This occurs because the model structure imbeds cross-correlations between parameters (i.e., a change in one paraemter can be compensated by a change in another parameter so as to not alter the model output). Clearly, if the parameters can vary greatly without altering the quality of fit to a given data set, one must refrain from physiological interpretation of the specific estimated parameters themselves. Assuming that the data is not overly noisy, large parameter uncertainty indicates that the model is over defined (too complicated) for the data fit. The solutions are to reduce the complexity of the model or to acquire new data (outside the range of existing data) which are sufficiently sensitive to the parameters that are currently poorly determined.

ACKNOWLEDGMENTS

The authors' research work contributing to this chapter has been supported by NIH HL-50515 and NSF BCS-93094276.

REFERENCES

1. Barnas, G.M., D.N. Campbell, C.F. Mackenzie, J.E. Mendham, B.G. Fahy, C.J. Runcie, and G.E. Mendham. Lung, chest wall, and total respiratory system resistance and elastance in the normal range of breathing. *Amer. Rev. Resp. Dis.* 143, 240-244, 1991.
2. Barnas, G.M., K. Yoshino, S.H. Loring, and J. Mead. Impedance and relative displacement of relaxed chest wall up to 4 Hz. *J. Appl. Physiol.* 62, 71-81, 1987.
3. Barnas, G.M., C. Mackenzie, M. Skacel, S. Hempleman, K. Wicke, C. Skacel, and S.H. Loring. Amplitude dependency of regional chest wall resistance and elastance at normal breathing frequencies. *Amer. Rev. Respir. Dis.* 140, 25-30, 1989.
4. Barnas, G., D. Stamenovic, and K.R. Lutchen. Lung and chest wall impedance in the dog in the normal range of breathing: effects of pulmonary edema. *J. Appl. Physiol.* 73: 1049-1046, 1992.
5. Barnas, G.M., D. Stamenovic, K.R. Lutchen, and C.F. Mackenzie. Lung and chest wall impedances in the dog in normal range of breathing: effects of frequency and tidal volume. *J. Appl. Physiol.* 72:87-93, 1992.
6. Bard, Y. *Nonlinear Parameter Esrtimation* New York, Academic Press., 1974.
7. Bates, J.H.T. , F. Shardonofsky, and D.E. Stewert. The low-frequency dependence of respiratory system resistance and elastance in normal dogs. *Resp. Physiol.* 78, 369-389, 1989.
8. Bates, J.H.T. Stochastic model of the pulmonary airway tree and its implication for bronchial responsiveness. *J. Appl. Physiol* 75(6): 2493-2499, 1993.
9. Bates, J.H.T., A.-M. Lauzon, G.S. Dechman, G.N. Maksym, and T.F. Schuessler. Temporal dynamics of pulmonary response to intravenous histamine in dogs: effects of dos and lung volume, *J. Appl. Physiol.* 76(2): 616-626, 1994.
10. Benade, A.H. On the propagation of sund waves in a cylindrical conduit. *J. Accoust. Soc. Am.* 44:616-623, 1968.
11. Bhansali, C.G. Irvin, J.A. Dempsey, R. Bush, and J.G. Webster. Human pulmonary resistance: effect of frequency and gas physical properties. *J. Appl. Physiol.* 47, 161-168, 1979.
12. Capper, W., R. W. Guelke, and A. Bunn. The estimation of tube wall compliance using acoustic input impedance. *IEEE Trans. Biom. Eng.* 18, 544-550, 1991.

13. Chalker, R., B. Celli, R. Habib, and A.C. Jackson. Respiratory system input impedance from 4-256 Hz in normals and patients with chronic airflow obstruction: comparison and correlation with spirometry. *Am. Rev. Respir. Dis.* 146:570-576, 1992.
14. Csendes, T. Nonlinear parameter estimation by global optimization-efficiency and reliability. *Acta Cybernetica.* 8,361-370. 1988.
15. Daroczy, B. and Z. Hantos. Generation of optimum pseudorandom signals for respiratory impedance measurements. *Int. J. Biomed. Comput.* 25: 21-31, 1990.
16. Davis, K.A.; Lutchen, K.R. Respiratory impedance spectral estimation for digitally created random noise. *Annls Biomed. Eng.* 19, 179-195, 1991.
17. Dorkin, H.L., K.R. Lutchen, and A.C. Jackson. Human respiratory input impedance from 4 to 200 Hz: physiological and modeling considerations. *J. Appl. Physiol.* 64:823-831, 1988.
18. DuBois, A.B., A.W. Brody, D.H. Lewis, and B.F. Burgess, Jr. Oscillation mechanics of lungs and chest in man. *J. Appl. Physiol.* 8:587-594, 1956.
19. Farre, R., R. Peslin, E. Oostveen, B. Suki, C. Duvivier, and D. Navajas. Human respiratory impedance from 8 to 256 Hz corrected for upper airway shunt. *J. Appl. Physiol.* 67:1973-1981, 1989.
20. Fredberg, J.J. and D. Stamenovic. On the imperfect elasticity of lung tissue. *J. Appl. Physiol* . 67, 2408-2419, 1989.
21. Fredberg, J.J. and A. Hoenig. Mechanical response of the lungs at high frequencies. *ASME J. Bimomech. Eng.* 44:616-623, 1978.
22. Fredberg, J.J., D.H. Keefe, G.M. Glass, R.G. Castille, and I.D. Frantz III. Alveolar pressure nonhomogeneity during small amplitude high-frequency oscillation. *J. Appl. Physiol* . 57: 788-800, 1984.
23. Fredberg, J.J., R.H. Ingram, Jr., R.G. Castile, G.M. Glass, and J.M. Drazen Nonhomogeneity of lung response to inhaled histamine assessed with alveolar capsules. *J. Appl. Physiol* . 58, 1914-1922, 1985.
24. Fredberg, J.J. Airway dynamics: recursiveness, randomness, and reciprocity in linear system simulation and parameter estimation. In *Respiratory Physiology: An analytical approach,* ed. H.K. Chang, and M. Piava, vol 40, Marcel Dekker, 167-194, 1989.
25. Fredberg, J.J. D. Bunk, E. Ingenito, and S. Shore. Tissue resistance and the contractile state of lung parenchyma. *J. Appl. Physiol* 74(3): 1387-1397, 1993.
26. Fuller, S.D. and A.N. Freed. Partitioning of pulmonary function in rabbits during cholinergic stimulation. *J. Appl. Physiol,* 78(4): 1242-1249, 1995.
27. Fung, Y.C. *Biomechanics: Mechanical Properties of Living Tissues.* New York, Springer, 1981.
28. Guelke, R.W. and A.E. Bunn. Transmission line therory applied to sound wave propagation in tubes with compliant walls. *Acustica.* 48:101-106, 1981.
29. Habib, R.H., B. Suki, J.H.T. Bates and A.C. Jackson. Seriel distribution of airway mechanical properties in dogs: effects of histamine. *J. Appl. Physiol* 77(2):554-566, 1994.
30. Hantos, Z., B. Daroczy, B. Suki, G. Galgoczy, and T. Csendes. Forced oscillatory impedance of the respiratory system at low frequencies. *J. Appl. Physiol* 60: 123-132, 1986.
31. Hantos, Z., B. Daroczy, T. Csendes, B. Suki, and S. Nagy. Modeling of low frequency pulmonary Impedance in Dogs. *J. Appl. Physiol* , 68, 849-860, 1990.
32. Hantos, Z , B. Daroczy, B. Suki, S. Nagy, and J.J. Fredberg. Input impedance and peripheral inhomogeneity of dog lungs. *J. Appl. Physiol* , 72, 168-178, 1992.
33. Hantos, Z. A.Adamicza, E. Govaerts, and B. Daroczy. Mechanical impedance of the lungs and chest wall in the cat. *J. Appl. Physiol* , 73, 427-433, 1992.
34. Hantos, Z. F. Petak, A. Adamicza, B. Daroczy, B. Suki, and K.R. Lutchen. Optimum ventilator waveform for the estimation of respiratory impedance: an animal study. *Eur. Respir. Rev.* 19:191-197, 1994.
35. Hildebrandt, J. Comparison of mathematical models for cat lung and viscoelastic ballooon derived by Laplace transform methods from pressure-volume data. *Bull. Math. Biophys.* 31:651-657, 1969.
36. Hildebrandt, J. Pressure-volume data of cat lung interpreted by a plastoelastic linear viscoelastic model. *J. Appl. Physiol* 28, 365-372, 1970.
37. Horsfield, K. Pulmonary airways and blood vessels considered as confluent trees. *The Lung: Scientificic Foundations,* edited by R.G. Crystal, J.B. West et. al. Raven Press, New York, 1991.
38. Horsfield, K., G Dart, D.E. Olson, and G. Cumming. Models of the human bronchial tree. *J. Appl. Physiol.* 31, 207-217, 1971.
39. Horsfield, K. W. Kemp, and S. Phillips. An asymmetrical model of the airway of the dog lung. *J. Appl. Physiol.* 52:21-26, 1982.
40. Ingenito, E.P., B. Davison, and J.J. Fredberg. Tissue resistance in the guinea pig at baseline and during methacholine constriction. *J. Appl. Physiol.* 75(6):2541-2548, 1993.
41. Jackson, A.C., C.A. Giurdanella, and H.L. Dorkin. Density dependence of respiratory system impedances between 5 and 320 Hz in humans. *J. Appl. Physiol.* 67:2323-2330.1989.

42. Jackson, A.C., K.R. Lutchen, and H.L. Dorkin. Inverse modeling of dog airways and respiratory system impedances. *J. Appl. Physiol.* 62:2273-2282, 1987.
43. Jackson, A.C. and K.R. Lutchen. Modeling of respiratory system impedances in dogs. *J. Appl. Physiol.* 62:414-420, 1987.
44. Jackson, A. and K.R. Lutchen. Physiological basis for resonant frequencies in respiratory system impedances in dogs. *J. Appl. Physiol.* 70:1051-1058, 1991.
45. James, A.L., P.D. Pare, and J.C. Hogg. The mechanics of airway narrowing in asthma. *Am. Rev. Respir. Dis.* 139: 242-246, 1989.
46. Kariya, S. L.M. Thompson, E. P. Ingenito, and R.H. Ingram, Jr. Effects of lung volume, volume history and methacholine on lung tissue viscance. *J. Appl. Physiol* . 66(2), 977-982, 1989.
47. Lauzon, A.-M., and J.H.T. Bates. estimation of time-varying respiratory mechanical parameters by recursive least squares. *J. Appl. Physiol.* 71: 1159-1165, 1991.
48. Larsen, G. L. Experimental models for reversible airway obstruction. in *The Lung: Scientific Foundations,* edited by R.G. Crystal, J.B. West et. al. Raven Press, New York, 1991.
49. Ludwig, M.S., F.M. Robatto, P.D. Sly, M. Browman, J.H.T. Bates, and P.V. Romero. Histamine-induced constriction of canine peripheral lung: an airway or tissue response. *J. Appl. Physiol.* 71, 287-293, 1991.
50. Ludwig, M.S., P.V. Romero, J.H.T. Bates. A comparison of the dose-response behavior of canine airways and parenchyma. *J. Appl. Physiol.* 67, 1220-1225, 1989.
51. Ludwig, M.S., F.M. Robatto, S. Simard, D. Stamenovic, and J.J. Fredberg. Lung tissue resistance during contractile stimulation: structural damping decomposition. *J. Appl. Physiol.* 72, 1332-1337, 1992.
52. Lutchen, K.R. and A.C. Jackson. Statistical measures of parameter estimates from models fit to respiratory impedance data: emphasis on joint variabilities. *IEEE Trans. Biomed. Eng.* 33, 1000-1010, 1986.
53. Lutchen, K.R. and A.C. Jackson. Reliability of parameter estimates from models applied to respiratory impedance data. *J. Appl. Physiol.* 62:403 413, 1987.
54. Lutchen, K.R., Z. Hantos, and A.C. Jackson. Importance of low-frequency impedance data for reliably quantifying parallel inhomogeneities of respiratory mechanics. *IEEE Trans. Biomed. Eng.* 35:472-481, 1988.
55. Lutchen, K.R.; Costa, K.D.; Physiological behavior of lumped parameters estimated from respiratory impedance data: use of forward/inverse modeling. *IEEE Trans. Biomed. Eng.* 11:1076-1086, 1990.
56. Lutchen, K.R., R.H. Habib, H.L. Dorkin, and M.A. Wall. Relation of respiratory impedance to a multibreath nitrogen washout in healthy, asthmatic, and cystic fibrosis subjects. *J. Appl. Physiol.* 68:2139-2149, 1990.
57. Lutchen, K.R. and A.C. Jackson. Effects of tidal volume and methacholine on low-frequency total respiratory impedance in dogs. *J. Appl. Physiol.* 68:2128-2138, 1990.
58. Lutchen, K.R. D. W. Kaczka, B. Suki, G.M. Barnas, P. Barbini, and G. Cevenini. Low frequency respiratory mechanics using ventilator-driven oscillations. *J. Appl. Physiol.* 75 (6): 2549-2560, 1993.
59. Lutchen, K.R. Sensitivity analysis of respiratory parameter uncertainties: Impact of criterion function form and constraints. *J. Appl. Physiol.* 69:766-775, 1990.
60. Lutchen, K.R. and A.C. Jackson. Confidence bounds on respiratory mechanical properties estimated from tranasfer fersus input impedance in humans versus dogs. *IEEE Trans. Biomed. Eng.* (39) 6, 644-651, 1992.
61. Lutchen, K.R.; Guirdenella,C; and Jackson, A.C.. Inability to separate airway from tissue properties using input impedance in humans. *J. Appl. Physiol.* 68, 2403-2412, 1990.
62. Lutchen, K.R., J.R. Everett and A.C. Jackson. Influence of frequency range and input impedance on interpreting the airways tissue separation implied from transfer impedance *J. Appl. Physiol.* 73(3), 1089-1099, 1993.
63. Lutchen K.R., K. Yang, D.W. Kaczka, and B. Suki. Optimal ventilator waveform for estimating low frequency mechanical impedance *J. Appl. Physiol.* 75(1): 478-488, 1993.
64. Lutchen, K.R., B. Suki, Q. Zhang, F. Petak, B. Daroczy, and Z. Hantos. Airway and tissue mechanics during physiological breathing and bronchoconstriction in dogs. *J. Appl. Physiol.* 77(1), 373-385, 1994.
65. Lutchen, K.R., J.L. Greenstein, B. Suki. How inhomogeneities and airway walls affect frequency dependence and seperation of airway and tissue properties. *J. Appl. Physiol.* 80(5), 1696-1707, 1996.
66. Lutchen, K.R., Z. Hantos, F. Petak, A. Adamicza, B. Suki. Airway inhomogeneities contribute to apparent lung tissue resistance during constriction. *J. Appl. Physiol* 80(5):1841-1849, 1996.
67. Lutchen, K.R. A. Sullivan, F.T. Arbogast, B.R. Celli, and A.C. Jackson. Use of transfer impedance measurements for clinical assessment of lung mechanics. *Amer. J. of Resp. and Crit. Care Medicine.* (under revision).
68. Maki, B.E. Interpretation of the coherence function when using pseudorandom inputs to identifynonlinear systems. *IEEE Trans. Biomed. Eng.* 33, 775-779, 1986 and Addendum, 35, 279-280, 1988.

69. Marshall, R. and A.B. Dubois. The measurement of viscous resistance of the lung tissues in normal man. *Clinical Sci.* 15:161-170, 1956
70. Marshall, R. and A.B. Dubois. The viscous resistance of lung tissues in patients with pulmonary disease. *Clinical Sci.* 15: 473-483, 1956
71. McIlroy, M.B., J. Mead, N.J. Selverstone, and E.P. Radford. Measurement of lung tissue viscous resistance using gases of equal kinematic viscosity. *J. Appl. Physiol.* 7:485, 1955.
72. Mead, J. Contribution of compliance of airways to frequency dependent behavior of the lungs. *J. Appl. Physiol.* 26: 670-673, 1969.
73. Michaelson, E.D., E.D. Grassman, and W. R. Peters. Pulmonary mechanics by spectral analysis of forced random noise. *J. Clin. Invest.* 56, 1210-1230, 1975.
74. Mishima, M., Z. Balassy, and J.H.T. Bates. Acute pulmonary response to intravenous histamine using forced oscillations through alvoelar capsules of dogs. *J. Appl. Physiol* . 77(5), 2140-2148, 1994.
75. Mitzner, W. , S. Blosser, D. Yager, E. Wagner,. Effect of bronchial smooth muscle contraction on lung compliance. *J. Appl. Physiol.* 72, 158-167, 1992.
76. Nagase, T,, J.G. Martin, and M.S. Ludwig. Comparative study of mechanical interdependence: effect of lung volume on Raw during induced constriction.*J. Appl. Physiol* 75(6):2500-2505, 1993.
77. Nagels, J., F. J. Landser, L. Van der Linden, J. Clement, and K.P. Van de Woestijne. Mechanical properties of lungs and chest wall during spontaneous breathing. *J. Appl. Physiol.*49, 408-416, 1980.
78. Navajas, D., R. Farre, M. Rotger, and J. Canet. Recording pressue at the distal end of the endotracheal tube to measure respiratory impedance. *Eur. Respir. J.* 2, 178-184, 1988.
79. Navajas, D., R. Farre, J. Canet, M. Rotger, and J. Sanchis. Respiratory input impedance in anesthesized paralyzed patients. *J. Appl. Physiol.* 69, 1372-1379, 1990.
80. Oostveen, E., R. Peslin, C. Gallina, and A. Zwart. Flow and volume dependence of respiratory mechanical properties studied by forced oscillation. *J. Appl. Physiol.* 67, 2212-2218, 1989.
81. Otis, A.B., C.B. McKerrow, R.A. Bartlett, J. Mead, M.B. McElroy, N.J. Selverstone, and E.P. Radford. Mechanical factors in the distribution of ventilation. *J. Appl. Physiol.* 8, 427-443, 1956.
82. Peslin, R., C. Duvivier, J. Didelon, and C. Gallina. Respiratory impedance measured with head generator to minimize upper airway shunt. *J. Appl. Physiol.* 59:1790-1795, 1985.
83. Peslin, R., C. Gallina, D. Teculescu, and Q.T. Pham. Respiratory input and transfer impedances in children 9-13 years old. *Bull. Eur. Physiopathol. Respir.* 23, 107-112, 1987.
84. Peslin, R. J. Papon, C. Duvivier, and J. Richalet. Frequency response of the chest: modeling and parameter estimation. *J. Appl. Physiol.* 39, 523-534, 1975.
85. Peslin, C. Duvivier, and C. Gallina. Total respiratory input and transfer impedance in humans. *J. Appl. Physiol.* 59, 492-501, 1985.
86. Peslin, R. C. Saunier, and M. Marchand. Analysis of low-frequency lung impedance in rabbits with nonlinear models. *J. Appl. Physiol.* 79: 771-780, 1995.
87. Petak, F., Z. Hantos, A. Adamicza, and B. Daroczy. Partitioning of pulmonary impedance: modeling vs alveolar capsule approach. *J. Appl. Physiol.* 75(2): 513-521, 1993.
88. Renzi, P.E., C.A. Giurdanella, and A. C. Jackson. Improved frequency response of pneumotachometers by digital compensation. *J. Appl. Physiol.* 68:382-386, 1990.
89. Robatto, S. Simard, and M.S. Ludwig. How changes in the serial distribution of bronchoconstriction affect lung mechanics. *J. Appl. Physiol.* 74(6): 2838-2847, 1993.
90. Romero, P.V., and M.S. Ludwig. Maximal methacholine-induced constriction in rabbit lung: interactions between airways and tissue? *J. Appl. Physiol.* 70, 1044-1050, 1991.
91. Romero, P.V., F.M. Robatto, S. Simard, and M.S. Ludwig. Lung tissue behavior during methacholine challange in rabbits in vivo. *J. Appl. Physiol.* 73, 207-212, 1992.
92. Rotger, M., R. Peslin, C. Duvivier, D. Navajas, and C. Gallina. Density dependence of respiratory input and transfer impedances in humans. *J. Appl. Physiol.*65, 928-933, 1988.
93. Rotger, M., R. Peslin, E. Oostveen, and C. Gallina. Confidence intervals of respiratory mechanical properties derived from transfer impedance. *J. Appl. Physiol.* 70, 2432-2438, 1991.
94. Rotger, M. R. Peslin, D. Navajas, and R. Farre. Lung and respiratory impedance at low frequency during mechanical ventilation in rabbits. *J. Appl. Physiol.* 78: 2153-2160, 1995.
95. Sato, J., B.L.K. Davey, F. Shardonofsky, and J.H.T. Bates. Low-frequency respiratory resistance in the normal dog during mechanical ventilation. *J. Appl. Physiol.* 70: 1536-1543,1991.
96. Sato, J., B. Suki, L.K. Davey, and J.H.T. Bates. Effect of methacholine on low frequency mechanics of canine airways and lung tissue. *J. Appl. Physiol.* 75(1): 55-62, 1993.
97. Shardonofsky, F.R. J.M. McDonough, and M.M. Grunstein. Effects of positive end-expiratory pressure on lung tissue mechanics in rabbits. *J. Appl. Physiol.* 75(6):2506-2513, 1993.

98. Sly, P.D. and C. J. Lanteri. Differential responses of the airways and pulmonary tissues to inhaled histamine in young dogs. *J. Appl. Physiol.* 68, 1562-1567, 1990.
99. Suki, B. R. H. Habib, and A.C. Jackson. Wave propagation, input impedance, and wall mechanics of the calf trachea from 16-1,600 Hz. *J. Appl. Physiol.* 75(6):2755-2766, 1993.
100. Suki, B. and K.R. Lutchen. Pseudorandom signals to estimate apparent transfer and coherence functions of nonlinear systems: Applications to respiratory mechanics. *IEEE Trans. Biomed. Eng.* 39(1), 1142-1151, 1992.
101. Suki, B., Z. Hantos, B. Daroczy, G. Alkaysi, and S. Nagy. Nonlinearity and harmonic distortion of dog lungs measured by low frequency forced oscillations. *J. Appl. Physiol.* 71, 69-75, 1991.
102. Suki, B. and J.H.T. Bates. A nonlinear viscoelastic model of lung tissue mechanics. *J. Appl. Physiol.* 71. 826-833, 1991.
103. Suki, B. Nonlinear phenomena in respiratory mechanical measurments. *J. Appl. Physiol.*. 74(5): 2574-2584, 1993.
104. Suki, B., A-L Barabasi, and K.R. Lutchen. Lung tissue viscoelasticity: a mathematical framework and its molecular basis. *J. Appl. Physiol.* 76(6), 2749-2759, 1994.
105. Suki, B., Q. Zhang, and K.R. Lutchen. Relationship between frequency dependence and amplitude dependence in the lung: a nonlinear block-structured modeling approach. *J. Appl. Physiol.* 79(2): 660-671, 1995.
106. Suki, B., F. Petak, A. Adamicza, Z. Hantos, and K.R. Lutchen. Partitioning of airway and lung tissue properties from lung input impedance: comparison of in situ and open chest conditions. *J. Appl. Physiol.* 79(2): 660-671, 1995.
107. Suki, B., F. Petak, A. Adamicza, B. Daroczy, K.R. Lutchen, and Z. Hantos. Airways and lung tissues are more sensitive to methacholine in closed chest than in open chest dogs. *J. Appl. Physiol.* (submitted).
108. Tepper, R. J. Sato, B. Suki, J.G. Martin, and J.H.T. Bates. Low frequency pulmonary impedance in rabbits and its response to inhaled methacholine. *J. Appl. Physiol.*. 73, 290-295, 1992.
109. Van Noord, J.A., J. Clement, K.P. Van de Woestijne, and M. Demedts. Total respiratory resistance as a measure of response to bronchial challenge with histamine. *Amer. Rev. Resp. Dis.* 139, 921-926, 1989.
110. Van Noord, J.A., J. Clement, M. Cauberghs, and K.P. Van de Woestijne. Effects of rib cage and abdominal restriction on total respiratory resistance and reactance. *J. Appl. Physiol.* 61, 1736-1740.
111. Wiggs, B.R., R. Moreno, J.C. Hogg, C. Hillian, and P.D. Pare. A model of the mechanics of airway narrowing. *J. Appl. Physiol.* 69: 849-860, 1990.
112. Wiggs, B.R., C. Bosken, P.D. Pare, A. James, and J.C. Hogg. A model of the airway narrowing in asthma and chronic obstructive pulmonary disease. *Am. Rev. Resp. Dis.* 145:1251-1258, 1992.
113. Ying, Y., R. Peslin, C. Duvivier, C. Gallina, and J. Felicia da Silava. Respiratory input and transfer impedances in patients with chronic obstructive pulmonary disease. *Eur. Respir. J.*3, 1186-1192, 1990.
114. Zwart, A. and R. Peslin, editors. Mechancial Respiratory Impedance: the forced oscillation method. Contributions to the workshop held for the European Commission. *Europ. Resp. Rev.* vol 1, rev. 3, 1991.
115. Zhang, Q., B. Suki, and K.R. Lutchen. An extended harmonic distortion index to quantify system nonlinearities from broadband inputs: application to lung mechanics. *Ann. Biomed. Eng.* 23: 672-681, 1995.

INDEX

MIX
Papier aus verantwortungsvollen Quellen
Paper from responsible sources
FSC® C105338

If you have any concerns about our products,
you can contact us on
ProductSafety@springernature.com

In case Publisher is established outside the EU,
the EU authorized representative is:
Springer Nature Customer Service Center GmbH
Europaplatz 3, 69115 Heidelberg, Germany

Printed by Libri Plureos GmbH
in Hamburg, Germany